RESEARCH IN HYBRID MATERIALS

RESEARCH IN HYBRID MATERIALS

SIMON J. BRUNNER AND JULIAN W. EGGER
EDITORS

Nova Science Publishers, Inc.
New York

For permission to use material from this book please contact us:
Telephone 631-231-7269; Fax 631-231-8175
Web Site: http://www.novapublishers.com

LIBRARY OF CONGRESS CATALOGING-IN-PUBLICATION DATA

Research in hybrid materials / Simon J. Brunner and Julian W. Egger (editors).
p. cm.
Includes bibliographical references and index.
ISBN 978-1-60456-718-2 (hardcover)
1. Composite materials. 2. Nanostructured materials. I. Brunner, Simon J. II. Egger, Julian W.
TA418.9.C6R47 2008
620.1'18--dc22

2008017036

Published by Nova Science Publishers, Inc. ≃ New York

CONTENTS

PREFACE

This book provides diverse articles from around the globe in the field of hybrid materials. Hybrid materials have currently an great impact on numerous future developments including nanotechnology. Hybrid materials not only represent a creative approach to design of new materials, but their improved or unusual properties also allow the development of innovative industrial applications. This interdisciplinary research field will open a land of promising applications in many areas including optics, electronics, ionics, mechanics, energy, environment, biology, medicine.

Chapter 1 - The state of the art survey of research findings on materials with nanoscale structure, including nonionic organic-inorganic hybrids, revealed that there is a large number of method producing such structures, that they posses a number of unique properties as a result of the synergy of their both organic and inorganic counterparts properties. This review describes the main general and specific sol-gel synthesis routes and also the classification of nonionic silica hybrids following several criteria. A special attention will be paid on the influence of the precursors and synthesis conditions to design nanostructured hybrid materials and hybrid silica nanocomposites ordered architecture or/and porosity. Classic and new investigation methods and several possible applications with significant technological importance including sensors and smart materials will be presented. At the present time, this very interdisciplinary scientific area is open for new ideas from all disciplines and therefore a promising future can be foreseen.

Chapter 2 - In general, artificial materials implanted into bone defects are encapsulated by a fibrous tissue. Some ceramics, such as Bioglass®, sintered hydroxyapatite and glass-ceramic A-W, however, form a bone-like apatite layer on their surfaces in the living body and bond to living bone through this apatite layer, i.e. they show bioactivity. Although these bioactive ceramics are used clinically as important bone-repairing materials, they are essentially brittle and hence limited in their applications. It is desirable to develop new types of deformable bioactive materials. Organic–inorganic hybrids prepared by a sol-gel method are expected to exhibit bioactivity as well as deformability. Polydimethylsiloxane (PDMS)-CaO-SiO_2-based hybrids were prepared. To evaluate their potential of bone-bonding property, their ability of apatite formation was examined in a simulated body fluid (SBF). Since the apatite formation on bioactive materials in the body can be reproduced even in SBF, we can estimate the bioactivity of the material by using SBF. Mechanical properties were also examined. Some of them showed apatite-forming ability and mechanical properties analogous to those of human cancellous bone. These CaO-containing hybrids, however, showed a decrease in mechanical strength in SBF. These hybrids containing no CaO do not form apatite on their surfaces in SBF. The authors then prepared CaO-free PDMS-TiO_2 and poly(tetramethylene oxide) (PTMO)-TiO_2 hybrids in which anatase was precipitated by hot-water treatments. These hybrids also showed apatite-forming ability on their surfaces in SBF and flexibility, and they showed little decrease in mechanical strength in SBF. In conclusion, it was revealed that we can design new bioactive materials for bone-repair by controlling the compositions and structures in organic-inorganic hybrid systems.

Chapter 3 - Organic-inorganic hybrid materials based on simple two-dimensional layers of tungsten oxide separated by organic molecules are merely one example of the huge variety of systems classed as "organic-inorganic hybrids". The materials in question have the potential to combine the high-electron mobility of the metal oxide layers with the flexibility of the organic layers. More importantly they identify the primary elements that might enable the formation of more complex nanostructures through self-assembling processes and the templating effect of the organic component. One such example that will be discussed here is the possibility of including magnetic ions within the inorganic layers, which is observed to have profound effects upon the magnetic behaviour of the resulting hybrids.

Tungsten oxide-organic amine hybrid materials have been synthesised via a range of methods and characterised using conventional materials characterisation techniques, such as X-ray diffraction, infrared and Raman spectroscopy, UV-visible spectroscopy and electron microscopy. The work is also complemented by ab initio calculations of the physical and electronic structure of the tungsten oxide hybrids. For compounds incorporating transition metal elements, the magnetisation was also measured. The effect of the low-dimensionality of the inorganic component is clearly seen in materials that include certain transition metals, in the form of magnetic transitions that indicate a crossover from 2-dimensional to 3-dimensional behaviour.

Chapter 4 - The chemistry of organoimido derivatives of polyoxometalates (POMs), an innovative emerging organic chemistry of polyoxometalates, has received increasing interest in the supramolecular chemistry and chemistry of materials recently. Organoimido derivatives of POMs are valuable building blocks for constructing novel nanostructured organic-inorganic hybrid molecular materials with the so-called ''value-adding properties'' and possible synergistic effects, which could result in unique catalytic, photo and electronic properties dramatically different from the corresponding parent materials, since the π electrons in the organic component may extend their conjugation to the inorganic framework and dramatically modify the electronic structure and redox properties of the corresponding parent POMs. The surface modification of POMs, that is, the substitution of one or more terminal or bridged oxo groups with organoimido ligand, is the fundamental method to synthesize organoimido derivatives of POMs via common organic chemistry reactions. In this chapter of the book, the synthetic chemistry of organoimido derivatives of POMs using different organic reagents will be reviewed, of which the novel DCC-dehydrating protocol to prepare organoimido derivatives of POMs recently developed by us and co-authors will be concentrated. As it can be expected, the reaction chemistry of organoimido derivatives of POMs stands for the fascinating future of the chemistry of organoimido derivatives of POMs since it opens not only a new road to the chemical modification of POMs, but also an exciting research arena where a variety of hybrid materials containing covalently bonded POM clusters and organic conjugated segments can be prepared in a more controllable and rational manner. Hence, in the rest of this chapter, Pd-catalyzed coupling reactions of the functionalized organoimido derivatives of POMs will be introduced in detail for this purpose. Especially, different rationally assembled hybrid molecules and polymers where the organic molecules and inorganic clusters are bonded together by an imido nitrogen atom will then be highlighted.

Chapter 5 - A general analysis is given of hybrid systems consisting of different combinations of 4 devices frequently employed for renewable energy utilization: Photovoltaic Solar Panel (PV), Solar Thermal Plane Collector (ST), Wind Generator (WG) and Heat-to-Electric/Mechanic Energy Convertor (HE); some of the combinations include radiation energy flux concentrators of different degrees. The main result of the consideration made is that the hybrid systems are more efficient than the sum of the constituents and more stable in relation to spontaneous variations of the renewable energy source potential (like wind velocity, insolation, etc.). However, to realize the possibilities mentioned, all the elements of a given hybrid system have to be especially designed and made for this specific system. For example, the PV panel for the hybrid PV/Thermal system ought to have a substrate with high thermal conductivity, to allow for heat extraction from the panel by the adjacent Solar Thermal Plane Collector, and practically no commercial panels with these characteristics are available. Besides, the PV panel as a part of the hybrid system will demand a special choice of semiconductor material and surface treatment which could be different from those for conventional panels. The limiting efficencies for some hybrid systems are estimated; these efficiencies exceed the efficiencies of separate use of the devices discussed. The most promising hybrid system is the PV panel made as spectrum splitter in combination with HE converter, of which total efficency could be around 50 %.

Chapter 6 - Distributed actuation and sensing are the key elements in the development of active structural control methodology. Piezoelectric materials are popularly considered as active elements (actuators or sensors) due to their good frequency bandwidth, low cost and fast energy conversion nature. As actuators, they develop isotropic or directional actuation strains, which are governed by mainly five piezoelectric constants (d_{31}, d_{32}, d_{33}, d_{15}, d_{24}). The longitudinal (d_{33}) and extension (d_{31}, d_{32}) actuations have been thoroughly studied; however shear actuation (d_{15}) is relatively a new concept but shows promising feature. It is a novel idea to combine the extension and shear actuations to develop a hybrid actuation mode for active vibration control applications, exploiting the benefits of both. The hybrid active laminate can be built, employing a transversely polarized (d_{31}) lamina and an axially polarized (d_{15}) lamina. Appropriate constitutive models are derived with an assumption that each lamina behaves as elastically orthotropic and electro-mechanically orthorhombic crystal class mm2. A two node sandwich beam element is developed using the isoparametric FE procedures to conduct numerical experiments. Active control analysis is performed using a modal control approach and the procedure is outlined to obtain the reduced order models without loosing the dynamic information of the vibrating systems.

Active stiffening (piezoelectric straining) and active damping (piezoelectric resistive force) are the two active effects systematically analyzed by numerical studies. Collocated and non-collocated actuator configurations are considered, employing extension and shear actuators in sandwich beam architectures to evaluate the performance of above mentioned active effects. In the vibration amplitude control, the shear actuation has been found very effective, as it develops locally shear strain. Also, a sine wave actuation mode is observed when a shear actuator is activated in a Clamped-Clamped construction. Interesting deflection behaviours are observed under hybrid actuation mode for various boundary effects. The mode shape control concept using piezoelectric stiffening has been introduced, where a Clamped-Free laminated beam is taken as an illustration. It is a useful technique, as the mode shapes influence significantly the dynamic instability of thin walled composite structures.

Chapter 7 - Hybrid molecules comprised of [60]fullerene and various π-conjugated oligomers have gained much attention owing to their promising potentials in molecular optoelectronics and photovoltaics. In this article, the authors intend to describe our recent study on a series of dumbbell-shaped C_{60}-oligomer-C_{60} triad molecules, primarily aimed at better understanding the roles of conjugation motif and dimensionality of the oligomer bridge in affecting their electronic and photophysical properties. Two types of conjugated oligo(phenylene ethynylene)/oligo (phenylene vinylene) (OPE/OPV) hybrid oligomers with linear or cruciform molecular topologies were designed and prepared via sequential Pd-catalyzed cross-coupling and Wittig-Horner reactions. These oligomers were readily endcapped with fullerenyl groups via an in situ ethynylation protocol, affording the desired C_{60}-oligomer-C_{60} triads. Electrochemical and photophysical properties of these novel bis(fullerene)-OPE/OPV hybrids were characterized by cyclic voltammetry, UV-vis and emission spectroscopic methods. Our results clearly indicate that properties of these triad molecules are subject to the influence of the π-bridge to varied degrees at the ground and the excited states. Moreover, evidence for rapid photoinduced electron and/or energy transfer was found in our excited state characterization. The photophysical behavior revealed suggests that these C_{60}-oligomer-C_{60} molecular dumbbells offer not only a new model system for in-depth understanding of the structure-property relationship but also a class of potentially useful candidates for advanced C_{60} based optoelectronic materials.

Chapter 8 - Organic-inorganic hybrid materials are nanostructured materials whose inorganic phase is synthesised by sol-gel process. Polymer based hybrid materials contain inorganic phase dispersed in polymer matrix on a nanometric level. Sol-gel process is particularly suitable for preparation of these materials since the inorganic phase is formed at low temperatures, preserving polymeric matrix from degradation during synthesis. Similarly to other polymeric nanocomposites, organic-inorganic hybrid materials show improved thermal stability compared to unmodified polymeric materials. This can be ascribed to poorer diffusion of degradation products through the stable inorganic phase, and also to stabilising effect of silicon.

Organic-inorganic hybrid materials based on diglycidyl ether of bisphenol A (DGEBA) and 3-glycidyloxypropyltrimethoxysilane (GLYMO) were prepared, with a poly(oxypropylene) diamine, Jeffamine D230, as a crosslinking agent. Previous research on these hybrid materials has shown that epoxy groups of GLYMO are able to react only with primary hydrogen atoms in amine groups of D230. Therefore two sets of hybrids were prepared, one with stoichiometric quantity of amine (1 mole of amine groups per 2 moles of epoxy groups) and the other with doubled quantity of amine in relation to epoxy groups of GLYMO. Thermal degradation of all prepared hybrid materials, as well as unmodified epoxy resin, was investigated by combined differential thermal analysis and thermogravimetric analysis (DTA-TGA) in inert atmosphere of nitrogen and oxidative atmosphere of air. It was determined that although increased content of amine improves conversion of epoxy groups and therefore degree of organic crosslinking in the hybrid materials, their overall thermal stability is decreased. Hybrid materials with increased amine content have lower degradation onset temperatures and lower residue at 1000 °C. This can be attributed to unreacted amine groups that remain in the material, and to lower density of inorganic network. Atmosphere has very little influence on the main degradation step, which probably proceeds in self-generated atmosphere.

Chapter 9 - Some systems for investigation of the biosynthesis, biotransformation and biodestruction processes with a simultaneous chemical or photochemical reaction have been developed. It is expected that such hybrid processes will improve biotechnological performances and a variety of the biotechnological methods for pollutant's destruction.

Chapter 10 - Oxidation behavior of various copolymers and hybrid materials is discussed on the basis of recent references in the field. Structural and functional changes induced by oxidation ageing are discussed in some significant cases. The influences of chemical structure, additives or impurities are also discussed as possible factors influencing the photo-oxidation susceptibility of nanocomposite materials. The study revealed that the oxidation and ageing of nanocomposites and hybrids is tentatively explored, in spite of the interest regarding long term exploitation of these materials.

In: Research in Hybrid Materials
Editors: Simon J. Brunner and Julian W. Egger
ISBN 978-1-60456-718-2

Chapter 1

NONIONIC ORGANIC-INORGANIC HYBRID MATERIALS

Cecilia Savii*[1], *Claudia Adriana Ionescu, Mihaela Ioana Popovici, Corina Stefania Enache and Ramona Radu
Institute of Chemistry Timisoara,
24 Mihai Viteazul Blvd., 300223 Timisoara, Romania

ABSTRACT

The state of the art survey of research findings on materials with nanoscale structure, including nonionic organic-inorganic hybrids, revealed that there is a large number of method producing such structures, that they posses a number of unique properties as a result of the synergy of their both organic and inorganic counterparts properties. This review describes the main general and specific sol-gel synthesis routes and also the classification of nonionic silica hybrids following several criteria. A special attention will be paid on the influence of the precursors and synthesis conditions to design nanostructured hybrid materials and hybrid silica nanocomposites ordered architecture or/and porosity. Classic and new investigation methods and several possible applications with significant technological importance including sensors and smart materials will be presented. At the present time, this very interdisciplinary scientific area is open for new ideas from all disciplines and therefore a promising future can be foreseen.

Keywords: Nonionic silica hybrid nanomaterials, nonionic silica hybrid nanocomposites, organic, inorganic polymers, sol-gel, xerogels, aerogels, sonogels, hydrolytic and non-hydrolytic sol-gel routes, hydrolysis ratio, network modifiers, inteligent (smart) materials, fractal character, Small angle neutron scattering techniques, rheology, X-ray diffraction, MALDI-TOF, NMR, XANES, EXAFS, EPR, SEM, TEM, AFM.

[1] e.mail: cecilias@acad-icht.tm.edu.ro,Tel.: +40.256.491818; Fax: +40.256.491824

1. Introduction

Due to the exciting range of properties demonstrated by hybrid materials (belonging to nanomaterials field) in recent years, much attention has focused on the novel combination of organic species with inorganic minerals and networks [1-2].

The organic functional groups incorporated into the inorganic silicate structure, both of these counterparts being neutral (uncharged from electrical point of view), results in the fact that the obtained non-ionic organic-inorganic hybrids possess both mechanical strength and glassy optical transparency. In fact, the organic-inorganic hybrid solids may be considered as a generic concept that comprises all materials within it can be found at least one association between an organic group and an inorganic network. From the beginning, it is important to establish that inorganic polymer which will be referred in the present work is Si-O-Si network.. Silicon-based hybrid solids is a category of materials of growing interest because of their great potential in synthesis and possible applications, but it is important to make a distinction between polyphase nanocomposites and monophase nanostructured hybrid materials [1-11].

The sol-gel nanostructured hybrid materials usually are obtained by hydrolytic or non hydrolytic polycondensation of a molecular precursor in which n-functionalised (with respect to hydrolysis reaction) silica precursors ($n \leq 4$) are covalently bound to the organic moiety [1]. In other words, these solids correspond to monophase hybrid inorganic–organic hybrid materials.

The hybrid nanocomposites are obtained by hydrolytic polymerization of a metal alkoxide previously mixed with organic molecules (or polymers). These solids can be separated into two phases by washing with an organic solvent and they correspond to the heterophase hybrid materials. After Mauritz and coworkers [12] there are 4 distinct areas in which these materials, organic-inorganic nanocomposites, can be classified: Organic-Inorganic Monomers, Organic-Inorganic Polymer Associations, Organic-Inorganic Interpenetrating Networks, and Organic-Inorganic Telechelic Polymers.

Concerning the synthesis methods, used in order to obtain non-ionic organic-inorganic hybrid solids, they can be classified as: General Methods and Morphologically Specific Methods [1].

The relatively new concept of "Chimie Douce" or "Soft Chemistry" has been introduced in the field of Material Science by Rouxel and Livage [13,14].This method of obtaining solids represents a breaking down of the preparation of materials and this concept, which is based on inorganic polymerisation, has opened the field of Materials Science to molecular chemistry. This methodology can be considered a General Method in the mean time encompassing numerous specific routes or methods based on kinetic control of the solid synthesis. One commonly used approach of „Soft Chemistry" is the sol-gel methodology [15,16]. It corresponds to an inorganic polymerisation, exhibiting a great complexity of the hydrolytic or non hydrolytic polycondensation of alkoxide precursors such as $Si(OR)_4$ or derivate functionalised compounds by showing possible competitive reactions occurring at the same time.

Beyond the compositional and microstructure complexity, an important aspect of the nanophase materials is the large fraction of their atoms that reside in grain boundaries, which results from their ultrafine grain size. Therefore the properties of nanophase materials can be expected to be strongly influenced by the nature (atomic and electronic structure) of their internal boundaries, because of the high number density of these boundaries and the significant fraction of atoms associated with them [17]. As a consequence, the complex and unique properties of such materials, and their functionalities will be a challenge to be investigated and mathematically simulated in order to be able finally to design and develop materials with desired characteristics. Fortunately, nanoscale structured solids can be studied by using almost all classical method of investigations. Nevertheless unusual methods became more and more of actuality, especially those destined to provide information about the size, shape and orientation, peculiar structure (like fractal character) [18,19] or magnetic (monodomain, superparamagnetism, giant coercitivity) behavior [20-25] of some component of the sample at nanoscale level of microstructure, in 10-1000 Å range. An example is the small angle scattering (SAS) representing collective name given to the techniques of small angle neutron (SANS), X-ray (SAXS) and light scattering (SALS, or just LS). In each of these techniques, radiation is elastically scattered by a sample and the resulting scattering pattern is analyzed. Transmission electron microscopy, TEM [26], provide also information concerning the mass density distribution of aggregates at different length scales and changes resulting from aggregate growth or aging [27]. The interpretation of scattering data, however, is not a straightforward process. Computer simulation is a valuable tool in studying the effect of changes in different process parameters on both the aggregation and aging processes in sol-gel systems. Eventually the combination of experimental and simulation gives an insight into the effect of process parameters on the chemistry and physics of these important materials.

The hybrid materials are important by themselves and they also can be used as host for other compounds having many applications: magnetic, optics, electrooptics, environmental sensors, smart materials.The possible applications of nonionic organic–inorganic hybrid materials are manifold and the future of these materials appears bright. Especially, the tremendous possibilities of different combinations between inorganic and organic species, in form of both monophasic and polyphasic hybrid solids, allow inducing properties which cannot be obtained by traditional composite materials [28]. In this review chapter, only a limited amount of presently available information can be discussed. But one cannot forget about the most exciting and almost unexplored application of hybrids in designing smart materials. Corriu pointed out that in this field the long term target is to reach intelligent (smart) materials [29,30]. In the present chapter it is envisaged (even if briefly and partially) the great effort of the material science scientists dedicated in the last decade to synthesis, characterization and modeling of nonionic hybrid materials based on silica inorganic counterpart.

1.1. Definitions

Numerous silica or/and siloxane based hybrid organic-inorganic materials have been intensively developed in the past few years because they present several advantages for designing materials for a plethora of applications (versatile and relatively facile chemistry, easy shaping and patterning, materials having good mechanical integrity and excellent optical quality). The hybrid organic-inorganic materials which are composed from neutral (electrically uncharged) counterparts and there are no electrostatic attraction between them often are called nonionic organic-inorganic hybrid materials [31]. Saegusa defines hybrid materials as blends at molecular level dispersion between an organic polymer and metal oxide (mainly silica). These blends are called "organic-inorganic polymers hybrids". The dimension of dispersion of "hybrids" is much smaller than that of the so-called "composites". They are called by a general term of "composites", which are usually heterogeneous materials. The "composites" and "hybrids" differ from each other in respect to the dimension of dispersion [32]. Functional hybrids are defined also as nanocomposite materials lying at the interface of organic and inorganic realms, whose high versatility offers a wide range of possibilities to elaborate tailor-made materials in terms of chemical and physical properties [33]. In fact the term of "hybrid" is generally used to describe organic/inorganic materials [34].

Sanchez and collab. defined a hybrid material as any organic-inorganic or bio-mineral system in which at least one of the components, organic or inorganic, is present with a size scaling within tens of nanometers. Components making up the hybrids could be molecules, oligomers or polymers, aggregates and even particles [35]. The organic/inorganic hybrid materials combine at least two components, organic and inorganic, respectively, that bring not only their specific properties to the hybrid, but also the combination results in synergetic properties, difficult to reach by each component material as a single part [36]. Although such materials are macroscopically homogeneous, their properties reflect the nature of the chemical building blocks from which they are composed [37]. To a certain extent, these hybrid materials combine the most important properties of their constituents, like high transparency (glass-like), low processing temperatures (polymer-like), sufficient thermal stability (silicone-like), and are easily accessible because of an unique availability of the respective precursors (commercially available metal alkoxides and organo(alkoxy)silanes as well as nanoparticles) [34].

1.2. Classification

Many attempts were made to classify organic/inorganic materials synthesized by sol–gel process. The chemical structures of the hybrids are considered the main criteria for the basic classification. There can be find many types of organic/inorganic nonionic hybrids considering also other criteria of classiffication.

Surveying the literature results one of the most relevant classification of hybrid materials is that introduced by Sanchez and Ribot [3,38] referring to:

a) The nature of chemical interactions between the constituents (organic-inorganic counterparts) in other words in function of the involved chemical bonds;
b) The organic/inorganic components ratio;
c) The effects of polymer association types related to organic-inorganic moieties
d) Compatibilities using certain preparation methods;
e) The porosity type involved within solid network;
f) The morphological combinations and destination for use (e.g. hybrids coatings).

a) The nature of chemical bonding between organic and inorganic species

Following this criterion the different nonionic organic-inorganic hybrids can be classified in two broad families:

Class I corresponds to hybrid systems in which weak interactions (such as, van der Waals forces, hydrogen bonds and hydrophilic–hydrophobic balance) are created between organic and inorganic phases. In this cathegory are included also the hybrid nanocomposite without chemical interactions between the two components [5,28,39]. Class I includes hybrid systems where one of the component (organic, biologic or inorganic), which can be molecules, oligomers or polymers, is entrapped within a network of the other component. These materials are generaly synthesized by carrying out the hydrolysis and condensation of the inorganic compound, i.e., the formation of the inorganic network, in the presence of the organic compound or by polymerizing organic monomers in porous inorganic hosts.

"Nanocomposite" bi- or multiphasic hybrid materials [1] fall in this category, because they often refer to materials where organic and inorganic constituents are not covalently bound. The forces acting between the constituents vary from weak to relatively strong. The concept of "nanocomposite" was adopted for ceramic materials by Roy and Komarneni and coworkers about two decades ago [40,41]. A nanocomposite is a material that contains a reinforcement component in the form of one or more ultrafine phase with dimensions less than 100 nm. In mechanical field, it is worth recalling that Yajima work [42] preceded the concept of nanomaterials from some decades.

Class II corresponds to hybrid organic-inorganic compounds where both organic and inorganic components are bonded through strong covalent chemical bonds. This approach requires molecular precursors that contain a hydrolytically stable chemical bond between the elements that will develop the inorganic and the organic moieties during sol-gel processing [5,29,34,43]. In this class are included "ORMOSILs" (organically modified silicates) [44,45] or "ORMOCERs" (organically modified ceramics) [46]. Another given name is "CERAMERS" (ceramics modified with polymers) [47]. All of these terms refer to organic/inorganic materials where the constituents are covalently bonded to each other [48].

Corriu considers that it is an important task to make a distiction between "hybrid nanocomposites" (Class I) and the "nanostructured hybrid materials" (Clas II) [1]. In fact, the frontier between the two classes is not always simple and one can eventually have hybrids systems with class I and class II characteristics. Typical examples of such case are hybrids for optical applications made by encapsulation of organic chromophores within a hybrid matrix which, according this clasification criterion belongs to class II [49].

b) The components ratio

The organic/inorganic component ratio can be an important criterion since its variation leads to hybrid materials with distinctive physico-chemical behaviors and profoundly different properties. Two kind of hybrid materials can be distinguished :

- *Organics or inorganic doped systems.*They are usually based on one major phase which contains a second one in relatively low amounts (generally less than 1%).
- *Organic-inorganic systems or hybrids.* Here the fraction of each component in the system is of the same order of magnitude (eg. higher than 10%).

Following the same criterion, it was also found in ref. [50] a classification of hybrid materials according to Allock [51]:

- Inorganic-Organic Polymers (IOP). They have inorganic elements in the main chain/network and may also have organic side groups. When the organic groups are reactive, additional organic crosslinking/polymerization is possible. As an example of sol-gel derived such a material is indicated ORMOCERs.
- Organic-Inorganic Polymers (OIP). These materials have C-atoms in the main chain/network and may contain inorganic elements in the side groups connected to organic network. An example of sol-gel derived such a material is indicated as CERAMERS [52].

Sanchez concludes that components ratio, by itself, is not a pertinent criterion for classification [49].

c) Effects of polymer association types related to inorganic-organic moieties compatibilities using certain preparation methods

Considering this criterion of classification, four hybrids cathegories can be identified: Organic-Inorganic Monomers, Organic-Inorganic Polymer Associations, Organic-Inorganic Interpenetrating Networks (IPNs) and Organic-Inorganic Telechelic Polymer [12].

Organic-Inorganic Monomers are molecules containing a metal core such as a silicon bound to reactive alkoxy groups and/or organic groups. These monomers, when assembled together, form organic-inorganic siloxane materials. There are two distinctive types of organic-inorganic monomers [12]:

- Organically Modified Silicate monomers (e.g. ORMOSILs);
- Silsesquioxane-Type monomers.

ORMOSILs are derived from tetrafunctional (toward hydrolysis) silicone alkoxides such as tetraethylorthosilicate (TEOS) by organic modification at any one of the reactive alkoxide arms [12,44,45].

In the case of silsesquioxane-type monomers, the organic substituent behaves as a spacer unit (bridging functionality) that resides between two reactive di- or tri-alkoxysilane structures. Such hybrid materials can be synthesized to have almost any kind and number of organic spacer functionalities [53,54]. By removing the spacer units after the network formation, well defined pores corresponding to the organic spacer are built into the sol-gel network [12].

Organic-Inorganic Polymer Associations are those in which there is no covalent bonding between the organic polymer and the inorganic network but the systems exhibit characteristics similar to systems which are covalently bound. This synergistic relationship can occur through two types of associative behavior: hydrogen bonding, and association/coupling equilibrium. Various polyesters, polycarbonates, polyacrylates, and polyamides possess moieties such as carbonyls, ethers, and amine groups which have free electrons available for hydrogen bonding [12]. Hydrogen bonding of the sol-gel with these polymers retards severe phase separation during network densification. There are also copolymers, which does not have the capability of interacting with an inorganic phase. In that case the hydrogen bonding polymer, in small quantity is used as a compatibilizer with the other polymer system which is of interest for inorganic modification. An association/coupling equilibrium occurs between polymer systems which implies hydrogen bonds. Here there is a dynamic equilibrium between hydrogen bound groups and a covalently bound inorganic network, producing -C-O-Si- bonds [55].

Organic-Inorganic Interpenetrating Networks (IPNs) is any material containing two polymers, each in network form. The three conditions for eligibility as an IPNs [12,56] are presented:

a) the two polymers are synthesized and/or crosslinked in the presence of the other;
b) the two polymers have similar kinetics;
c) the two polymers are not dramatically phase separated.

IPNs that have only one polymer crosslinked (where the polymers are synthesized separately) or where the polymers have vastly different kinetics are still considered to be IPNs. IPNs are distinguishable from blends, block copolymers, and graft copolymers in two ways: (1) an IPN swells but does not dissolve in solvents, and (2) creep and flow are suppressed [56]. These systems differ mainly because of the number and types of crosslinks existing in the system, which results in several kinds of IPNs architectures:

- A non-covalent semi-IPNs is one in which only one of the polymer systems is crosslinked;
- A non-covalent full IPNs is one in which the two separate polymers are independently crosslinked;
- A covalent semi-IPN (similar to a non-covalent IPN) contains two separate polymer systems that are crosslinked to form a single polymer network. The covalent semi-IPNs are developed with organic-inorganic composite materials [12]

Organic-Inorganic Telechelic Polymers. A telechelic polymer is one containing one or more functional end groups (hydroxyl or carboxyl) that have the capacity for selective reaction to form bonds with another molecule. Composite materials or block copolymers can then be made through these endgroups by post-reactions [12,57].

d) The porosity type involved within solid network

Accordingly to the IUPAC classification, if hybrids are considered as members of the porous materials familly, they can be: *(1) microporous*, having pore sizes below 2.0 nm, *(2) macroporous*, with pore sizes exceeding 50.0 nm, and *(3) mesoporous*, with intermediate pore sizes between 2.0 and 50.0 nm [58,59].

e) The morphological combinations and destination for use

The final form of hybrid material can be as bulk monoliths, thin films (coatings), fibers or powders. After Keijman, organic-inorganic hybrids can be distinguished in different morphological combinations, which include the following categories:

- *category 1*, the inorganic matrix, where organic materials are embedded in an inorganic polymer. An example of this category is obtained by the incorporation of small quantities of organic products into an inorganic matrix.

- *category 2*, the organic matrix, where inorganic materials are embedded in an organic polymer. Some of the oldest and well-known organic – inorganic hybrids of this category are used in the paint industry by the dispersion of inorganic pigments into organic binders;
- *category 3*, the interpenetrating network, where inorganic and organic polymeric networks are independently formed without mutual chemical bonds. The incorporation of inorganic macro-molecular sol-gel networks into organic polymer structures is an example of this category;
- *category 4*, true hybrids, where inorganic and organic polymeric systems with mutual chemical bonds are formed. Specifically these true hybrids have been explored successfully for the development of new binders for use in high technology organic-inorganic polysiloxane hybrid based coatings.

Depending on the paint market segments involved, different technologies are used. For the protective coatings industry, for example, the interpenetrating networks of inorganic and organic polymers and the chemically reacted hybrids networks are of major importance [60].

With respect to organic-inorganic materials having specific optical functions, the chemical structure of the hybrids is considered also by Wojcik and Klein as a very important criterion for classification, in order to distinguish different types of organic/inorganic gels. The authors [48] are also mentioning several other criteria of classification, such as preparation methods, and the nature of the links between organic and inorganic constituents.

2. Synthesis Methods

The morphologies and properties of the sol-gel resulting materials are controlled by the reaction conditions as well as the used precursors [28]. Because of their importance and for more clarity of the description, the precursors section was lightly extended within the nonionic hybrids chapter, before the synthesis methods presentation.

2.1. Precursors

Soluble oligomers of silicic acid are often used as a *silica source* (inorganic precursor) to obtain hybrids in sol-gel process [61,48,62,63]. Chemists currently use alkoxide precursors $Si(OR)_4$, where R is an alkyl group, rather than silicic acid $Si(OH)_4$ as silica polymeric network source [64,65]. Starting from silica alkoxides, silicic acid is first formed upon hydrolysis, by adding water:

$$Si(OR)_4 + 4H_2O \leftrightarrow Si(OH)_4 + 4ROH \quad (1)$$

Tetrafunctionalized silica alkoxides (with respect to hydrolysis reaction), Figure 1, and/or n-functionalized alkoxides ($n<4$), Figure 2, are generally used as precursors in a sol-gel process.

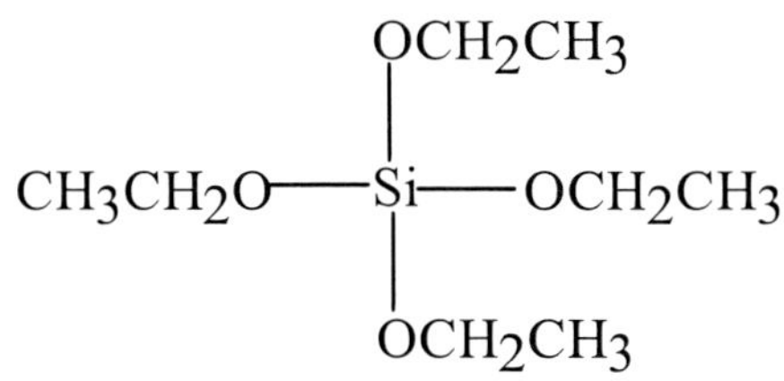

Figure 1. Tetraethylorthosilicate (TEOS)

Tetrafunctionality of alkoxide precursor can be modified by replacing an organic radical (C-Si) at any one of the reactive alkoxide arms as shown in Figure 2. Here, *n* is representative of the number of organic moiety connected to the silicon molecule and *f* is representative of the number of reactive alkoxy groups connected to the silicon molecule.

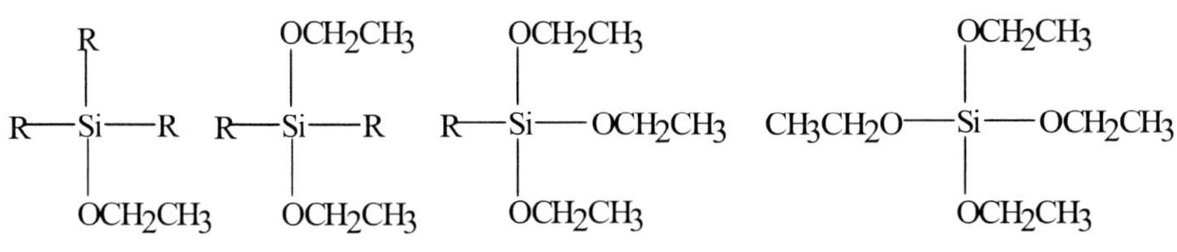

Figure 2. The $R'_nSi\ (OEt)_f$ Functionality Spectrum

For a mono-functional silicon alkoxide (f = 1, n = 3), the monomer „terminates" the polymer chain since there is only a single reaction site on the molecule. A di-functional silicon alkoxide (f = 2, n = 2) behaves as a „bridging" agent, connecting molecules in a linear fashion. A tri-functional silicon alkoxide (f = 3, n = 1) behaves as an "crosslinker", allowing for branching in the network. A tetra-functional silicon alkoxide (f = 4, n = 0) behave as a "networking" agent, allowing for complete connectivity between all functional groups of the molecule.

The *organic group* R' can be a simple organic group such as alkyl or aryl groups. R' groups can have different functions and generally acts as a network modifier. Few examples are given, as follows:

Methyl alkoxysilane, $CH_3Si(OR)_3$ is often used to provide some hydrophobic character to a silica matrix or to modify its mechanical properties. For example, the use of a linear ''rigid'' precursor, as $R_n'Si(OR)_{4-n}$ where R' = phenylene or a ''flexible'' one $R_n''Si(OR)_{4-n}$ where R"= saturated aliphatic long chain, has a drastic influence on the properties of the hybrid solids [66].

Phenyl siloxanes are sometimes preferred to methyl siloxanes, to increase the thermal stability, glass transition temperature, or refractive index. Moreover, they can also act as a prestructured starting material for further chemical modifications [67].

Alkyltrimethoxysilanes with various alkyl chain lengths (C_nH_{2n+1}-$Si(OMe)_3$; C_nTMS, n = 6,8,10,12,14,18) have been known to form spontaneously organized assemblies based on the amphiphilic nature of hydrolyzed species containing both hydrophilic silanol groups and hydrophobic alkyl groups, by cohydrolysis and polycondensation of the C_nTMS and TMOS where M = CH_3 [68].

It opens the possibility of structural and morphological variations of hybrids by utilizing both the *molecular-assembling* property of long-chain *alkyltrialkoxysilanes* and the network-forming ability of *tetraalkyloxysilanes.* The functionality of the trialkoxysilane can be used for the formation of a covalent linkage between the polymer and the silica species. For example, aminophenyl-trimethoxysilane was used as a compatibilizer which reacts with the phthaloyl chloride end-capped polyamide [62]. R' can also be a functional group used to bring some new properties to the oxide matrix. R' being an organic chromophore are now currently used to bring optical properties into hybrid photonics materials. Several studies show that the optical response increases significantly when the organic dye is covalently bound in the class II of hybrid sol-gel matrix rather than simply embedded within an oxide network [66]. Organic polymerization of R' can be induced by photochemical or thermal curing before or after the sol-gel reaction.

Bifunctional silanes R'R''$Si(OR)_2$ can also be used as precursors. Upon polymerization, these organoalkoxysilanes form cyclic species or linear chains, and therefore are not able to crosslink and form networks themselves. They can be used with tetrafunctional silanes (TEOS) to incorporate desirable functionality into the hybrid network, such as hydrophobicity (R', R'': methyl, phenyl or fluoroalkyl chain) [69]. The organic R can form a bridge between two Si atoms. Various aliphatic or aromatic molecules have been used as organic spacer between two silicon atoms [48].

Metal alkoxides modified by unsaturated ligands. Homo-or copolymerisation reactions involving a polymerizable moiety Z are required for extended organic arrays and access to hybrid materials. If covalent association between the networks is a goal, the M-Z linkage should resist to processing and be thermodynamically and kinetically stable. The approach used for silanes is valid in certain cases, but in some cases the reactivity of metal-carbon bonds makes it useless for certain metals.

A better strategy is to link the polymerizable functionalities to the metal via O-donor ligands forming M-Z bonds which are less susceptible to hydrolysis than metal-alkoxide ones. Chelating or bridging-chelating ligation is preferable. The hydrolytic stability of the metal to β -diketonate bond is generally higher than that of the metal carboxylate one but functional β -diketones are less available than functional carboxylic acids. Carboxylic acids (acrylic or methacrylic), β-diketones such as vinylacetylacetone, represent reasonable choices for unsaturated O-donor moieties. Accessibility of the polymerisation sites, the nature of the unsaturated functionality (acrylate ones are among the best for polymerisation) are also of importance for reactivity [70].

New classes of organically functionalised precursors are often mentioned in literature, such as, organically modified silicate monomers and silsesquioxane-type monomers. These monomers are typically used in the synthesis of controlled porosity materials. *ORMOSILs* when assembled together, form organic-inorganic siloxane polymeric materials [50]. In the mid -1980s started the new developpments in the field of ORMOSILs chemistry [44,71-77].

ORMOSILs of general formula $(RO)_4$-xSiZ_x are derived from tetrafunctional silicone alkoxides. They allow tailoring of porosity, of surface properties, of refractive index, of coatings thickness, to improve mechanical properties, etc. ORMOSILs are based on the stability of the Si-C bonds (Z is a functional alkyl group) versus the hydrolysable Si-OR ones.

Many reviews have emphasized the use of those derivatives and these aspects will not be developed here [70].

CERAMERS monomers represent also a hybrid system in which several precursor type are combined [16]. Schmidt and coworkers [78] have developed silicate products with unique properties by combining tetraalkoxsilanes with alkyl-substituted and organofunctional alkoxysilanes: $Si(OR)_4 + R_2Si(OR)_2 + YR'Si(OC_2H_5)_3$. R is an alkyl group, R' is an alkylene, and Y is an organofunctional group such as $-(CH_2)_3NH_2$, $-(CH_2)_3S(CH_2)_2CHO$, etc. If Y is a polymerizable ligand such as an epoxide, it is possible to form an organic network in addition to the inorganic one [16,79]. In principle such hybrid systems allow an unlimited number of chemical and structural modifications to be performed. Choice of specific precursors may be made on the basis of solubility or the thermal stability of the organofunctional substituents.

Silsesquioxane-Type monomers are a class of silicon-oxygen compounds based on a fundamental $Si_8O_{12}T_8$ cube unit. Silsesquioxanes can be made by the hydrolysis of alkyltrichlorosilanes in acetone [80]. Klemperer and coworkers [81,82] have pioneered the use of specific oligomers as "molecular buildings blocks" for the construction of silicate networks with precisely controlled architectures, such as: hexamethoxydisiloxane, octamethoxy-trisiloxane and a methoxylated, cubic octamer $(Si_8O_{12})(OCH_3)_8$ [83]. Later, different mesoporous silica oxide with ordered structure in the form of hexagonal (denoted as MCM-41 type architecture) [84-91] cubic (denoted as MCM-48-type architecture) [86-88], and lamella (denoted as MCM-50 type architecture) [88,89] were obtained and used as precursors building blocks for desired sofisticated architectures. For example, an ordered organo-SiO_2-surfactant mesophase containing a covalently-linked chromophore was synthesized with MCM-41-type architecture by template-directed co-condensation of TEOS and 3-(2,4-dinitrophenylamino) propyl(triethoxy) silane [91].

Since the pioneering research of Schmidt [44] and Wilkes [92] in 1985, hundreds publications on this type of ORMOSILs have appeared [93]. The reason for such a level of activity is presumably twofold. The first reason is that, since materials constitute a new family, the synthesis of more and more new materials and the understanding of the reactions in solution and structures of the new hybrids are sufficient justification for the research. The other reason is these new materials are predicted or expected to have unique useful properties and/or to be less expensive than existing materials to produce [66].

A variety of polymerizable alcohols and polymerizable cosolvents can be used as well to produce composites with different mechanical and physical properties [94]. Organic functionalization of silicates permits precise control over the surface properties, modification of the hydrophilic/hydrophobic balance of the surface, alteration of the surface reactivity, protection of the surface from attack of undesired external chemical agents, modification of the bulk properties of the materials and, at the same time, stabilizing the materials towards hydrolysis.

The organic groups introduced into the sol–gel precursor can fulfill two functions:

I. modification of the inorganic network, resulting in improved compatibility between the two phases, and/or
II. the formation of a strong linkage between the organic and the inorganic phase via covalent bonds.

The amount of incorporated organic as well as inorganic moiety can be varied over a wide range. In addition to the sol–gel processing parameters, the morphology of the hybrid materials obtained by the interpenetration of polymers and inorganic moieties is dependent on polymer variables such as the molecular weight, the presence and extent of polymer functionality and shape, and the solubility of the polymer in the sol solution. A typical application area for hybrid materials are optical systems that require homogeneous and transparent materials which cannot be obtained from mixtures that are already macroscopically phase separated. This is often the case if hydrophobic polymers are mixed with sol–gel solutions. Therefore, organic polymers with functional groups that have specific interactions with compounds created in the sol–gel process, e.g. hydrogen bonding to residual silanol groups on the formed silica, have to be chosen. A broad variety of organic polymers with hydrogen bonding ability such as poly(2MO$_x$), poly(VP), poly(DMAA), poly(VA), poly(MMA), poly(VAc), polyamides, polyethersulfones [95-97] and hydroxy end-capped polysiloxanes were incorporated into sol–gel reaction mixtures [98-102].

Due to the interactions between polymer and silanol groups generated during the sol–gel process, a macrophase separation is avoided, and the resulting materials have a high degree of homogeneity and optical transparency [103,104]. In a hybrid material formed from PVAc and phenyltriethoxysilane it was observed that the glass transition temperature (Tg) of the polymer increases in a similar way as it does with TEOS as precursor [28,105]. This behavior probably has its origin in the different extend of network formation with the two silicate precursors. In the case of tetraalkoxysilane, depending on the reaction conditions, a highly crosslinked silica network can be formed, while in the case of trialkoxysilanes a lower degree of crosslinking is usually observed. Additionally, the functionality of the trialkoxysilane can be used for the formation of a covalent linkage between the polymer and the silica species. For example, aminophenyltrimethoxysilane was used as a compatibilizer which reacts with the phthaloyl chloride end-capped polyamide [101].

In other studies, the interfacial bonding between the phases of hydroxypolybenzoxazole and sulfopolybenzobisthiazole polymers and the silica, was improved by the use of isocyanatopropyltriethoxysilane and N,N-diethylaminopropyltrimethoxysilane, respectively [106,107].

Another way to enhance the homogeneity of the hybrid materials is the use of *the polymeric compatibilizers* such as copolymers of PSt and P4VPh which help to reduce the interfacial tension in incompatible systems [108]. Applying this method the particle size of the silicate species can be significantly reduced leading also to a better adhesion between the organic polymer and the inorganic moieties [50]. Hybrid polymers can be composed of *structural units* which span the whole range of more inorganic glass like up to the composition with a high content of organic networks. If all structural units are connected via strong covalent bonds, the building principle of these hybrid polymers is based on different types of precursor molecules. These precursors can be classified by their network forming or modifying roles [109]. Various polymerization, polycondensation or crosslinking reactions are possible.

Haas and Wolter [109] present four different types of precursors which can be used for the synthesis of hybrid polymers (Class 2):

type 1: inorganic network formers (alkoxides);

type 2: inorganic network formers with organic nonreactive modification. Due to the hydrolytic stability of the Si-C bond a wide variety of organic groups are available for the modification of the inorganic networks;

type 3: forming inorganic networks with reactive organic modification for crosslinking/polymerization reactions or organic monomers/polymers with silylated end groups for crosslinking/co-condensation via inorganic Si-O-Si bonds. These precursors are key elements for hybrid polymers;

type 4: organic monomers which can react via chemical crosslinking or polymerization reactions with the modified metal alkoxides.

Beside the importance of precursors selection it is important to mention once again that the parameters employed for the sol–gel process have a crucial influence on the morphology of the inorganic species in the resulting material which directly influences its properties.

2.2. General Sol-Gel Synthesis Methods

Over the past few years, there has been growing interest in ''Chimie Douce'' or "Soft Chemistry" [13-15,110-113] including sol-gel processing [16]. The sol-gel process has been known for 150 years, but only in the last 15 years it has been exploited for the formation of organic-inorganic hybrids, where in many cases the phases co-exist on a nanometer size scale [12,114-116]. Sol-gel chemistry has been investigated extensively since the mid-1970's, when sol-gel reactions were shown to produce a variety of inorganic networks from metal alkoxide sols and solutions [16,117]. A sol is a colloidal dispersion of particles in a liquid. A gel is usually a substance composed of a continuous solid network encompassing a continuous liquid phase [16,75,118].

According to Corriu the sol-gel methodology is an alternative to the classical thermodynamic routes for the synthesis of solids. It allows materials to be obtained by ''step-by-step'' chemistry starting from molecular precursors at low temperatures. The preparation of the solids through ''Chimie Douce'' processes is the natural bridge between molecular and solid state chemistry [1]. This method makes possible the preparation of both nanocomposite and nanostructured hybrid organic–inorganic materials (Class I and Class II) under smooth conditions [6,11,119,120].

The combination of organic species and inorganic minerals and networks has attracted much interest in recent years [121-123] for a number of good reasons. As a consequence, much work has been focused on the preparation of sol–gel hybrids, such as the combination of inorganic oxides (particularly silica) with polymers [123,124] and the organic modification of silica (ORMOSIL's) starting from alkylalkoxysilanes [8,125,126]. The majority of these sol-gel methods use the well established *hydrolytic sol-gel route* to the oxide. Corriu's group [127,128] and other researchers [113,129] have reported the synthesis of inorganic oxides also via *non-hydrolytic sol–gel route* which involves the reaction of a „metal" halide with an oxygen donor such as an alkoxide, ether, alcohol, etc.

As a *general synthetic process* the sol-gel can undergo via hydrolytic or nonhydrolytic routes, which briefly will be presented as follow.

2.2.1. Sol-Gel Hydrolytic Route

Hydrolytic sol-gel route can be described as a versatile process initially used for the preparation of inorganic materials such as glasses and ceramics of high purity and homogeneity. It involves the transition of a system from a liquid sol (or solution) into a solid gel phase and has been the subject of several books and reviews [16,130-132].

The sol-gel process can ordinarily be divided into the following steps:

- forming a solution, or sol,
- gelation,
- aging,
- drying, and
- densification

The hydrolythic sol-gel route generally starts with an appropriate tetrafunctional alkoxide, which is mixed with water and a mutual solvent, such as an alcohol, to form a solution. Hydrolysis leads to the formation of silanol groups Si-OH. These species are only intermediates as they react further, i.e., condense, to form siloxane Si-O-Si groups. As the hydrolysis and condensation reactions continue, viscosity increases until the sol ceases to flow and form the gel. The overall reactions leading to the formation of a silica gel starting from a tetrafunctional alkoxide, $Si\text{-}(OR)_4$, are:

$$nSi(OR)_4 + 4nH_2O \leftrightarrow nSi(OH)_4 + 4nROH \tag{2}$$

$$nSi(OH)_4 \leftrightarrow 4nSiO_2 + 2nH_2O \tag{3}$$

From chemical point of view, the sol-gel process consists in two main reactions, hydrolysis and condensation. The parameters which naturally controll these reactions, from the beginning to the end, will mainly control sol to gel transition process. The several parameters which influence the hydrolytic sol-gel process, including the precursors activity, water/alkoxide ratio, solution pH, temperature, and nature of the catalyst, solvent and additive (type of organic group(s) used as network modifier(s) [133,16,71,75] will be very briefly considered first.

Precursors. The principal alkoxide precursors, cited as general source of sol-gel derived materials are the n-functionalised alkoxides ($n \leq 4$) as it has been described in the previous section.

Water-to-Alkoxide Ratio. The hydrolysis ratio h ($h = [H_2O]/[Si(OR)_n]$ allows to control the extent of hydrolysis. The effect of the water-to-alkoxide ratio for the sol-gel process is such that as the ratio increases, so does the SiO_2 content of the gel; for complete hydrolysis to occur, there must be at least one mole of water for every alkoxide group. Some researchers have gone further to state that if more than one mole of water per alkoxide group is used, the reverse reaction, reesterification, will occur faster than the forward reaction [16,75]. However, the experiments of McCormick and coworkers [134] conducted over a wide range of water-to-TEOS ratios, indicates no correlation between the water:alkoxide and complete hydrolysis. Another problem consists of deciding how much water should be initialy added given that water is generated in situ through condensation reactions. It is possible, from different reasons, that the finally resulted water content to be improper. This makes calculation of any specific amount of water for addition difficult.

Type and Amount of Catalyst. Hydrolysis and condensation reactions of most inorganic alkoxides can occur over a short period of time without catalysis because of the very quick rates of reaction. Alkoxysilanes, however, hydrolyze slowly, making catalysis an important reaction factor. Catalysis is typically accomplished through the addition of acid or base. Relative rates of reaction for various pH are very different for hydrolysis and condensation reactions. It is the variability of the rates of the two reactions that must be taken into account when trying to control the outcome of the reaction. pH-rate profiles for the hydrolysis and condensation reactions have been reported [16,75,135]. Acid catalysis, promotes the rate of the hydrolysis reaction. In the rate-determining step, oxygen from the water attacks the silicon atom which has low electron density. This results in the possible formation of a pentacoordinate transition state in which partial positive charges are developed. Due to this small charge development, steric effects, such as bulkiness and/or long alkyl substituents, will determine the rate of the reaction. At pH >>> 7, molecular hydrolysis occurs at a slow rate, while molecular condensation occurs at a fast rate. The basic conditions, (eg. alkaline) produce strong nucleophiles via protonation of hydroxoligands [16].

Type of Network Modifier. The hydrolysis rate has to be controled also in order to slow the reaction kinetics when there are some reasons for that. To slow down hydrolysis rates, several strategies were given by Hubert-Pfalzgraf [70]:

- Changing the nature of the organic group R: alkoxides with primary organic groups such as n-butoxides are less sensitive to hydrolysis than secondary ones such as isopropoxides;
- Increase of the metal coordination number thus hindering attack of water and formation of the metal hydroxyl bond, M-OH, necessary for the development of the network;
- Decreasing the functionality of the precursor by partial substitution of the OR ligands by anionic ligands such as carboxylates or β-diketonates leading to M-Z bonds less susceptible to hydrolysis.

These different approaches controlling hydrolysis are often interdependent, for instance replacing R by a functional group increases also the coordination number of the metal. Differential hydrolysis is observed for $M(OR)_{n-x}Z_x$ species (IR data show the retention after hydrolysis of the less hydrolysable M-Z bond). Anisotropy of the network can be promoted as well as porosity. Carboxylates are usually more labile than β-diketonates but hydrogen-bonding can assist the elimination of the latter and thus modify the behavior. Trialkylsiloxide groups R_3SiO are also less susceptible to hydrolysis than alkoxide ones such as butoxides, isopropoxides etc. When the sol is cast into a mold, a wet gel will form, after which the reactions do not stop [16]. Many direct and/or reverse reactions occur during ageing and also a lot of bonds are broken and others rebuilt. Aging is also an important and complex step of the process and it has a significant influence upon resulted materials, but it will not be detalied here. The extensive problem of drying sol-gel materials has really plagued scientists trying to study sol-gel processing [16,136]. In a nutshell, the problem arises from the fact that there is a very large portion of the precursor materials that ends up as a liquid in the small pores inside the inorganic network. During removal of this liquid by evaporation, large stresses build up on the inorganic network due to capillary forces in the gel pores. The capillary pressure is $p_c \propto 1/d$ (d = pore diameter). So, very small pores produce high stresses. As portions of the piece are dried then other portions will crack under stress unless great care is taken to control the evaporation rate. A few techniques to solve the drying problem have been developed, but all still take relatively long duration for drying. The step that involves drying and the products that are produced are:

Xerogels. Xerogels are gels that are dried to the atmosphere at ambient pressure. The gel collapses on itself during this kind of drying. This causes small pores to form that develop very large stresses. These pores can be on the order of 2 nm. The temperature is controlled to accelerate drying but cannot be raised too much because the vapor pressure of the solvent will increase to unacceptable levels and break the gel. These gels are denser then aerogels with about 40 to 60% of the fired density. This process increases the drying time but is the simpler way of drying.

Aerogels. If the liquid is extracted, under a supercritical condition, from wet gel, a highly porous and extremely low density aerogel is obtained. Aerogels are produced by the hypercritical drying of solutions. This is a technique where drying is done at increased pressure and temperature to dry the part above the critical point and therefore avoid a liquid-vapor interface. This is because the liquid and vapor having the same densities. This process, when finished, leaves a structure that usually is 95% porous. One problem with this technique is the expensive high pressure equipment that is used. These materials have a very open structure that has a large shrinkage when sintered to full density. The drying time is on the order of days for this technique.

Cryogels. These are gels that are dried below the triple point of water. This freezes the water and the water sublimes off as vapor. The unfortunate result in this process is that the part cracks into a fine powder since water expands upon freezing.

Alcogels. These are gels that, before drying, pore water content is extracted (exchanged) using alcohols, in ordinary conditions.

Sonogels. Sonogels are gels that before drying are subjected to an ultrasound treatment of about 20 kHz, which cavitates the gel. This treatment is done to ensure better retention of other components into the gel. Ultrasonically assisted sol-gel process will be described more detailed later, in the sol-gel specific processing methods and strategies section.

By varying the processing parameters, materials with different microstructure and surface chemistry can be obtained. Further processing of the sol enables one to make sol- gel materials in different forms. Thin films can be produced on a piece of substrate by spin-coating or dip-coating. With further drying and heat-treatment, the gel is converted into porous or dense ceramic or glass particles. As the viscosity of a sol is adjusted into a proper viscosity range, ceramic fibers can be drawn from the sol.

2.2.2. Sol-Gel Non Hydrolytic Route

In 1956, Andrianov et al. [137] synthesized some alkyl- and arylmodified silicates (and linear polyorganosiloxanes) by reacting various combinations of dimethyldichlorosilane, methylphenyldichlorosilane, phenyltrichloro-silane, phenyltriethoxysilane and phenylethyldiethoxysilane in the presence of iron(III) chloride or aluminium(III) chloride catalyst at 95-100 °C. Examination of the experimental data revealed these reactions to proceed via a heterofunctional stepwise polycondensation to form insoluble, branched organically-modified silicates, when di- and tri-functional precursor combinations were used. No further analyses were performed on these hybrids.

The nonhydrolytic method has been extended [138-140] and further works in the field [141] revealed that the nonhydrolytic sol-gel method is particularly useful in the preparation of luminescent gels. Conventional inorganic gels are notoriously inadequate matrices for trivalent, rare-earth luminescence because of hydroxyl quenching. Methyl-modified silicates with low hydroxyl contents were found to be critical in maintaining the luminescent properties of erbium in these hybrids.

Recently, several groups [114,115,127,128,142] have reported the synthesis of inorganic oxides via a nonhydrolytic sol-gel route. This involves the reaction of a “metal” halide with an oxygen donor such as an alkoxide, ether, alcohol, etc. under nonaqueous conditions. The by-product of this reaction is commonly an alkyl halide.

Sol-gel synthesis of silica hybrids, using a nonhydrolytic pathway, was represented by five general equations:

$$RSiCl_3 + 0{:}75Si(OR')_4 \rightarrow [RSi_{1:75}O_3]_n + 3R'Cl \tag{4}$$

$$0{:}75SiCl_4 + RSi(OR')_3 \rightarrow [RSi_{1:75}O_3]_n + 3R'Cl \tag{5}$$

$$RSiCl_3 + RSi(OR')_3 \rightarrow [R_2Si_2O_3]_n + 3R'Cl \tag{6}$$

$$RSiCl_3 + Si(OR')_4 \rightarrow [RSi_2O_3\text{-}OR']_n + 3R'Cl \tag{7}$$

$$SiCl_4 + RSi(OR')_3 \rightarrow [RSi_2O_3\text{-}Cl]_n + 3R'Cl \tag{8}$$

The use of a single organically-modified precursor leads to an ORMOSIL with only one type of organic modifying group. If both precursors are organically modified, the resulting hybrid can contain two different types of organic modification. The organic groups can be alkyl (groups of variable length from methyl to decyl have been used), vinyl or aryl [121].

The nonhydrolytic sol-gel approach has a number of associated advantages, including:

- the potential to avoid the use of solvents, with their various drawbacks;
- reduction or elimination of residual silanol groups in the final product resulting from the different mechanistic course of the reaction compared with the hydrolytic sol-gel route [143].

Nonhydrolytic condensation reactions can be alternatives to control the hydrolysis [1]. Building-up of the M-O-M network can be achieved by condensation reactions between species with different ligands. Metal alkoxides and carboxylates (elimination of ester), metal halides MX_n and alkoxides (formation of alkylhalide) or elimination of dialkylether as the source of the oxo ligand is usual given examples [70].

The “nonaqueous" synthesis strategy proved to be a real versatile approach to rapidly produce patterned porous or nanocomposite materials with morphologies of films, membranes, fibres and monoliths [144-146] which offer large potential applications such as sensors, separation, supports for heterogeneous catalysis and optoelectronic devices [147]. As it was shown, optically transparent composite films are impossible to be obtained if during synyhesis a phase separation occurred. If the time of the sol–gel network formation is fast enough, the macroscopic phase separation is avoided [28]. This was proved applying the very fast non-aqueous sol–gel process where formic acid was used together with silicon alkoxides. In this process the water necessary for the condensation of the silicate species is produced by an esterification reaction of the alcohols released by the silica precursor and formic acid. In these systems, a TEOS/PMMA hybrid was obtained in 15 minutes compared to 200 hours by the usually employed systems [148].

The synthesis of ORMOSIL's has been carried out using both the hydrolytic and non-hydrolytic sol-gel routes. Using the nonhydrolytic sol-gel route in order to obtain ORMOSIL's particles may lead to products of different morphologies because of the different polarity of the reaction medium. Different reaction conditions were studied in order to optimize the size and shape of the particles, including choice of solvent, use of surfactants and addition of polystyrene. Dimethylsulfoxide acts as a novel oxygen donor for the catalyst-free formation of colourless silsesquioxanes [149].

A new nonhydrolytic sol-gel method was reported for the preparation of PI/silica hybrid materials using silicic acid oligomer extracted from water glass, and it has been proved that transparent PI/silica nanohybrids with silica content up to 40 wt.-% can be obtained by the addition of aminopropyltriethoxysilane (APTES) as coupling agent [150]. The films of PI/silica nanohybrids remained transparent even at high silica content due to formation of nanometer-scaled SiO_2 particles at the addition of APTES. In comparison with pure PI, PI/silica hybrids showed improved thermal stability and mechanical properties, and lower linear coefficients of thermal expansion (CTEs). The glass transition temperatures of the hybrids were increased with increasing content of silica or APTES as a result of the increasing limitation to the movements of the PI backbone [151].

The mesoporous materials synthesized under „nonaqueous" media always contain less ordered or more distorted domains than those obtained by conventional aqueous synthesis. Besides the topological defects induced by the stress or epitaxial mismatch, the solvent may be another important factor to cause the less-ordered domains in mesostructured materials. For the mesoporous materials prepared from „nonaqueous" media, the solvents with a less polarity than H_2O will increase the solubility of hydrophobes, namely diminish the hydrophilic (solvophilic) and hydrophobic (solvophobic) contrast. Thus, the inorganic-organic hybrid mesostructures resulted from self-assembly of inorganic silica species and organic amphiphiles under „nonaqueous" conditions may be less ordered [152].

Summarizing, the main advantages and disadvantages of a general sol-gel processing can be pointed out as follows [136]:

Advantages of Sol-Gel Processing

- The wet and dried product has a very high surface energy that is ideal for further processing;
- The thermal treated products can be obtained below their traditional temperatures. Temperatures can be reduced 25-50% from conventional ones.
- Raw materials such as alkoxides and other precursors can be readily distilled down to *ppb* level impurities through simple standard organic chemistry methods. This leads to an ultra high purity material suitable for many destinations including optical applications.

- Mixing the constituents at molecular level leads to excellent homogeneity of synthesized material; sol–gel also allows creating proper conditions in order to homogenize immiscible precursors, avoiding their phase separation.
- Lower temperatures of the sol-gel process raise possibility for forming amorphous inert phase materials (especially porous silica or hybrid matrices), and very important, it offer a good protection for associated organic materials against thermal alteration; sol-gel system offers also, the possibility for supersaturation of different phases. Room temperature forming allows for net shape capability. It can form complex shapes as bulk, thin films or fibers.
- Sol-gel processing allows obtaining new materials that cannot be made otherwise;
- There is a relatively low environmental impact, relative to other known processes.

Disadvantages of Sol-Gel Processing

- Structure has high percentage of liquid solvent that must be removed (drying). Very long gelation and aging times with special equipments are needed. Also a large amount of carbon based material remains along with H2O. These species can cause optical problems if they are not removed. Furthermore, removal of these species too quickly can cause cracking, which is a major problem.
- Compared to conventional materials, metal functionalized alkoxides are relatively expensive, but often, due to the new and valuable properties of sol-gel derived materials, this inconvenient is compensated.
- Long drying times and multiple steps increase costs. This prevents sol-gel from wide scale applications as bulk materials.
- It is worth to mention also that sol-gel represents a relative easy way to obtain a great diversity of nanostructured materials including nanosized powders but major difficulties arise during manipulation and further processing. It may cause seriously human and environmental damages.

Nevertheless, the huge versatility and other many advantages overcome the mentioned disadvantages which eventually can be avoided by a proper exploitation of the sol-gel processing steps.

2.3. Specific Sol-Gel Synthesis Methods and Strategies

A critical challenge in the design of inorganic-organic composites is the control over the mixing between the two chemically dissimilar phases [94]. Tremendous progress in both fundamental understanding of the sol-gel process and the development of new nonionic organic-inorganic hybrid materials have been made and extensively reviewed [16].

The *specific sol-gel methods* can be such routes or strategies which are used in order to design target specimens both from Class 1 and Class 2 [85,87]. Kaddami and coworkers presents several strategies to be used for synthesizing polymer-silica nanocomposites by sol-gel chemistry of metal alkoxydes such as alkoxysilanes and organic precursors [153]:

Route 1/ The hydrolysis and condensation of alkoxysilanes are performed in the presence of a polymer.

Route 2/ The hydrolysis and condensation reactions of the organic precursor are done at the same time with the polymerisation of an organic monomer.

Route 3/ Alkoxy-terminated oligomers of various natures can be synthesised and condensed after the hydrolysis of the alkoxysilane groups.

Wen and Wilkes propose the use of different synthetic strategies to generate organic/inorganic hybrid materials by the sol-gel process such as by incorporating various starting inorganic and organic components with varied molecular structure [52]. Basically they give a more detalied classification of hybrid networks (class II) synthesis routes, as follows:

(1) Hybrid networks can be synthesized by using *low molecular weight organoalkoxysilanes* as one or more of the precursors for the sol-gel reaction in which organic groups are introduced within an inorganic network through the = Si – C - bond [73,78,135];
(2) Organic/inorganic hybrid network materials can be formed via the *co-condensation of functionalized oligomers or polymers with metal alkoxides* in which chemical bonding is established between inorganic and organic phases [92];
(3) A hybrid material can be synthesized through the *in situ formation of inorganic species within a polymer matrix* [154].
(4) Hybrids can be obtained by infiltration of previously formed oxide gels with polymerizable organic monomers or the mixing of polymers with a single or mixture of metal alkoxides in a common solvent;
(5) Organics can be simply impregnated or entrapped as a guest within inorganic gel matrixes (as a host);
(6) Hybrid networks can be formed by interpenetrating networks and simultaneous formation of inorganic and organic phases.

Other new synthesis strategies have also been developed in recent years. Cubic and hexagonal ordered porous silicas MCM-41 and MCM-48 synthesis under acidic or alkaline conditions at room temperature by using symmetric (10–2–10 and 12–2–12) and dissymmetric (8–2–16) gemini surfactants as templates was recently reported [155].

Kickelbick believe that probably the most prominent strategy for the synthesis of IPNs materials is the incorporation of inorganic building blocks in organic polymers. The most commonly employed preparation procedures for these materials are the use of the sol–gel process for the formation of the inorganic network, the incorporation of preformed dense inorganic structures, such as clusters and particles, and the formation of hybrids using porous or layered inorganic materials either by interpenetration of the organic polymers into the voids, or by exfoliation of the inorganic material [28]. The formation of IPNs between inorganic and organic moieties can be performed by either a sequential two-step process, wherein a secondary network is formed in a primary one, or the simultaneous formation of the two networks. The resulting materials (from Class I) are microscopically phase separated, but macroscopically uniform. Examples of the two-step process include the creation of an organic polymer penetrating into the pores of an inorganic porous material, such as mesoporous amorphous materials, or between the sheets of a layered lattice. In these cases the inorganic moiety is a rigid, preformed natural or synthetic host matrix, with a regular pore or channel structure in the nanoscale.

In another process, an inorganic (crosslinked) moiety is formed by a polycondensation reaction, and interpenetrates during this process into an organic polymer. Difficulties of such an approach are potential incompatibilities between the moieties, leading to phase separation, and the challenge to find reactions for the formation of the second network which can be carried out in the presence of the first one [28]. Another major problem arises from the different stabilities of the materials. While inorganic systems are thermally quite stable and are often formed at high temperature, most organic polymers have an upper temperature limit of around 250 °C. Therefore, the synthesis of hybrid systems requires a strategy wherein the formation of the components is well-suited to each other, e.g. the use of a low temperature formation procedure like sol-gel [16].

Sumarizing, the frequently used strategies for the formation of IPN hybrid materials applying the sol–gel approach cited by Kickelbick [28] are:

- mixing of the precursor for the sol–gel process with the organic polymer followed by an inorganic polycondensation reaction;
- polymerization of an organic monomer in the inorganic gel;
- simultaneous formation of the inorganic and organic network, and
- dual network materials in which the molecular precursors for the formation of the two networks are covalently linked to each other (telechelics, etc.).

The resulting materials show strong structural differences depending on whether a one step or a two-step procedure was used to form the IPNs [103,104].

Surface-modification with polymerizable groups. Research in this area spans the modification of commonly used inorganic fillers or particles, up to the recent commercial availability of a variety of polyhedral oligomeric silsesquioxanes (POSS), $R_8Si_8O_{12}$ nanoclusters There are two main possibilities to functionalize the inorganic species with organic groups: either by grafting functional organic groups onto the surface of preformed clusters or particles, or by *in situ* functionalization of the clusters or particles during their formation. Another possibility for surface-modification is the attachment of organic groups by ionic interactions which is not intends to be treated in this section. An important parameter which has an influence on the properties of the final hybrid materials is the number of reactive functional groups available on the surface. If only one group is present, the particulate systems can be used as monomers to form linear polymers. More than one functional group leads to crosslinked systems [156,157].

Hybrid organic-inorganic materials from sonogels. The application of ultrasound (sonocatalysis) to the precursors gives rise to the materials with new properties, known as sonogels [158-161]. The absence of additional solvent and, mainly, the effects of ultrasonic cavitation create a unique environment for sol-gel reaction leading to particular features in the resulting gels: high density, fine texture, homogenous structure, etc. These properties determine the evolution of sonogels on further processing and the final material structure. One of the most relevant features of the ultrasonically assisted sol-gel process (sonosynthesis) is also low temperature requiered. This offers a huge number of possibilities in the field of organic-inorganic composites [44,47,78,92]. A proposed route to these materials has been described by several authors [158-161]. When both phases chemically bound [79,162,163] a sono-ORMOSIL results after gelation [164], while pure physical interactions between the two phases [165,166] lead to an organic-doped sonogel.

The preparation and properties of sono-ORMOSILs were first investigated by Mackenzie's group [167-169]. Kinetic studies were carried out in order to elucidate the influence of ultrasounds on the reactions involved. Once sonocatalysis starts, the reacting liquid temperature increases sharply to a plateau at 78 °C, probably due to the evaporation of ethanol produced by the hydrolysis. The gelation time is also reduced by ultrasounds with a dependence on the energy applied, i.e. radiation time, which exhibits the typical crossover around 450 Jcm^{-3} [199].

Ultrasound-assisted reactions carried out for different irradiation times have been monitored by NMR [168,169]. The spectra reveal that hydrolysis and condensation of TEOS take place within several minutes of irradiation and further condensation occurs as time passes. Ultrasounds not only cause mechanical mixing but also gives rise to chains breaks, which lead to the appearance of free radicals on the end of the broken chains [158-162]. Mackenzie *et al.* report also other effects of reaction parameters on gelation process in sono-ORMOSIL [170,171]. The final properties of the ORMOSIL are affected by ultrasound irradiation. They are denser and less porous, with better thermal stability too, than the ones prepared by the classic procedure. Specific surface areas, lower than 1 m^2g^{-1}, have been measured [168] on samples of the so-called hard-ORMOSILs that have been dried at 150 °C for 24 h to remove all the residuals from the pores. However, it has been shown that if these hybrid samples are heated at 100 °C their pore size distribution and volume are strongly modified with respect to non-heat-treated sample. The experiments on nitrogen adsorption on sono-ORMOSIL samples, previously evacuated at 60 °C, show an enlargement of the pores with increasing organic content as well as a decrease in surface area wich can range to hundreds of m^2g^{-1}).

One of the most interesting features of ORMOSILs is the rubber-like behavior that samples with high PDMS content (>10%) exhibit [163,172]. Certainly, the mechanical properties of these materials also have attracted considerable attention. In this respect, the increased density of sono-ORMOSILs can enhance mechanical properties such as hardness [169]. When compared with other materials, sono-ORMOSILs are much harder than the hardest organic plastics with Vickers hardness values close to those of soft glasses (~180 $kg \cdot mm^{-2}$). The dynamic elastic behavior of sono-ORMOSILs has been examined as a function of the structural features by using Brillouin scattering [173]. From the measured dynamic and static hypersonic properties, the elastic constant, and the related hypersonic attenuation were inferred. Calculated values range from 1 to 17 GPa. Refraction indices of sono-ORMOSILs were also obtained from the Brillouin frequencies corresponding to the $\pi/2$ rad and backscattering geometries. All three parameters exhibit behavioral changes when the monomer molar ratio reaches a defined value, close to 50 mol %. This fact pointed to the existence of two different PDMS morphologies in the samples: segment like and quasi-globular (liquid-like) forms, as it was clearly shown by NMR studies of dried samples [163,172].

Organic – doped sonogels. Ultrasonic cavitation has also been applied to the preparation of the aforementioned second type (class II) of hybrid materials. For effective entrapment of organic molecules, a stable and transparent sol-gel matrix is required with a narrow pore size distribution. This minimizes light dispersion effects and induces a uniform cluster size distribution of the guest [174]. Several organic dyes were effectively trapped in the sono-xerogel pores, by adding the optically active phase in the sol step, leading to very stable composites for optical applications [173-175].

Alternative synthesis strategy. The inclusion of functionalities precisely located in the channel pores can be performed by *grafting*. Groups of ordered silica was very often used [176-178] after removal of the surfactant. This method, in two steps, is very interesting in that it is general and allows the grafting of even bulky organic groups by choosing a mesoporous ordered silica with a convenient pore size.

An alternative approach in one step, overcoming the main restrictions of the post-synthesis method, has been developed for the preparation of ordered hybrid materials with functionalization within the channel pores. It consists in the co-hydrolysis and polycondensation of TEOS and an organotrialkoxysilane $R'Si(OR)_3$ in the presence of a *structure directing agent* [179-182]. This route provides a more regular distribution of organic groups inside the channel pores than by grafting and allows the control of the organic content. These hybrids prepared in the presence of structure directing agent are of great interest for nanoscience, very high surface (sometimes higher than 1000 $m^2\ g^{-1}$) enabling the linkage at the surface of an amount of functionalities sufficient for an easy characterization using directly classical methods (solid state NMR, IR, etc).

In literature are mentioned the recent developments concerning hybrid materials prepared in the *presence of a surfactant* [183,184]. These materials were novel in that the organic fragments are located within the framework leaving the unoccupied pores. That renders this class of materials called Periodic Mesoporous Organosilica (PMOs) of great interest, the functionalization of channel pores leading to bifunctional materials [185].

Synthesis strategy of the sol-gel derived non-Euclidean objects, based on fractal character. Using polymeric sols, the pore size is established by the structure of the gel and its extent of collapse during drying. The gel structure depends on the polymer size, structure (if fractal or not), and condensation rate. The extent of collapse during drying depends on the magnitude of the capillary pressure and the strength of the gel network. Under conditions where polymer interpenetration and collapse are promoted, it is possible to create sufficiently small pores as molecular sieving effects to be observed. When the capillary pressure is minimized by either elimination of surface tension or surface modification, little shrinkage occurs, resulting in highly porous aerogels. In order to independently control pore size and pore volume, it is possible to introduce sacrificial organic templates into a sol-gel matrix. The oxidative removal of these templates creates pores that appear to reproduce the size and volume fraction of the added template phase.

An alternate composite approach to produce microporous films is to embed structurally microporous templates type material within a dense sol-gel matrix [16]. As the polymeric silica and/or hybrid matrices are obtained as porous materials, valuable by themselves or as a host for a plethora of guests, in our opinion, the strategies to control porosity is a crucial task. Some information will be briefly presented here in order to lightening the less known concept of the peculiar fractal behavior of non-Euclidean objects correlated with tailoring the porosity. The cited literature [186-195] can be visited to accomplish the knowledge concerning the existent data focused on the subject.

Sol-gel derived materials often exhibit a randomly structure, with fractal character within a certain space domain, in other words they remain similar to themselves at any length scale of observation within this space range. Brinker and coworkers [195] proposed a strategy, which deserves a special attention, in order to control porosity based on the physics and chemistry principle of sol-gel processing [16]. The distinction between polymers and particles can be made by ^{29}Si-NMR spectroscopy. There is a considerable body of evidence based on ^{29}Si-NMR [186,187]. Sol-gel chemistry studied by ^{1}H and ^{29}Si-NMR and SAXS [188,189] exhibits data indicating that, depending on the synthesis conditions, a wide spectrum of structures ranging from weakly branched polymers to fully condensed particles may be prepared from silicon alkoxides.

The distinction between polymers and particles can also be made from SAS (SAXS, SANS, etc.) studies that probe structure on the ~0.1-50 nm length scale. The Porod slope, P, of a plot of log scattered intensity versus log scattering wave vector, Q

$$Q = (4\pi/\lambda)\sin(\theta/2) \quad (1)$$

is related to the mass fractal dimension, D_f, by equation [231]:

$$P = D_f - 2D \quad (2)$$

where D_f is fractal dimension and D = 3 is the dimension of space.

For uniform (non-fractal) objects such as particles, (mass fractal dimension) D_m = 3 and (surface fractal dimension) D_s = 2, P reduces to – 4. For mass fractal objects, D = D_m and P = - D. For surface fractal objects, D = 3 and P = D_s - 6.

The mass fractal dimension D_f relates an object's mass M to its radius r_c [190]:

$$M \sim r_c^{D_f} \quad (3)$$

where for mass fractal objects, D_f is less than the embedding dimension of space D. The surface fractal dimension D_s relates an object's area, A, to its size [190]:

$$A \propto r_c^{Ds} \quad (4)$$

where for D = 3 and $2<D_s<3$. Since in three dimensions, D=3, the density of a mass fractal cluster decreases with distance from its center of mass:

$$\rho \propto 1/r_c^{(3-D)} \quad (5)$$

Because density is inversely related to porosity, this relationship requires that, unlike Euclidean objects, fractal objects become more porous as their size increases. It was shown that this property may be exploited to tailor the pore structure of films deposited from fractal precursors. For both bulk and thin film specimens, the final porous microstructure is established by drying [191]. The drying process is composed of two successive stages. In the first stage, referred to as the constant rate period, the loss of pore fluid is accommodated by shrinkage of the gel, causing the liquid-vapor interface to remain at the exterior gel surface. The driving force for this shrinkage is the capillary tension P developed in the liquid by the radius of curvature of the meniscus r_m:

$$P = -2\gamma_{LV}/r_m \tag{6}$$

where γ_{LV} is the liquid vapor interfacial energy. Assuming for simplicity cylindrical pores, r_m is related to the cylindrical pore radius r_p by:

$$r_m = -r_p / \cos(\theta) \tag{7}$$

where θ is the contact angle. Combining eqs. (6) and (7) we arrive at the familiar Kelvin equation:

$$P = 2\gamma_{LV}\cos(\theta) / r_p \tag{8}$$

For wetting fluids ($\theta < 90°$), the meniscus is concave, causing the liquid to be in tension (positive P); in essence, the reduced volume of liquid must stretch to cover the solid surface and avoid the creation of higher energy solid-vapor interfaces. This tension in the liquid is supported by the gel network causing it to contact [16]. Initially the gel network is quite compliant. Shrinkage easily occurs, maintaining a relative flat meniscus. However, the contraction of the network promotes further condensation as reactive terminal hydroxyl species are brought into closer contact. Eventually, the contracting network becomes sufficiently stiff and can resist to any further compaction due to the capillary tension. At this point, referred to as the critical point, shrinkage ceases, and the capillary tension is maximized. Any further evaporation of liquid causes the meniscus to recede into the gel interior. This constitutes the beginning of the falling rate period of drying [191].

The shrinkage of the solid network at the critical point establishes the final pore volume, pore size, and surface area of the dry gel or xerogel. Therefore, many strategies designed to tailor the porosity of either bulk or thin film xerogels rely on controlling the extent of shrinkage. The extent of drying shrinkage depends on the balance between the magnitude of the capillary pressure that collapses the gel and the stiffness of the gel that provides resistance to collapse. This balance depends in turn on the characteristic time scale of the drying process. For bulk gels, long pathways for flow and diffusion require drying to occur slowly (hours to days), especially if cracking is to be avoided [16,191]. For films, vigorous evaporation overlaps the complete deposition process. Within seconds the entrained sol is concentrated and dried to form an extremely thin solid film (typically 10 to 300 nm).

Compared to bulk gels, films are less highly condensed prior to the critical point and hence suffer greater collapse during drying. This has the synergistic effect of reducing the pore size, increasing P (eq. 10) and promoting greater collapse of the network. Consequently, films are characterized by lower pore sizes, pore volumes and surface areas than their bulk counterparts. In the case of porous or randomly structured gels, porosity generally cannot be defined by the same rules for a spatially extensive crystalline lattice. Thus strategies designed to control porosity often rely on managing the packing efficiency of polymers or particles during gelation and aging and their extents of collapse upon drying.

For a better analysis of porosity generation process, both the solid particle packing principle and sol-gel derived (fractal) polmeric primary particles agregation process will be presented as follow.

Dense Particle (Euclidean Objects) Packing

Virtually all particle packing concepts utilize dense particles that are packed together to create pores of a size related to the primary particle size [196]. Ideally, monosized (spherical) particles could be assembled into a colloidal crystalline lattice. For example, random dense packing of monosized spherical particles always results in about maximum 33% porosity (theoretically calculated which do not depends on sphere diameter).

Aggregation of Fractals (Non –Euclidean Objects)

A strategy that is generally applicable to the wide range of polymeric sols, characterized by a mass fractal dimension, is aggregation. This strategy depends on the scaling relationship of size and mass of a fractal object, eq. (3), causing the density of an individual fractal cluster to decrease (porosity to increase) with cluster size r_c (eq. 5) [16]. When the individual fractal clusters comprising the sol are concentrated during slipcasting or evaporation, this porosity is incorporated into the film or membrane provided that: (1) the clusters do not completely interpenetrate and (2) that there exists no monomer or low molecular weight species that are able to fill-in the gaps of the fractal clusters [197].

The clusters interpenetration depends on their mean number of intersections: the fewer the number of intersections, the greater the ease of interpenetration. Mandelbrot [192] has shown that for two clusters with fractal dimensions D_{f1} and D_{f2} and size r_c, placed in the same region of space, the mean number of intersections $M_{1,2}$ is expressed as:

$$M_{1,2} \propto r_c^{Df1+Df2-D} \tag{9}$$

where D is the dimension of space (for our purposes, D = 3). Thus, if each structure has a fractal dimension less than 1.5, the probability of intersection decreases indefinitely as r_c, increases.

One refers to these structures as *mutually transparent:* during film or membrane formation they should freely interpenetrate one another as they are forced into close proximity, first by the increasing concentration, and then by the capillary pressure accompanying drying. Alternatively, if the mass fractal dimensions exceed 1.5, the probability of intersection increases with r_c. These structures are *mutually opaque.* During film or membrane formation, they do not interpenetrate as they are concentrated, so, for *opaque fractals*, we expect both the percent porosity and mean pore size of the membrane to increase with r_c.

Based on *fractal aggregation* principle, it can be predicted the efficiency of aggregation to design a certain porosity. The efficacy of the aggregation approach shows that for films formed from polymeric sols, characterized by D_m = 2.4, the porosity volume fraction increases monotonically with the average size r_c of the polymers comprising the sol. In addition to the volume fraction porosity, the surface area and pore size also increase with the aging time employed to grow the fractal clusters to different sizes prior to film deposition. The porosity of a film depends on both the porosity of the individual clusters and their extents of interpenetration during film formation. Although a reduction in D increases the porosity of a cluster according to eq. (7) it also facilitates interpenetration according to eq. (9). Thus, highly porous films require an intermediate value of D_f: generally $2.0 < D_f < 3.0$ [16].

According to eq. (5), low values of D promote interpenetration of polymers during film formation. Interpenetration can completely mask the porosity of the individual polymers. For example, for $D < 2$, quite dense films (5 vol % porosity) with small pores (< 1.0 nm) are prepared over a wide range of aging times [16], i.e., there is no dependence of the film porosity on polymer size. This situation is beneficial for the preparation of ultrathin membranes on porous supports, because aging can be employed to grow polymers large enough to be trapped onto the support surface without suffering an increase in pore size [193,195].

Influence of condensation on rate for tailoring fractals porosity. The criterion for opaque or transparent fractals established by eq. (9) is based on the assumption that instantaneous and irreversible reaction or *sticking* occurs at each point of intersection. Thus, as the points of intersection are reduced, the polymers more easily interpenetrate one to another - they become transparent. Polymer transparency is also promoted by reducing the probability of sticking at each point of intersection. Chemically, this is achieved by reducing the condensation rate. Instantaneous and irreversible sticking is chemically equivalent to an infinite reaction (condensation) rate. Of course, in reality the probability of sticking is much less than unity and is influenced by catalyst concentration, temperature, and reactivity of the terminal ligands. For example, for silicate polymers prepared from TEOS by a sol-gel process, the condensation rate is strongly dependent on the sol pH as estimated by the reciprocal gel times [194]. If the condensation rate is reduced, the probability of sticking at each point of intersection is reduced. This promotes polymer interpenetration, leading to denser films with smaller pore sizes. An additional (and perhaps more influential) consequence of the condensation rate is that it largely dictates the stiffness of the network and thus its extent of collapse during drying. A reduction in the condensation rate causes the network to be more compliant fostering its collapse in response to the capillary pressure created at the final stage of drying. It seems that the primary effect of the condensation rate is not to alter the polymer structure but rather to influence the extent of polymer interpenetration and collapse during membrane deposition.

From the above discussion it is clear that a reduction in the condensation rate promotes the collapse of the structure during drying. For films, it is possible to achieve pore sizes in the range appropriate for separations based on size exclusion. Such extensive collapse of the gel network has not been observed for bulk specimens. Although the condensation rate can be minimized, the considerable longer processing times associated with bulk xerogel formation allow considerable more condensation to take place, leading to stiffer structures that resist collapse.

In conclusion, according to Brinker and coworkers [195], sol-gel process affords many strategies for the preparation of inorganic and/or a hybrid material, with pore sizes in the range appropriate for membranes and sensors, and allows thin films to be fabricated using simple procedures. The porosity (i.e., pore volume, pore size, and surface area) of ceramic materials prepared by sol-gel processing depends on the size and structure of primary particles or polymers formed by condensation reactions, the organization of these structures, often by aggregation, to form a gel, and the collapse of the gel by drying. Among specific *strategies designed to control pore sizes* in the range appropriate for many applications the study of *aggregation of fractals* is a challenge [195]. It can be pointed aut that the fractal microstructure could be a starting point and a driving force for developing new theories and models for designing peculiar propeties and functionalities exhibited by a large cathegory of nanomaterials derived from sol-gel.

3. Properties and Investigation Methods

Typical properties of organic polymers are flexibility, low density, toughness and formability whereas ceramics have excellent mecanical and optical properties such as surface hardness, modulus, strength, transparency, and high refractive index. If this two kind of materials can be combined effectively, a new class of high performance functional organic-inorganic hybrid materials may be achieved. Some hybrids containing small concentrations of organic polymer are porous. More typically, hybrids are designed to have low porosity. With high concentrations of the organic component, hybrids have low densities, indicating high free volumes characteristic to the organic polymers [198]. Generally, it is claimed that covalently bound hybrids combine the flexibility and mechanical properties of the organic polymer chains with the hardness, stiffness and transparency of silica. In detail, the properties of the hybrids are governed by the chemistry, presence of side chains or pendant groups, and length of polymeric chains [48,199-200]

Basically, the accepted and used methods of characterization hybrid materials are classical techniques, suitable for both characterization of each component, (organic part and inorganic part), and resulted hybrids [44,201]. Nevertheless, new nonconventional investigation techniques are required in order to monitore and identify the new peculiar properties arised from synergy. Sol to gel transition process can be monitored using different investigation tehniques, such as: Raman and infrared spectroscopy (FT-IR), NMR, SAS tehniques- (SANS, SAXS or SALS) reology, magneto-reology, etc.

Both dry gels and thermal treated samples can be investigated by using thermo-analitical methods, DSC (TG, DTA), specific methods for porous media characterization like, surface area, porosity (BET), electron microscopy, SEM and TEM, Atomic Force Microscopy (AFM), spectroscopic methods (NMR, FTIR), diffraction methods, (SANS, SAXS), mass spectrometry MALDI TOF, etc. [202].

Eventually, methods for specific mechanical, magnetical (Mossbauer, VS-magnetometry, etc.), optical, catalysys, biochemical properties caracterisation, can be added to obtain an image of the studied hybrid material. In a brief description the investigation methods, properties and applications of hybrid nanomaterials hardly can be sharply separated [203].

Small angle scattering tehniques. Small angle scattering (SAS) investigations utilizing SANS, SAXS, or SALS have been employed to investigate the growth and topology of macromolecular networks, that precede gelation, the aggregation of colloids, and the structure of porous gels and aerogels [16,204]. Among other methods of investigations, SANS technique provides an appropriate way to study the evolution of inorganic polymer random microstructures obtained by sol-gel route because of averaging in reciprocal space [205-207]. SANS is a powerful tool for quantifying porosity in materials. A unique feature of the SANS technique is the ability to distinguish between open and closed porosity, allowing for more accurate measurements of total porosity and surface area.

Characterization of the microstructure using SANS. As microstructure plays a significant role in determining the sensitivity of an energetic material [208], detalied structural information are vital for advanced computer modeling of the properties of material and hence the predictability of performance. This ability allows for quantitative assessment of microstructural features and is of particular utility in studying porosity in complex materials [18,19,209,210]. In addition, the penetrating nature of neutrons makes this technique much less invasive than other techniques (e.g., scanning electron microscopy). This allows for more direct correlation between insult and microstructural changes, a feature that is highly advantageous for future studies of material.

Spatial range. Essentially, the scattered intensity I(Q) is measured as a function of the scattering vector, Q using a multidetector to record the profile for an incident monochromatic beam; most SAS instruments cover a range of 10^{-3} to 0.1 $Å^{-1}$, corresponding to a spatial range of 50 Å to 1μ. The intensity may vary over many decades in magnitude. Special techniques of ultra-low small-angle scattering [USAXS] may be used to study the profile at lower Q-values corresponding to larger distributions in real space [211]. SANS is a useful tool in order to study and evaluate certain microstructure parameters such as, the size and the distribution of particles and pores, and the aggregation tendency during ageing.

Within certain length scale range SANS measurements provide, direct information on the type of microstructure fractal behavior (mass or surface fractal). Both mass and surface fractals cause a power law behavior in the SANS intensity. For mass fractals this is $I(Q) \propto Q^{-D_m}$ with $1 \leq D_m \leq 3$. The scattering from a surface fractal follows the power-low: $I(Q) \propto Q^{Ds-2d}$ with $(d-1) \leq Ds \leq d$ (d = 3 is the space dimension). For a dense or homogeneous object with well defined smooth, non-fractal, fractal surface $D_s = 2$ and the usual Porod low $I(Q) \propto Q^{-4}$ is effective. The log-log plot of the scattering intensity curve I(Q) exhibits a straight line section, whose slope value (P) lies between –1 and –3 for mass fractals and between –3 and –4 in case of the surface fractals.

In our group porous silica-PVA hybrid xerogels, were obtained by sol-gel processing. The sample microstructure evolution during the thermal treatment at 300 °C, and 1000 °C was observed and qualitatively evaluated using SANS technique [18,19,210].

Silica-PVA nanocomposites have been prepared using acid catalysed sol-gel. As precursors were employed TEOS (Fluka, 98%), and PVA having two different molecular masses (M1 = 18000 g/mol, Austranal, p.a.) and M2 = 49000 g/mol, Fluka, p. a).

Both series of hybrid xerogels and hybrid nanocompozite microstructure behavior obtained by annealing up to 1000 °C were characterised by SANS and presented here for the first time. The obtained SiO_2/PVA sol-gel derived monoliths are homogenous and present high transparency. After the precursor mixture was stirred for 2 hours, the resulted homogeneous, clear sol was kept in covered vessels at room temperature. After gelation of sols and aging (3 weeks) hybrid silica-PVA transparent monolith wet gels have been obtained. Wet gels were dried at 60 °C for 48 hours. The obtained SiO_2/PVA sol-gel derived monoliths were homogenous and presented high transparency. Further thermal treatment of hybrid silica/PVA matrix xerogels, was performed in laboratory owen, at 300 °C and 1000 °C for 3 hours, in moderate oxidising conditions (air).

The SANS measurements on silica (S_0) and silica hybrid xerogels (V1S10), (V2S10) and heat treated samples respectively (V2S10 300 and V2S10 1000), were overdone at the facility in function, the IBR-2 pulsed reactor, in Dubna. The Q (scattering vector) values covered in these measurements, 0.005-0.37 $Å^{-1}$ range, providing information, at length scales (2π/Q) situated approximately between 1 and 100 nm. The observed scattered intensities of both series of samples were monotonous decreasing functions versus Q with no evidence of any peak. The lack of peak position in the measured I = f(Q) spectra, (data not shown) might be due to the larger size of the oxide regions, to an unknown value as $Q_{peak} < Q_{min}$ [207]. The log-log plots of the neutron scattering intensities I(Q) versus Q of the measured samples are shown in Figures 3 to 7. As Figures 3 to 7 show, in certain cases the intensity decreases with Q until a value Q_0, the first cross-over. At this point a plateau begins, extending to the second cross-over at the value Q_1, after which a linear descent is observed. Two linear parts in the SANS profile of the S_0 and V1S10 can be interpreted by two scale size of fractal behavior. In small Q range the silica xerogel sample shows a mass fractal dimension of 1.76. The observed mass fractal dimension is consistent with those reported in literature for open structures formed by diffusion- limited cluster-cluster aggregation (DLCCA), in acidic conditions. The reported $D_m = 1.78$ [16].

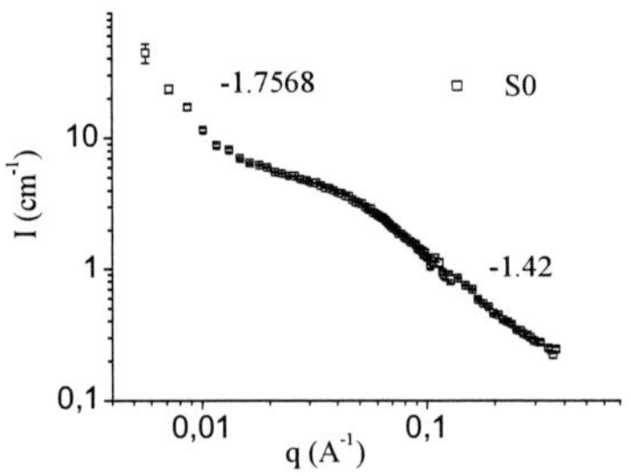

Figure 3. Log-log plot I=F(Q) of silica xerogel sample, S_0

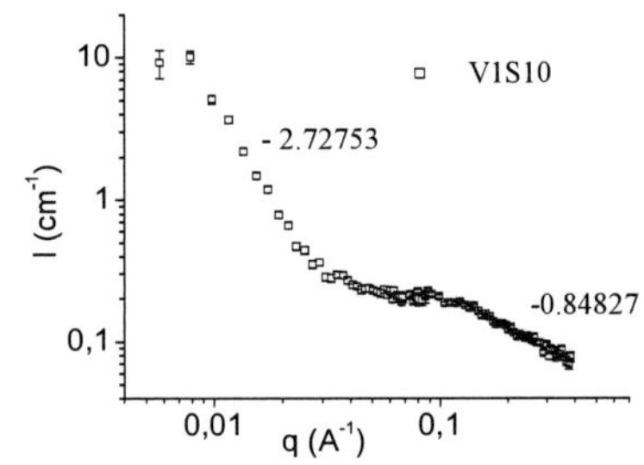

Figure 4. Log-log plot I=F(Q) of silica/PVA (M1) hybrid xerogel sample, V1S10

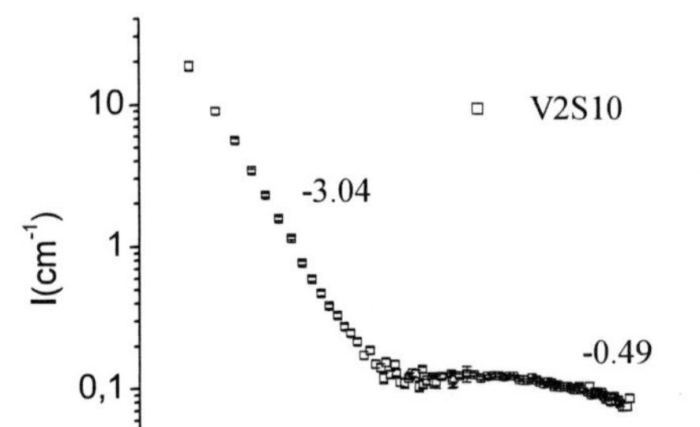

Figure 5. Log-log plot I=F(Q) of silica/PVA(M2) hybrid xerogel sample, V2S10

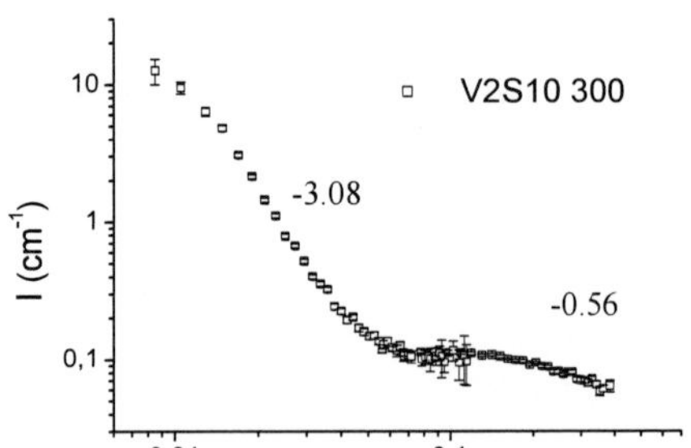

Figure 6. Log-log plot I=F(Q) of silica/PVA(M2) hybrid sample, annealed at 300 °C, V2S10 300

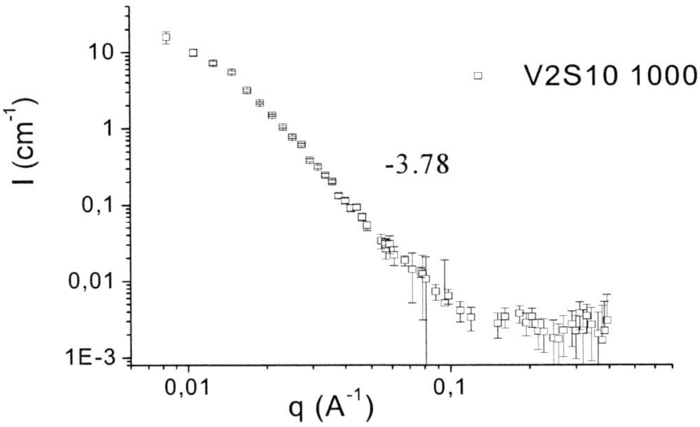

Figure 7. Log-log plot of silica/PVA (M2) hybrid sample annealed at 1000 °C, V2S10 1000

At high Q value, the mass fractal dimension is smaller, D_m = 1.42. Lower fractal dimension compared with the classic aggregation model can be explained by tip-to-tip cluster-cluster aggregation process [214], which involves polarisation of the clusters when they approach each other. This process results in more linear structures, and consequently a lower fractal dimension occurs. A more general model includes the difference in reactivity of species which have already been reacted, which is the reason that a wide variation in fractal dimensions can be obtained. In small Q range, the V1S10 sample shows a mass fractal dimension, D_m = 2.73 (D_m = -P, where P is slope) revealing a quite compact silica structure [215]. In the high Q region, the curve slope, P, tends toward unity, denoting the silica polymer tendency to extend its chains [216]. In the small Q range a significant change, surface fractal behaviour of V2S10 hybrid xerogel sample, D_s ~2.96, ((D_s = 6- (-P)) was observed. In the high Q range, the slope P = -0.49. The increased mass fractal dimension, observed for V1S10 sample, compared to D_m of silica matrix, S_0, can be attributed to the steric restriction, to the growth of the siloxane structure exerted by the rigid organic moiety [216]. The surface fractal behaviour, implied by higher molecular mass of PVA in V2S10 sample, can be the expression of more effective steric influence due to the extended chains of organic polymer. In the mean time, if their physical dimensions fit, the PVA oligomers may fill the silica polymeric network pores [217,218].

At high Q values, the plot of V2S10 (PVA with a higher M) sample is almost flat, probably indicating the absence, at this length scale, of the fractal behavior. D_s = 2.92, characteristic for V2S10 300 sample, Figure 6, is not essentially different comparing the xerogel, indicating however a slight smoother surface due probably to temperature influence. It can be attributed to the elimination of organics, shrinkage of the gel backbone without a significant pores collapse [218].

More dramatic changes, caused by the higher firing temperature (1000 °C), were obvious in the case of V2S10 1000 sample. The slope of the scattered intensity P ~ -3.77, from which diminished value of D_s ~ 2.23 results, in the case of the sample annealed at 1000 °C. It could mean that the particles are almost well defined, exhibiting smoother surface. The surface fractal dimensions series ranging from 2.23 (1000 °C), 2.92 (300 °C) and 2.96 (hybrid xerogel) shows without doubt that the lower temperature of 300 °C do not drastically damage the porosity; the higher the temperature of thermal treatment the smoother the particle surfaces.

Sumarizing, all the synthesised samples (xerogels and thermal treated samples up to 300 °C) present mass or surface fractal characteristics, most of them in both long and short length scale range. The significant conclusion was that both molecular mass and temperature of thermal treatment magnification cause a dramatic change in microstructure fractal behavior from mass fractal to surface fractal. The 1000 °C fired sample shows the smoothest surface.

EXAFS, XANES, ESR, UV–Vis spectroscopies showed that the active species from which there is oxygen uptake, can be mixed valence complex of some included metals cations bridged by an anion, making possible to evaluate the meta-metal distance [219].

Among the important morphological properties of porous materials are: porosity, specific surface area, permeability, breakthrough capillary pressure, diffusion properties of liquids in pores, pore size distribution, radial function, etc. Porosity measurement techniques on porous media are:

- Optical which involve filling the pores with a material, sectioning and inspecting with SEM ;
- Gas adsorbtion which measures the change in pressure as a gas is adsorbed by the sample;
- Mercury Intrusion which measures the volume of mercury forced into the sample as a function of pressure;
- Thermoporosimetry which is based on filling the pores with a liquid, freeze it, then measure the heat evolved as the sample is warmed, until all the liquid is melted;
- NMR Cryoporometry implies also to fill the pores with a liquid, freeze it, but then measure the amplitude of NMR signal from the liquid component as the sample is warmed, until all the liquid is melted.

Porosity, was defined as the ratio of the void space volume to the bulk volume of a sample (expressed as a fraction or percentage) [220]. It was revealed that a sol-gel sample, heated at certain temperatures remain mesoporous and the N_2 adsorbtion isotherm permits to be assigned after IUPAC regulations in the field [221]. Dore and coworkers have been studied porous media by a variety of NMR and neutron scattering techniques. These give complementary information [222]. One may generate model porous structures and study them by analytic and numerical techniques, and then compare the predicted properties with the measured ones [223].

Mechanical durability and reliability belong to the most important features of these materials. The relevant mechanical parameters are: hardness, elastic modulus, residual stress, fracture toughness, and interfacial fracture toughness (or adhesion energy) Thus, it is practical important to characterize and, if possible, control and improve these properties [224]. Surface roughness is of crucial importance for applications in many fields [225]. Most of these materials must have a specific surface roughness in order to guarantee optimal adherence for retaining various active agents (e.g., inks) [226]. The roughness scaling characteristics can be determined by statistical analysis of the AFM surface images using scaling analysis software. The surface texture is defined as the local deviation of a surface from its ideal shape. The deviations may be repetitive or random and may result from roughness, waviness, lay, and defects. Roughness comes from the finer irregularities of the surface, usually including those that result from the inherent action of the production process [48].

4. APPLICATIONS

Nanostructured nonionic hybrids can yield altogether new multifunctional materials where the optical and other macroscopic properties of the continuous phase (when such definition applies) are basically unchanged, while others sought for their application relevance are drastically modified. The incorporation of organic/oligomeric/polymeric materials into organic/inorganic networks by the sol-gel process makes it possible to optimize selected properties independently [34]. Specifically, the introduction of organic groups into an inorganic network leads to new structure property variation, thereby promoting new potential applications for the resulting composite materials, as follows:

- flexibility can be introduced by the incorporation of organic/oligomeric/polymeric materials into the inorganic networks [66];
- new electronic properties, such as conductivity, redox properties, etc., can be achieved by introducing conductive polymers along with transition-metal alkoxides;
- by incorporating organic dyes or $\eth$-conjugated polymers into the inorganic networks, the optical properties can be systematically altered in both the linear as well as nonlinear optical properties (NLO).

A combination or totally new properties with respect to the two components of the hybrid can arise from nanostructuration, achieved by surface modification of nanostructures, self-assembling or simply heterophase dispersion. The mechanical [227], adhesive [228], cohesive, electrical, optical, photochemical, catalytic and magnetic properties of these new hybrid materials are often a synergistic combination or utter new ones with respect to those of the constituents. Thus, organic polymers or ceramic coatings with improved toughness [229], elasticity, low surface energy, [98,230], enhanced hardness [231], chemical, radiation and heat resistance [232] availability of reactive functional groups or catalytically active host sites [233-235], can be produced by interpenetration [236-238], inclusion [239] or dispersion [240] of an inorganic component, where hydrophilic–hydrophobic interaction, coordinative [241,242] or covalent binding provide the required stabilization of incompatible phases with large interface area. Entirely new materials and applications can be attained with respect to the traditional ones, as in the case of the combination of typical molecular properties, from electrical [243], photochemical [244], magnetic [245,246], catalytic [247], biological activity [248-253] fast photochromic response, very high and stable second order NLO response, or being original pH sensors and electroluminescent diodes, have been reported in the past five years [254].

Another application that demonstrates the interest of using multifunctional hybrids concerns sol–gel derived hybrid organic-inorganic coatings developed by several glass packaging or glass decoration industries in Japan and Europe. These hybrid coatings cannot only give access to a wide variety of colours enhancing the consumer appeal, but they also improve the mechanical properties of glass bottles [255,256]. Moreover, these dye coloured bottles are easy to recycle as uncoloured glass (these materials do not need colour classification recycling), because, contrary to conventional glass bottles, their colorations do not arise from transition metals that are very difficult to remove upon remelting.

In nonionic hybrids synthesis a number of organics have been used, including organic dyes and chromophores, [257], liquid crystals [258], and active biomolecules, enzymes and proteins, [259-262] in order to produce specific optically active or bioactive devices for a broad range of optic and electro-optic applications. Suitably doped gels exhibit highly efficient luminescence, photochemical hole burning, photochromism and tunable laser action [263]. Wojcik and Klein pointed out [48] that hybrid host was better than either pure organic polymer or pure silica [264].

Wide application in recent years of scanning probe microscopies (SPM), including scanning tunnelling microscopy (STM) and AFM, has provided new information at the atomic scale about the state of the surfaces of model catalysts in situ under conditions close to those of real catalytic reactions. X-ray absorption fine structure techniques (XAFS, EXAFS, XANES), applicable to both model systems and „real" catalysts are further tools extensively applied recently to the in situ study of the local structure of catalysts. Quantum chemical calculations are becoming almost routine in practically all fields of catalysis: a new impetus has been provided by the wide application of density functional theory (DFT) to the description of the electronic structure of clusters which model the real catalysts, to studies on the interaction of molecules with such models, as well as to the determination of the reaction pathways [265]. Materials of high porosity in the meso- and macroporous range are aerogels, as prepared by the sol-gel technique studied by Schneider and Baiker [266]. They provide promising systems of molecularly mixed oxides and organic-inorganic nanocomposites; however, their application on a large scale is at the moment rather limited, mainly due to their expensive method of preparation. In addition to microporous zeolites, which uses are well established in the electrochemical industry there has been an intense interest in the application of larger pore materials, in particular, in silicate and alumino-silicate based mesoporous molecular sieves, discovered in 1992 and designated as M41S materials.

The nanocomposite hybrid materials can find applications in many fields which are far beyond the scope of application of traditional composite materials and this is one important reason for the strong interest in the application of organic/inorganic hybrid materials such as:

scratch and abrasive-resistant hard coatings and special coatings for polymeric materials, metal, and glass surfaces [267-270], electrical and NLO materials [271,273], adhesives and contact lens materials [69,274], reinforcement of elastomers and plastics [275], catalyst and porous supports, adsorbents [276], tunable solid-state lasers [277] and chemical/biomedical sensors [278,279].

Silicon-based inorganic polysiloxane hybrid coatings have entered the Protective Coatings industry as alternatives to the traditionally used organic polymer-based coatings and their use and benefits in the Protective Coatings industry are discussed in ref. [60]. The uses of inorganic coatings have been limited to specific applications. Polysiloxane coatings like two component epoxy-polysiloxane hybrid as well as one component acrylic polysiloxane hybrid are frequently used for the protective coatings since their introduction in 1994 [280], based on the advanced study of the properties of nanotools [281] and hybrid nanomaterials containing these nanotools.

Intelligent (Smart) materials, in our opinion, represent the most exciting domain of applications and a real challenge opening new perspectives of scientific development. Corriu pointed out that it is only a part of the target to reach nanomaterials and, in the long term, Smart materials. Smart materials are one of the main challenges for nanosciences. They correspond to nanomaterials in which at least two physical properties will be interactively coupled together. Until now the chemistry has never be thought in terms of possible interactive coupling of properties. This research field is widely open.

The nanoporous hybrid materials prepared in the presence of structure-directing agent are certainly, at the moment, the best supports for exploring some ways to polyfunctional materials [29,30]. Movement, motion, and related dynamic responses are distinctive features of animate objects. Conversion of different forms of energy to mechanical energy is the key factor responsible for dynamic responses associated with movement and motion in living as well as artificial systems [30]. Consequently, materials that can convert physicochemical signals to a definite dynamic response in the form of a change in shape or size-the so-called "smart" or "intelligent" materials-are useful for many applications including design of shape memory systems, drug delivery devices, chemical valves, artificial muscle mimics, and actuators [282-285]. The hybrid sol-gel derived materials represent an initial example of stimuli-responsive silica-based sol-gels that exhibit bulk volume changes and active mechanical responses with respect to several environmental variables including temperature, pH, salt, and solvents. These glasses incorporate an optimum balance of hydrophobic, hydrophilic, and ionic moieties in the silica-based structure and are therefore capable of showing bulk volume changes with respect to applied physicochemical stimuli [30].

The advantage of using sol-gel-derived glasses is that the parent SiO_2 material is functional inactive, and operational nonresponsive. Therefore, by selective integration of specific response-active molecular units into the glass, by means of chemical modification of the precursor, it is possible to introduce desired functional and operational properties into the sol-gel product. Only a few exemples of potential operational responsive were found [285-287].

Solvent Response. The organically-modified silica sol-gels show changes in volume with respect to the interactions with solvents. In general, solvents with high polarity provide an environment conducive to swelling (or less shrinkage) and solvents with low polarity lead to enhanced shrinkage.

pH Response. The organosilica sol-gels containing ionizable functional groups such as NH or COOH exhibit responses to variations in pH of the surrounding medium. Due to the presence of ionizable groups in the materials, protonation/deprotonation as a result of change in pH alters the electrostatic interactions and leads to volume changes. In general, it is found that the organosilica sol-gels containing amino groups undergo shrinkage at high pH while at low pH they exhibit swelling.

Salt Response. The organically-modified silica sol-gels exhibit environmental responses in the form of swelling/deswelling upon exposure to different salt solutions. Overall, the environmental responses are found to be function of the ionic strength and the nature of salt. In some cases the swelling of the gels correlates linearly with the concentration of salts. This is possibly due to increased intake of ions along with water.

Thermal Response of Hybrid Sol-Gels. The most effective indicator of the environmental sensitivity of a material is its response to variation in temperature [30,288] by the thermally-regulated reproducible swelling and deswelling with respect to variation of temperature. The molecular mechanism of thermoresponsive behavior of the certain sol-gel samples was ascribed to thermally-induced changes in hydrophobicity and altered interactions between the organic groups present in the material which cause expulsion/intake of water [288].

Molecular Programming of Environmental Responses. It has to be mention that the concept of *intelligent (smart) material* represent a complex device composed from three components: *sensor*, *microprocesor* and *actuator*. Each of this components has precise functionality in the architecture of a smart material or device. The *functionality of sensor* is to sense the stimul from environment, to transform it in a signal (e.g. electric, mechanic, magnetic, etc.) and to transmit it to microprocessor. The *microprocessor functionality* is to amplify and than to compare the received signal to a pattern or refference single value and to compute the value and sens of deviation from the standard value. Further, the microprocessor transmits the result to actuator (eventually reconverting the signal). *Actuator has the function* to transmit and aply the correction back to the environment. The *real time of response* is a factor which play a crucial role in the smart material efficiency. It is important to be pointed that, for each component of an intelligent device, it can be found an apropriate material capable to accomplish the specific functionality. As sensor or as actuator it can be used shape memory aloys, piezoelectric ceramics, organic polymers, electroreologic fluids, etc., materials which usualy are able to play a single function. The only known materials which coresponds entierely to required functionalities of an intelligent material are nanocomposites including hybri ceramic and polymeric nanocomposites, and hybrid organic inorganic composites.

Vital criteria for generating stimuli-induced dynamic responses, in the case of hybrid materials, include the occurrence of a bulk volume transition, which is initiated by alteration of noncovalent interactions within a material, and subsequent expulsion/intake of water [363]. An important aspect of these materials is that they combine all the essential functions of an intelligent material in an one-component monolithic unit; they are able to sense the external stimulus, translate the stimulus to mechanical energy, and generate an active dynamic response. Since the sensing-actuating mechanism is an intrinsic property of the sol-gels, the resulting monolithic device elements are simple, self-sufficient, and self-sustaining under a given applied stimulus. Conversion of different forms of physicochemical energy to mechanical energy underlies the basis of actuation, and these mechanically stable sol-gel derived materials may find potential applications as technological compatible intelligent systems for practical devices [30].

Finally, these materials constitute a new family of environmentally-sensitive glasses. The suitable strategy offers a potential powerful approach for designing a diverse range of stimuli-responsive sol-gel materials whose properties may be tailored precisely with systematic control over both the composition of product materials and the extent of environmental sensitivity to ultimately be able to design materials with tunable active responses [289-292].

5. Conclusion

From the briefly description presented in this chapter results that non ionic hybrid materials can possibly be used both as *innovative advanced materials* and as *precursors* of novel inorganic solids, opening the way to promising applications in many fields, including optics, electronics, ionics, mechanics, membranes, functional and protective coatings, catalysis, sensors, smart materials and biology.

Compared to soft polymeric materials, highly crosslinked silicates are more chemically and mechanically robust as a matrix. The addition of organic species in the silicates makes them elastic and miscible with organic guest molecules, and the composite can be synthesized easily through sol-gel chemistry. Material properties can be further modified through introduction of functional groups in both templates and organosilicates.

The preparation, characterization, and applications of nonionic organic/inorganic hybrid materials have become a fast expanding area of research in materials science. The major driving forces behind the intense activities in this area are the new and different properties of the nanocomposites which the traditional macroscale composites and conventional materials do not have. Among other synthesis methods, the low temperature of sol-gel processing offers unique possibilities for creating new organic-inorganic nonionic hybrid materials and it offers some advantages over other systems.

Nanotools make up the elementary bricks of the future materials. They can be a molecule, an assembly of molecules such as a cluster, a metallic nanoparticle or the elementary stage of an inorganic combination (oxide, sulphide, etc.). However, many of them will correspond to a specifically designed molecule for obtaining a desired property. This one can be physical (optical, magnetic, electric, etc.) or chemical (reactivity, catalysis, selective separation, etc.), but sometimes also mechanical applications can be designed. In the near future, bio applications will also certainly be developed. In all cases, the property must be precise, able to be measured, and controlled.

Some outstanding issues of hybrids must be explored at the frontiers of science, technology and knowledge. During early studies, the cooking of complex mixtures leds to materials with unexpected features; many applications of these new sophisticated materials have been described. After the initial studies, a more critical point of view has been adopted and more systematic studies were carried out. A wide investigation of these materials began with the help of most of the available techniques (such as water titration, elementary analysis, chromatography, rheology, small angle neutron, light and X-ray scattering, X-ray diffraction, optical techniques) and spectroscopies (such as IR, UV-VIS, fluorescence, electrochemistry, electrospray, MALDI-TOF mass spectrometry techniques, multinuclear liquid and solid-state NMR, X-ray spectroscopies XANES and EXAFS, EPR) and electronic microscopies as SEM, TEM or AFM. They demonstrated that both structural and 'smart' hybrid materials may be achieved in the near future. These trends are supported by the research works published during the last few years.

The state of the art survey of research findings on materials with nanoscale structure, including nonionic hybrids, revealed that there is a large number of method producing such structures as well as a huge number of potential precursors and combination among them, that leads to the unique properties when compared to coarsed-scaled structures, and to many possible applications with significant technological importance. At the present time, this very interdisciplinary scientific area is open for new ideas from all disciplines and therefore a promising future can be foreseen.

6. REFERENCES

[1] Corriu, R. J. P. *Eur. J. Inorg. Chem.* 2001, 1109-1121.

[2] Edelstein, A. S.; Cammarata, R. C. *In Nanomaterials, Synthesis, Properties and Applications*, Edelstein, A. S.; Cammarata, R. C. Eds.; Inst. Phys. Pub., Bristol and Philadephia, 1998; pp 1-55.

[3] Sanchez, C.; Ribot, F. *New J. Chem*. 1994, *18*, 1007-1074.
[4] Boilot, J. P.; Chaput, F.; Gacoin, T.; Malier, L.; Canva, M.; Brun, A.; Levy, Y.; Galaup, J. P. C. R. *Acad. Sci. Paris*, t. 322 Serie IIb 1996, 27-43.
[5] Judenstein, P.; Sanchez, C. *J. Mater. Chem*. 1996, *6*, 511-525.
[6] Schubert, U.; Huesing, N.; Lorenz, A. *Chem. Mater*. 1995, *7*, 2010-2027.
[7] Hüsing, N.; Schubert, U. *Angew. Chem. Int. Ed.* 1998, *37*, 22-45.
[8] Loy, D. A.; Shea, K. J. *Chem. Rev.* 1995, *95*, 1431-1442.
[9] Corriu, R. *Polyhedron* 1998, *17*, 925-934.
[10] Cerveau, G.; Corriu, R. *Coord. Chem. Rev.* 1998, *178-180*, 1051-1071.
[11] Corriu, R. *Angew. Chem. Int. Ed.* 2000, *39*, 1376-1398.
[12] Young, S. (2005). Organic-Inorganic Nanocomposite Materials [*http://www.psrc.usm.edu/mauritz/nano.html*].
[13] Rouxel, J. *Chem. Scr.* 1988, *28*, 33-42.
[14] Livage, L. *Chem. Scr.* 1988, *28*, 9-18.
[15] Sakka, S.; Kamiya, K. *J. Non-Cryst. Solids* 1980, *42*, 403-422.
[16] Brinker, C. J.; Scherer, G. W. *Sol-Gel Science: The Physics and Chemistry of Sol-Gel Processing*, Academic Press, San Diego, 1990, pp 45-229.
[17] Siegel, R.W. *Annu. Rev. Mate. Sci.*1991, *21*, 559-578.
[18] Savii, C.; Balasoiu, M.; Ionescu, C.; Popovici, M.; Kuklin, A.; Islamov, A.; Kovalev, Y. *Ann. West Univ. Timisoara,* Ser. Chem. 2003, *12*, 1029-1039.
[19] Ionescu, C.; Savii, C.; Balasoiu, M.; Popovici, M.; Enache, C.; Kuklin, A.; Islamov, A.; Kovalev, Y.; Almásy, L. *Acta Per. Technol.* 2004, *1450–7188*, 95-101.
[20] Popovici, M.; Ghich, M.; Niznansky, D.; Roig, A.; Savii, C.; Casas, Ll.; Molins, E.; Zaveta, K.; Enache, C., Sort, J.; De Brion, S.; Chouteau, G.; Nogues, J. *Chem. Mater*, 2004, *16*, 5542-5548.
[21] Savii, C.; Popovici, M.; Enache, C.; Subrt, J.; Niznansky, D.; Bakardjieva, S.; Caizer, C.; Hrianca, I. *Solid State Ionics* 2002, *151*, 219-227.
[22] Popovici, M.; Gich, M.; Roig, A.; Casas, L.; Molins, E.; Savii, C.; Becherescu, D.; Sort, J.; Suriñach, S.; Muñoz, J. S.; Baró, M. D.; Nogués, J. *Langmuir* 2004, *20*, 1425-1429.
[23] Caizer, C.; Savii, C.; Popovici, M. *Mater. Sci. Eng.* 2003, *97*, 129-134.
[24] Caizer, C.; Popovici, M.; Savii, C. *Acta Mater*. 2003, *51*, 3607-3616.
[25] Popovici, M.; Savii, C.; Niznansky, D.; Subrt, J.; Bohacek, J.; Caizer, C.; Enache, C.; Ionescu, C. *J. Optoelect. Adv. Mater.* 2003, *5*, 251-257.
[26] Dobrescu, G.; Crisan, M.; Zaharescu, M.; Ionescu, N. I. *Mater. Chem. Phys*. 2004, *87*, 184-189.
[27] King, S. In *Modern Techniques for Polymer Characterisation*, Pethrick, R. A.; Dawkins, J. V.; Eds.; *Small-angle Neutron Scattering*, Chapter 7; J. Wiley, Chichester, England, 1999; pp 182-213.
[28] Kickelbick, G. *Prog. Polym. Sci*. 2003, *86*, 83–114.
[29] Corriu, R.; Mehdi, A.; Reye, C. *J. Organomet. Chem*. 2004, *689*, 4437–4450.
[30] Rao, M. S.; Gray, J.; Dave, B. C. *J. Sol-Gel Sci. Techn*. 2003, *26*, 553–560.
[31] Barthelemy, P.; Ameduri, B.; Chabaud, E.; Popot, J. L.; Pucci, B. *Org. Lett.* 1999, *2*, 1689-1692.
[32] Saegusa, T. *Pure Appl. Chem*. 1995, *67*, 1965-1970.
[33] Boury, B.; Corriu, R. J. P. *In Organic Silicon Compounds*; Rappoport, Z.; Apeloig, Y. Eds.; Series The Chemistry of Functional Groups; *Nanostructured Hybrid Organic-*

Inorganic Solids from Molecules to Materials, Chpt. 10, Wiley and Sons, Ltd., N-Y, 2003, Vol. 3, pp 565-640.

[34] Schottner, G. *Chem. Mater.* 2001, *13*, 3422-3435.

[35] Sanchez, C.; Lebeau , B.; Chaput, F.; Boilot, J.-P. In *Functional Hybrid Materials*; Gómez-Romero, P.; Sanchez , C.; Eds.; *Optical Properties of Functional Hybrid Organic-Inorganic Nanocomposites*; Chapter 6, Wiley-VCH: Weinheim, 2003; pp 135-180.

[36] José, N. M.; Prado S. A.; Quim, L. A. *Nova*, 2005, *28*, 281-288.

[37] Lin, Y.; Boker, A.; He, J.; Sill, K.; Xiang, H.; Abeto, C.; Li, X.; Wang, J.; Emrick, T.; Long, S.; Wang, Q.; Balazs, A.; Russell, T. P. *Nature* 2005, *434*, 55-59.

[38] Sanchez, C.; Ribot, F. *Comm. Inorg. Chem.* 1999, *20*, 327-371.

[39] Schubert, U.; Hüsing, N. *In Synthesis of Inorganic Materials*; VCH-Wiley, Weinheim, 2005; pp 20-113.

[40] Roy, R.; Roy, R. A.; Roy, D. M. *Mater. Lett.* 1986, *4*, 323-328.

[41] Komarneni, S. *J. Mater. Chem.* 1992, *2*, 1219-1225.

[42] Yajima, S.; Okamura, K.; Hayashi, J. *Chem. Lett.* 1975, *931*, 1209-1215.

[43] Shea, K. J.; Loy, D. A. *Chem. Mater.* 2001, *13*, 3306-3319.

[44] Schmidt, H. *J. Non-Cryst. Solids* 1985, *73*, 681-691.

[45] Ravaine, D.; Seminel, A.; Charbouillot, Y.; Vincens, M. *J. Non-Cryst. Solids* 1986, *82*, 210-219.

[46] Schmidt, H.; Wolter, H. *J. Non-Cryst. Solids* 1990, *121*, 428-435.

[47] Huang, H. H.; Orler, B.; Wilkes, G. L. *Polym. Bull.* 1985, *14*, 557-564.

[48] Wojcik, A. B.; Klein, L. C. *Appl. Organomet. Chem.* 1997, *11*, 129–135.

[49] Sanchez, C. (2005). The Sol-Gel Gateway. What are hybrids? [*http://www.solgel.com /hybrids/hybrids.htm*].

[50] Haas, K. H.; Rose, K. *Adv. Mater. Sci.* 2003, *5*, 47-52.

[51] Allcock, H. R. *Adv. Mater.* 1994, *6*, 106-115.

[52] Wen, J.; Wilkes, G. L. *Chem. Mater.* 1996, *8*, 1667-1681.

[53] Shea, K. D.; Sasaki, Y. *J. Am. Chem. Soc.* 1989, *111*, 3442-3444.

[54] Shea, K. J.; Loy, D. A.; Webster, O. *J. Am. Chem. Soc.* 1992, *114*, 6700-6710.

[55] Landry, C. J. T.; Coltrain, B. K.; Teegarden, D. M.; Long, T. E.; Long, V. A. *Macromolecules* 1996, *29*, 4712-4721.

[56] Sperling, L. H. *In Interpenetrating Polymer Networks and Related Materials*; Chpt. 1. Plenum Press: New York, 1981; pp 1-6.

[57] Odian, G. *In Principles of Polymerization* ; *Step polymerization* Chpt. 2, John Wiley and Sons, Inc.: New York, 1991; pp 40-197.

[58] Sing, K. S. W. *Adv. Colloid Interface Sci.* 1998, *76-77*, 3-11.

[59] Kresge, C. T.; Leonowicz, M. E.; Roth, W. J.; Vartuli, J. C.; Beck, J. S. *Nature*, 1992, *359*, 710-712.

[60] Keijman, J. M. (2005). Properties and use of inorganic polysiloxane hybrid coatings for the protective coatings industry [*http://www.psx700.com/pdf/lecture_portugal_22_11_ 00.PDF*].

[61] Raileanu, M.; Stanciu, L.; Tanase, G.; Trandafir, I.; Zaharescu, M. *Rev. Roum. Chim.* 2002, *47*, 455-460.

[62] Ellsworth, M. W.; Novak, B. M. *Polym. Prepr.* 1992, *33*, 1088-1096.

[63] Hoebbel, D.; Endres, K.; Reinert, T.; Pitsch, J. *J. Non-Cryst. Solids* 1994, *176*, 179-188.

[64] Braun, S.; Rappoport, S.; Zusman, R.; Avnir, D.; Ottolenghi, M. *Mater. Lett.* 1990, *10*, 1-5.
[65] Ellerby, L. M.; Nishida, C. R.; Nishida, F.; Yamanaka, S. A.; Dunn. B.; Valentine, S. J.; Zink, J. I. *Science* 1992, *255*, 1113-1115.
[66] Cerveau, G.; Corriu, R. J. P.; Framery, E. *J. Mater. Chem.* 2001, *11*, 713–717.
[67] Makarova, L. I.; Sergienko, N. V.; Trankina, E. S.; Zhdanov, A. A.; Ahmeteva, E. A. *Russ. Chem. Bull.* 2003, *52*, 170-172.
[68] Katz, E.; Shipway, A. N.; Willner, I. *In Nanoscale Materials*; Liz-Marzan, L. M.; Kamat, P.; Eds.; *Organization Of Layer-By-Layer Assembled Nanocomposites: Functional Effects In Photonics And Biomaterials*; Chapter 13, Kluwer Academic Publisher: 2003, pp 273-297.
[69] Schmidt, H.; Scholze, H.; Tunker, G. *J. Non-Cryst. Solids* 1986, *80*, 557-563.
[70] Hubert-Pfalzgraf, L. G. *Coord. Chem. Rev.* 1998, *178-180*, 967-997.
[71] Yoldas, B. E. *J. Non-Cryst. Solids* 1984, *63*, 145-154.
[72] Scholze, H. *J. Non-Cryst. Solids* 1985, *73*, 669-680.
[73] Interrante, L. V.; Moraes, K.; Liu, Q.; Lu, N.; Puerta, A.; Sneddon, L. G. *Pure Appl. Chem.* 2002, *74*, 2111–2117.
[74] Sakka, S.; Tanaka, Y.; Kokubo, T. *J. Non-Cryst. Solids* 1986, *82*, 24-30.
[75] Brinker, C. J. *J. Non-Cryst. Solids* 1988, *100*, 31-50.
[76] Schmidt, H. *J. Non-Cryst. Solids* 1989, *112*, 419-423.
[77] Babonneau, F.; Thorne, K.; Mackenzie, J. D. *Chem. Mat.* 1989, *1*, 554-558.
[78] Philipp, G.; Schmidt, H. *J. Non-Cryst. Solids* 1984, *63*, 283-292.
[79] Huang, H.; Orler, B.; Wilkes, G. L. *Macromolecules* 1987, *20*, 1322-1330.
[80] Zhou, W.; Shephard, D, S.; Thomas, J. M.; Maschmeyer, T.; Johnson, B. F. G.; Bell, R. G.; Ge, Q. *Science* 1998, *280*, 705-708.
[81] Klemperer, W. G.; Mainz, V. V.; Millar, D. M. *In Better Ceramics Through Chemistry II*; Brinker, C. J.; Clark, D. E.; Ulrich, D. R. Mat. Res. Soc. Pittsburg, PA., 1986; pp 3-25.
[82] Day, V. W.; Klemperer, W. G.; Mainz, V. V.; Millar, D. M. *J. Am. Chem. Soc.* 1985, *107*, 8262-8264.
[83] Stein, A.; Melde, B. J.; Schroden, R. C. *Adv. Marter.* 2000, *12*, 1403-1407.
[84] Tanev, P. T.; Pinnavaia, T. J. *Science* 1995, *267*, 865-867.
[85] Raman, N. K.; Anderson, M. T.; Brinker, C. *J. Chem. Mater.* 1996, *8*, 1682-1701.
[86] Morey, M. S.; Davidson, A.; Stucky, G. D. *J. Porous Mater.* 1998, *5*, 195-204.
[87] Ying, J. Y.; Mehnert, C. P.; Wong, M. S. *Angew. Chem. Int. Edit.* 1999, *38*, 56-77.
[88] Huo, Q.; Margolese, D. I.; Stucky, G. D. *Chem. Mater.* 1996, *8*, 1147-1160.
[89] Vartuli, J. C.; Kresge, C. T.; Roth, W. J.; McCullen, S. B.; Beck, J. S.; Schmitt, K. D.; Leonowicz, M. E.; Lutner, J. D.; Sheppard, E. W. *Adv. Catal. Nanostruct. Mater.* 1996, 1-19.
[90] Ciesla, U.; Schuth, F. *Micropor. Mesopor. Mater.* 1999, *27*, 131-149.
[91] Hou, Z.; Liu, L.; Xu, L.; Xu, Z.; Wang, W.; Li, F.; Ye, M. *Chem. Mater.* 1999, *11*, 3177-3180.
[92] Wilkes, G. L.; Orler, B.; Huang, H. *Polymer Prepr.* 1985, *26*, 300-302.
[93] Mackenzie, J. D.; Eric, P. *J. Sol-Gel Sci. Techn.* 1998, *13*, 371–377.
[94] Ellsworth, M. W.; Novak, B. M. *Chem. Mater.* 1993, *5*, 839-844.
[95] Juangvanich, N.; Mauritz, K. A. *J. Appl. Polym. Sci.*1998, *67*, 1799–1810.

[96] Deng, Q.; Jarrett, W.; Moore, R. B.; Mauritz, K. A. *J. Sol–Gel Sci. Techn.* 1996, *7*, 177–190.
[97] Shao, P. L.; Mauritz, K. A.; Moore, R. B. *J. Polym. Sci, Part. B: Polym. Phys.* 1996, *34*, 873–82.
[98] Chujo, Y.; Saegusa, T. *Adv. Polym. Sci.* 1992, *100*, 11–29.
[99] Saegusa, T.; Chujo, Y. *Macromol. Chem, Macromol. Symp.* 1992, *64*, 1–9.
[100] Landry, C. J. T.; Coltrain, B. K.; Wesson, J. A.; Zumbulyadis, N.; Lippert, J. L. *Polymer* 1992, *33*, 1496–1506.
[101] Wang, S.; Ahmad, Z.; Mark, J. E. *Polym. Bull.* 1993, *31*, 323–330.
[102] Zhao, M. X.; Ning, Y. P.; Mark, J. E. *Ceram. Trans.* 1991, *19*, 891–898.
[103] Matejka, L.; Dusek, K.; Plestil, J.; Kriz, J.; Lednicky, F. *Polymer* 1998, *40*, 171–181.
[104] Matejka, L.; Plestil, J.; Dusek, K. *J. Non-Cryst. Solids* 1998, *226*, 114–121.
[105] Landry, C. J. T.; Coltrain, B. K. *J. Macromol. Sci, Pure Appl. Chem.* 1994, *A31*, 1965–1973.
[106] Mark, J. E.; Premachandra, J.; Kumudinie, C.; Zhao, W.; Dang, T. D.; Chen, J. P.; Arnold, F. E. *Mater. Res. Soc. Symp. Proc.* 1996, *435*, 93–104.
[107] Mauritz, K. A.; Jones, C. K. *J. Appl. Polym. Sci.* 1990, *40*, 1401–1420.
[108] Landry, C. J. T.; Coltrain, B. K.; Teegarden, D. M.; Long, T. E.; Long, V. K. *Macromolecules* 1996, *29*, 4712–4721.
[109] Haas, K. H.; Wolter, H. *Curr. Opin. Solid State Mater. Sci.* 1999, *4*, 571-580.
[110] Livage, J. *L'Act. Chim.* 1997, *10*, 4-10.
[111] Livage, J. *New J. Chem.* 2001, *25*, 1-7.
[112] Corriu, R. J. P.; Leclercq, D. *Comm. Inorg. Chem.* 1997, *19*, 245-262.
[113] Agarwala, R. P. (2005). Defect and Diffusion [*http://www.scientific.net/1*].
[114] Hay, J. N.; Raval, H. M. *J. Mater. Chem.* 1998, *8*, 1233-1239.
[115] Hay, J. N.; Raval, H. M. *J. Sol-Gel Sci. Techn.* 1998, *13*, 109-112.
[116] Adrianainarivelo, M.; Corriu, R. J. P.; Leclercq, D.; Mutin, P. H.; Vioux, A. *J. Sol–Gel Sci. Techn.* 1997, *8*, 89-93.
[117] Nogami, M.; Moriya, Y. *J. Non-Cryst. Solids* 1980, *37*, 191-201.
[118] Khimich, N. N. *Glass Phys. Chem.* 2004; *30*, pp 430-442.
[119] Corriu, R. *Polyhedron* 1998, *17*, 925-934.
[120] Chevalier, P.; Corriu, R. J. P.; Delord, P.; Moreau, J. J. E.; Man, M. W. C. *New J. Chem.* 1998, *22*, 423-433.
[121] Hay, J.; Porter, D.; Raval, H. *Chem. Commun.* 1999, *1*, 81-82.
[122] Avnir, D. *Acc. Chem. Res.* 1995, *28*, 328-334.
[123] Novak, B. M. *Adv. Mater.* 1993, *5*, 422-433.
[124] Wei, Y.; Yeh, J. M.; Jin, D.; Jia, X.; Wang, J.; Jang, G. W.; Chen, C.; Gumbs, R. W. *Chem. Mater.* 1995, *7*, 969-974.
[125] Ou, D. L.; Seddon, A. B. *J. Non-Cryst. Solids* 1997, *210*, 187-203.
[126] Ou, D. L.; Seddon, A. B. *J. Sol–Gel Sci. Techn.* 1997, *8*, 139-145.
[127] Acosta, S.; Arnal, P.; Corriu, R. J. P.; Leclercq, D.; Mutin, P. H.; Vioux, A. *Mater. Res. Soc. Symp. Proc.* 1994, *346*, 43-51.
[128] Corriu, R. J. P.; Leclercq, D.; Lefèvre, P.; Mutin, P. H.; Vioux, A. *J. Non-Cryst. Solids* 1992, *146*, 301-303.
[129] Schaudel, B.; Guermeur, C.; Sanchez, C.; Nakatani, K.; Delaire, J. *J. Mater. Chem.* 1997, *7*, 61-67.

[130] Hench, L. L.; West, J. K. *Chem. Rev*. 1990, *90*, 35-40.
[131] Livage, J.; Henry, M.; Sanchez, C. *Prog. Solid State Chem*. 1988, *18*, 259-341.
[132] Klein, L. C. *Sol-Gel Technology for Thin Films, Fibers, Preforms, Electronics and Specialty Shapes*, Noyes Publications: Park Ridge, NJ, 1988; pp 37-85.
[133] Setnescu, R.; Barcutean, C.; Jipa, S.; Setnescu, T.; Negoiu, M.; Mihalcea, I.; Dumitru, M.; Zaharescu, M. *Polym. Degrad. Stab*. 2004, *85*, 997-1001.
[134] McCormick, A.V.; Ng, L.V.; Thompson, P.; Sanchez, J.; Macosko, C.W. *Macromolecules* 1995, *28*, 6471-6476.
[135] Schmidt, H.; Scholze, H.; Kaiser, A. *J. Non-Cryst. Solids* 1984, *63*, 1-11.
[136] Finley, M. *Advanced Glass II* 1999, *4*, 12.
[137] Andrianov, K. A.; Ganina, T. N.; Sokolov, N. N. *J. Gen. Chem.* (English Translation), 1956, 26, 1897-1905.
[138] Corriu, R. J. P.; Leclercq, D.; Mutin, P. H.; Samson, H.; Vioux, A. *J. Organomet. Chem.* 1994, *466,* 43-50.
[139] Bourget, L.; Mutin, P. H.; Vioux, A.; Frances, J. M. *J. Polym. Sci. A: Polym. Chem.* 1998, *36*, 2415-2425.
[140] Calleja, G.; Carre, F.; Cerveau, G.; Corriu, R. J. P. *CR ACAD SCI II C,* 1998, *1*, 285-289.
[141] Yuh, S. K.; Bescher, E. P.; Babonneau, F.; Mackenzie, J. D. *Mat. Res. Soc. Symp. Proc.* 1994, *346*, 803-808.
[142] Porta, G. M.; Rancourt, J. D.; Taylor, L. T. *Chem. Mater*. 1989, *1*, 269-273.
[143] Hay, J. N.; Porter, D.; Raval, H. M. *J. Mater. Chem*. 2000, *10*, 1811-1818.
[144] Zhao, D.; Yang, P.; Melosh, N.; Feng, J.; Chmelka, B. F.; Strucky, G. D. *Adv. Mater.* 1998, *10*, 1380-1385.
[145] Feng, P.; Bu, X.; Stucky, G. D.; Pine, D. J. *J. Am. Chem. Soc.* 2000, *122*, 994-995.
[146] Melosh, N. A.; Davidson, P.; Chmelka, B. F. *J. Am. Chem. Soc.* 2000, *122*, 823-829.
[147] Tian, B.; Liu, X.; Zhang, Z.; Tu, B.; Zhao, D. *J. Solid State Chem*. 2002, *167*, 324-329.
[148] Sharp, K. G. *ACS Symp. Ser*. 1995, *585*, 163–180.
[149] Arkhireeva, A.; Hay, J. N.; Lane, J. M.; Manzano, M.; Masters, H.; Oware, W.; Shaw, S. J. *J. Sol-Gel Sci. Techn*. 2004, *31*, 31–36.
[150] Wang, H.; Zhong, W.; Du, Q.; Yang, Y.; Okamoto, H.; Inoue, S. *Polym. Bull.* 2003, *51*, 63-68.
[151] Wang, H.; Zhong, W.; Xu, P.; Du, Q. *Macromol. Mater. Eng*. 2004, *289*, 793–799.
[152] Melosh, N. A.; Lipic, P.; Bates, F. S.; Wudl, F.; Stucky, G. D.; Fredrickson, G. H.; Chmelka, B. F. *Macromolecules* 1999, *32*, 4332-4336.
[153] Kaddami, H.; Pascault, J. P.; Hajji, P.; David, L.; Vigier G.; Gerard, J. F. (2005). Relationships between Mechanical Properties and Morphologies of Hybrid Organic Inorganic Materials [*http://www.unipd.it/solgel2001/page13.html*].
[154] Mark, J. E. *In Frontiers of Macromolecular Science*; Saegusa, T.; Higashimura, H.; Abe, A.; Eds.; Blackwell Scientific Publishers: Oxford, 1989; pp 90-115.
[155] Romero, F. J.; Jimenez, C.; Huc, I.; Oda, R. *Micropor. Mesopor. Mater*. 2004, *69*, 43–48.
[156] Kickelbick, G.; Schubert, U. *Monatsh. Chem*. 2001, *132*, 13–30.
[157] Schubert, U. *Chem. Mater*. 2001, *13*, 3487–3494.
[158] Suslick, K. S.; Price, G. J. *Annu. Rev. Mater. Sci*. 1999, *29*, 295-326.

[159] Blanco, E.; Esquivias, L.; Litrán, R.; Piñero, M.; Ramirez-del-Solar, M.; de la Rosa-Fox, N. *Appl. Organomet. Chem.* 1999, *13*, 399-418.

[160] Mason, T. J.; Lorimer, J. P. *Sonochemistry: theory, applications and uses of ultrasound in chemistry*, Ellis Horwood Limited. 1988; pp 98-129.

[161] Sora, I.; Nicoara, D.; Muntean, N.; Vekas, L.; Bica, D.; Savii, C., *Electroultraacustic Equipments for Liquid Media Advanced Processing*, (Rom) Ed. Orizonturi Universitare Timisoara, Ro, 2002; Chpt.5, pp 115-140.

[162] Pope, E. J. A.; Mackenzie, J. D. *MRS Bull.* 1987, *12*, 29-31.

[163] Mackenzie, J. D.; Chung, Y. J.; Hu, Y. *J. Non-Cryst. Solids* 1992, *147-148*, 271-279.

[164] Zarzycki, J. *Heterogen. Chem. Rev.* 1994, *1*, 243-249.

[165] Avnir, D.; Levy, D.; Reisfeld, R. *J. Phys. Chem.* 1984, *88*, 5956-5959.

[166] Levy, D.; Avnir, D. *J. Phys. Chem.* 1988, *92*, 4734-4738.

[167] Morita, K.; Hu, Y.; Mackenzie, J. D. *Mater. Res. Soc. Symp. Proc.* 1992, *271*, 693–698.

[168] Morita, K.; Hu, Y.; Mackenzie, J. D. *J. Sol–Gel Sci. Techn.* 1994, *3*, 109-116.

[169] Iwamoto, T.; Mackenzie, J. D. *J. Sol-Gel Sci. Techn.* 1995, *4*, 141-150.

[170] Fegley, B.; White, P.; Bowen, H. K. *Am. Ceram. Soc. Bull.* 1985, *64*, 1115-1120.

[171] Sakka, S.; Kamiya, K. *J. Non-Cryst. Solids* 1982, *48*, 31-37.

[172] Hu, Y.; Mackenzie, J. D. *J. Mater. Sci.* 1992, *27*, 4415-4422.

[173] Jiménez-Riobóo, R.; García-Hernández, M.; Prieto, C.; Fuentes-Gallego, J. J.; Blanco, E.; Ramírez-del-Solar, M. *J. Appl. Phys.* 1997, *81*, 7739-7745.

[174] Donatti, D. A.; Ibañez Ruiz, A.; Gambardella De Moraes, F.; Vollet, D. R. *J. Sol-Gel Sci.Techn.* 2003, *28*, 31-35.

[175] Canva, M.; Georges, P.; Brun, A.; Larrue, D.; Zarzycki, J. *J. Non-Cryst. Solids* 1992, *147–148*, 627-632.

[176] Zhang, W.; Froba, M.; Wang, J.; Tanev, P. T.; Wong, J.; Pinnavaia, T. J. *J. Amer. Chem. Soc.* 1996, *118*, 9164-9171.

[177] Price, P. M.; Clark, J. H.; Macquarrie, D. J. *J. Chem. Soc.,Dalton Trans.* 2000, *2*, 101-110.

[178] Liu, A. M.; Hidajat, K.; Kawi, S.; Zhao, D. Y. *Chem. Commun.* 2000, *13*, 1145-1146.

[179] Lim, M. H.; Blanford, C. F.; Stein, A. *Chem. Mater.* 1998, *10*, 467-470.

[180] Hall, S. R.; Fowler, C. E.; Lebeau, B.; Mann, S. *Chem. Commun.* 1999, *2*, 201-202.

[181] Corriu, R. J. P.; Embert, F.; Guari, Y.; Mehdi, A.; Reye, C. *Chem. Commun.* 2001, *12*, 1116-1117.

[182] Melero, J. A.; Stucky, G. D.; Van Grieken, R.; Morales, G. *J. Mater. Chem.* 2002, *12*, 1664-1670.

[183] Inagaki, S.; Fukushima, Y.; Guan, S.; Ohsuna, T.; Terasaki, O. *J. Am. Chem. Soc.* 1999, *121*, 9611-9614.

[184] Asefa, T.; MacLachlan, M. J.; Coombs, N.; Ozin, G. A. *Nature* 1999, *402*, 867-871.

[185] Yang, Q.; Kapoor, M. P.; Inagaki, S. *J. Am. Chem. Soc.* 2002, *124*, 9694-9695.

[186] Lo Celso, F.; Triolo, A.; Negroni, F.; Hainbuchner, M.; Baron, M.; Strunz, P.; Rauch, H.; Triolo, R. *Appl. Phys. A* 2002, *74*, S1430–S1432.

[187] Lippert, J. L.; Melpolder, S. B.; Kelts, L. W. *J. Non-Cryst. Solids* 1988, *104*, 139-147.

[188] Martin, J. E.; Hurd, A. J. *J. Appl. Cryst.* 1987, *20*, 61-78.

[189] Schaefer, D. W.; Keefer, K. D. *In Structure of random silicates: polymers, colloids and porous solids*; Pietronero, L.; Tosatti, E.; Eds.; Fractals in Physics; Amsterdam, North-Holland; 1986; pp 39-45.

[190] Mandelbrot, B. B. *The Fractal Geometry of Nature*, Freeman, San Francisco, CA, 1982.

[191] Scherer, G. W. *J. Non-Cryst. Solids* 1992, *147-148*, 363-374.

[192] Mandelbrot, B. B. *In Fractals: Form, chance, and dimension*; Freeman, W. H: San Francisco, CA, 1977; pp 1-79.

[193] Brinker, C. J.; Ward, T. L.; Sehgal, R.; Raman, N. K.; Hietala, S. L.; Smith, D. M.; Hua, D.-W.; Headley, T. J. *J. Membrane Sci.* 1993, *77*, 165-179.

[194] Iler, R. K. *The Chemistry of Silica*, Wiley, New York, 1978.

[195] Brinker, C. J.; Sehgal, R.; Hietala, S. L.; Deshpande, R.; Smith, D. M.; Lop, D.; Ashley, C. S. *J. Membrane Sci.* 1994, *94*, 85-102.

[196] Leenars, A. F. M.; Keizer, K.; Burggraaf, A. J. *J. Mater. Sci.* 1984, *10*, 1077-1088.

[197] Logan, D. L.; Ashley, C. S.; Brinker, C. *J. Mater. Res. Sot. Symp. Proc.* 1992, *271*, 541-546.

[198] Capozzi, C. A.; Pye, L. D. *SPIE* 1988, *970*, 135-139.

[199] Wolter, H.; Glaubitt, W.; Rose, K. *MRS Symp. Proc.* 1992, *271*, 719-724.

[200] Sanchez, C.; Soler-Illia, G. J. A. A.; Ribot, F.; Mayer, C. R.; Cabuil, V.; Lalot, T. *Chem. Mater.* 2001, *13*, 3061-3083.

[201] *Polyvinyl Alcohol Developments*; Finch, C. A.; Ed.; J. Wiley and Sons.; N-Y, 1992; pp 1-31.

[202] Mauritz, J. (2002). Atomic Force Spectroscopy [*http://www.psrc.usm.edu/mauritz/maldi.html*].

[203] Boury, B.; Corriu, R. J. P.; Delord, P.; Le Strat, V. *J. Non-Cryst. Solids* 2000, *265*, 41-50.

[204] Ramsay, J. D. F.; Kallus, S. *J. Non-Cryst. Solids* 2001, *285*, 142-147.

[205] Margaca, M. A.; Miranda Salvado, I. M.; Teixeira, J. *J. Non-Cryst. Solids* 1999, *258*, 70-77.

[206] Falcao, A. N.; Sousa, J. S.; Carrapico, M.; Margaca, F. M. A.; Carvalho, F. G.; Salvado, I. M.; Teixeira, J. *J. Sol-Gel Sci. Techn.* 2003, *26*, 345-348.

[207] Miranda Salvado, M.; Santos Sousa, J.; Margaça, F. M. A.; Teixeira, J. *Physica B: Condensed Matter.* 2000, *276-278*, 388-389.

[208] Tarver, C. M.; Chidester, S. K.; Nichols, A. L. *J. Phys. Chem.* 1996, *100*, 5794-5799.

[209] Hjelm, R.; Mang, J.; Skidmore, C.; Gerspacher, M. (2005). Materials Research Using Cold Neutrons at Pulsed Sources [*http://www.pns.anl.gov/staff/loong/pdf/Cold-n_Mtg_CoverandContent.pdf*].

[210] Ionescu, C.; Savii, C.; Balasoiu, M.; Popovici, M.; Kuklin, A.; Islamov, A.; Kovalev, Y. Fractality of Sol-Gel Derived Inorganic and Hybrid Matrixes Structures Investigated by SANS Experiments Proceedings of the 2nd Central European Training School, 7-11 April 2003, Budapest, Hungary, KFKI Report, 2004, Applied Physics B 77, pp 14-21.

[211] Dore and Webber, (2005). Structural studies of sol-gel and related materials by SAS and diffraction [*http://www.ukc.ac.uk./physical-science/cmr/index.html*].

[212] Ionescu, C.; Savii, C.; Popovici, M.; Simon, M. *Ann. West Univ. Timisoara, Ser. Chim.* 2002, *11*, 89-97.

[213] Ionescu, C.; Simon, M.; Ionascu, C.; Savii, C.; The IIIth International Symposium Young People and Multidisciplinary Research, Romania-Yugoslavia-Hungary, Timisoara, ROMANIA, 8-9 November 2001, pp 282-290.

[214] Lange, S. A.; Hekkink, J. H. A.; Keizer, K.; Burggraaf, A. J. *J. Non-Cryst. Solids* 1995, *191*, 1-16.

[215] Matejka, L.; Dukh, O.; Kolarik, J. *Polymer* 2000, *41*, 1449-1459.
[216] Falcao, A. N.; Sousa, J. S.; Carrapico, M.; Margaca, F. M. A.; Carvalho, F. G.; Salvado, I. M.; Teixeira, J. *J. Sol-Gel Sci. Techn.* 2003, *26*, 345-348.
[217] Pallikari, F.; Li, X.; King, T. A. *J. Sol-Gel Sci. Techn.* 1996, *7*, 203-209.
[218] Huang, W. L.; Liang, K. M.; Gu, S. R.; Cui, S. H. *J. Mater. Sci. Lett.* 2000, *19*, 1681–1683.
[219] David, G. *PhD Thesis*, Universite de Bourgogne, Dijon, 2004.
[220] Gregg, S. J.; Sing, K. S. W. *Adsorption, Surface Area And Porosity*, Academic Press, Second Ed., London, 1982.
[221] Sing, K. S. Ed.; *Pure and Appl. Chem.* 1985, 57, 603-619.
[222] Dore, J. C.; Webber, J. B. W.; Strange, J. H. *Colloid Surface A: Physicochem. Eng. Aspects* 2004, *241*, 191–200.
[223] Webber, J. B. W. (2005) Characterising Porous Media, Ph. D. Thesis [http://www.kent.ac.uk/physicalsciences/publications/theses/jbww.html].
[224] Malzbender, J.; Den Toonder, J. M. J.; Balkenende, A. R.; De With, G. *Mater. Sci. Eng. R.* 2002, *36*, 47-103.
[225] Castro-Rodriguez, R.; Oliva, A. I.; Sosa, V.; Caballero-Briones, F.; Peña, J. L. *Appl. Surf. Sci.* 2000, *161*, 340-346.
[226] Larena, A.; Millan, F.; Verdu, M.; Pinto, G. *Appl. Surf. Sci.* 2001, *174*, 217-224.
[227] Rhee, S.H. *Key Eng. Mater.* 2003, *240-242*, 187-190.
[228] Gunji, T.; Makabe, Y.; Takamura, N.; Abe, Y. *Appl. Organomet. Chem.* 2001, *15*, 683-692.
[229] Bourgeat-Lami, E.; Espiard, P.; Guyot, A.; Gauthier, C.; David, L.; Vigier, G. *Angew. Makromol. Chem.* 1996, *242*, 105-122.
[230] Kazuyuki, K.; Morita, M. *Trans. Mater. Res. Soc. Jpn.* 2001, *26*, 925-928.
[231] Du, Y. J.; Damron, M.; Tang, G.; Zheng, H.; Chu, C.-J.; Osborne, J. H. *Prog. Org. Coat.* 2001, *41*, 226-232.
[232] Viala, P.; Bourgeat-Lami, E.; Guyot, A.; Legrand, P.; Lefebvre, D. *Macromol. Symp.* 2002, *187*, 651-662.
[233] Blanford, C. F.; Do, T. N.; Holland, B. T.; Stein, A. *Mater. Res. Soc. Symp. Proc.* 1999, *549*, 61-66.
[234] Imhof, A. *Langmuir* 2001, *17*, 3579-3585.
[235] Tissot, I.; Reymond, J. P.; Lefebvre, F.; Bourgeat-Lami, E. *Chem. Mater.* 2002, *14*, 1325-1331.
[236] Matejka, L.; Dusek, K.; Plestil, J.; Kriz, J.; Lednicky, F. *Polymer* 1999, *40*, 171-181.
[237] Aglietto, L. B. M.; Ruggeri, G.; Ciardelli, F. *Polymer* 1994, *35*, 3296-3304.
[238] Donescu, D. S. S.; Stanciu, L.; Braileanu, A.; Zaharescu, M. *J. Sol-Gel Sci. Techn.* 2000, *19*, 839-843.
[239] Morikawa, A.; Iyoku, Y.; Kakimoto, M.; Imai, Y. *J. Mater. Chem.* 1992, *2*, 679-683.
[240] Corcos, F.; Bourgeat-Lami, E.; Novat, C.; Lang, J. *Colloid. Polym. Sci.* 1999, *277*, 1142-1151.
[241] Gomez-Romero, P. *Adv. Mater.* 2001, *13*, 163-174.
[242] Watkins, J. J.; McCarthy, T. *J. Chem. Mater.* 1995, *7*, 1991-1994.
[243] Dirix, Y.; Bastiaansen, C.; Caseri, W.; Smith, P. *Adv. Mater.* 1999, *11*, 223-227.
[244] Wirnsberger, G.; Scott, B. J.; Chmelka, B. F.; Stucky, G. D. *Adv. Mater.* 2000, *12*, 1450-1454.

[245] Philipse, A. P.; Van Bruggen, M. P. B.; Pathmamanoharan, C. *Langmuir* 1994, *10*, 92-99.
[246] Caruso, F.; Susha, A. S.; Giersig, M.; Mohwald, H. *Adv. Mater.* 1999, *11*, 950-953.
[247] Hanprasopwattana, A.; Srinivasan, S.; Sault, A. G.; Datye, A. K. *Langmuir* 1996, *12*, 3173-3179.
[248] Reetz, M. T. *Adv. Mater.* 1997, *9*, 943-954.
[249] Wang, B.; Li, B.; Deng, Q.; Dong, S. *Anal. Chem.* 1998, *70*, 3170-3174.
[250] Preda, G.; Savii, C.; Chiriac, A.; Peter, F.; Dragomirescu, M.; Ionescu, C. VI International Symposium »Interdisciplinary regional research« (Hungary – Romania – Yugoslavia), Novi Sad, 3–4 October 2002, pp 0436-0442.
[251] Dragomirescu, M.; Preda, G.; Savii, C.; Ionescu, C. *SCIENTIFIC WORKS,* Biotechn. Faculty, XXXV, Ed. Agroprint, Timisoara, 2002, ISSN 1221-5287, pp 510-514.
[252] Arsene, M.; Stanciu, L.; Parlog, C.; Bordeianu, L.; Raileanu, M. *Rev. Roum. Chim.* 2002, *47*, 1267-1271.
[253] Raileanu, M.; Stanciu, L.; Badea, M.; Stanescu, M.; Bordeianu, L.; Sci. Bull – Univ. "Politehnica" Bucharest, Ser. B*: Chem. Mater. Sci*. 2002, 64, 55-62.
[254] Sanchez, C.; Lebeau, B. *MRS Bull.* 2001, *26*, 377-387.
[255] Yamada, N.; Yoshinaga, I.; Katayama, S. *J. Sol-Gel Sci. Techn.* 2000, *17*, 123–130.
[256] Soler-Illia, G. J. A. A.; Sanchez, C.; Lebeau, B.; Patarin, J. *Chem. Rev.* 2002, *102*, 4093-4138.
[257] Reisfeld, R.; Jorgensen, C. K. *Struct. Bonding* 1992, *77*, 207-265.
[258] Levy, D.; Pena, J. M. S.; Serna, C. J.; Oton, J. M. *J. Non-Cryst. Solids* 1992, *147-148*, 646-653.
[259] Avnir, D.; Braun, S.; Lev, O.; Ottolenghi, M. *Chem. Mater.* 1994, *6*, 1605-1614.
[260] Raileanu, M.; Stanciu, L.; Parlog, C.; Bordeianu, D. L.; Stanescu, M. D.; Badea, M. *Rev. Roum. Chim*. 2002, *47*, 533-538.
[261] Arsene, M. L.; Stanciu, L.; Parlog, C. *Rev. Roum. Chim*. 2001, *46*, 757-762.
[262] Stanciu, L.; Arsene, M. L.; Parlog, C. *South. Brazilian J. Chem*. 1998, *6*, 77-86.
[263] Reisfeld, R.; Brusilovsky, D.; Eyal, M.; Miron, E.; Burstein, Z.; Ivri, J. *Chem. Phys. Lett*. 1994, *160*, 43-44.
[264] Reisfeld, R.; Gvishi, R.; Burstein, Z. *J. Sol–Gel Sci. Techn*. 1995, *4*, 49-55.
[265] Barbara, G. S. *Annu. Rep. Prog. Chem*., Sect. C, 2000, *96*, 297-334.
[266] Schneider, M.; Baiker, A. *Catal. Rev. Sci. Eng*. 1995, *37*, 515-556.
[267] Wen, J.; Wilkes, G. L.; Vasudevan, V. J. *J. Sol-Gel Sci. Techn*. 1995, *5*, 115-126.
[268] Kasemann, R.; Schmidt, H. *New J. Chem*. 1994, *18*, 1117-1123.
[269] Tamami, B.; Betrabet, C.; Wilkes, G. L. *Polym. Bull.* 1993, *30*, 393-399.
[270] Lebeau, B.; Guermeur, C.; Sanchez, C. *Mater. Res. Soc. Symp. Proc*. 1994, *346*, 315-324.
[271] Riehl, D.; Chaput, F.; Roustamian, A.; Levy, Y., Boilot J. P. *Nonlinear Opt*. 1994, *8*, 141-148.
[272] Luo, J.; Hua, J.; Qin, J.; Cheng, J.; Shen, Y.; Lu, Z.; Wang, P.; Ye, C. *Chem. Commun.* 2001, *2*, 171-172.
[273] Le Bouder, T.; Maury, O.; Bondon, A.; Costuas, K.; Amouyal, E.; Ledoux I.; Zyss, J.; Le Bozec, H. *J. Am. Chem. Soc*. 2003, *125*, 12284-12299.
[274] Fouassier, O.; Menetrier, M.; Videau, J. J.; Duguet, E. *Mater. Lett*. 2000, *42*, 305-310.

[275] Usuki, A.; Kujima, Y.; Kawasumi, M.; Okada, A.; Fukushima, Y.; Kurauchi, T.; Kamigaito, O. *J. Mater. Res.* 1993, *8*, 1179-1184.
[276] Junges, U.; Jacobs, W.; Voigt-Martin, I.; Krutzsch, B.; Schueth, F. J. *Chem. Soc.Chem. Commun.* 1995, *22*, 2283-2289.
[277] Canva, M.; Georges, P.; Perelgritz, J.-F.; Brum, A.; Chaput, F.; Boilot, J.-P. *Appl. Opt.* 1995, 34, 428-431.
[278] Zusman, R.; Rottman, C.; Ottolenghi, M.; Avnir, D. *J. Non-Cryst. Solids* 1990, *122*, 107-109.
[279] Dave, B. C.; Dunn, B.; Valentine, J. S.; Zink, J. I. *Anal. Chem.* 1994, *66*, 1120-1127.
[280] Bourg, S.; Broudic, J.-C.; Conocar, O.; Moreau, J. J. E.; Meyer, D.; Wong Chi Man, M. *Chem. Mater.* 2001, *13*, 491-499.
[281] Joachim, C.; Gimzewski, J. K.; Aviram, A. *Nature* 2000, *408*, 541-548.
[282] Osada, Y.; Gong, J. P. *Adv. Mater.* 1998, *10*, 827-837.
[283] Hoffman, A. S. *Artificial Organs*. 1995, *19*, 458-466.
[284] Galaev, I. *Russ. Chem. Rev.* 1995, *64*, 471-489.
[285] Annaka, M.; Tanaka, T. *Nature* 1992, *355*, 430-432.
[286] Rao, M. S.; Dave, B. C. *J. Am. Chem. Soc.* 1998, *120*, 13270-13271.
[287] Dave, B. C. *Mater. Technol.* 1999, *14*, 115-118.
[288] Tang, Y.; Dave, B. C. *Adv. Mater.* 1998, *10*, 1536-1540.
[289] Rao, M. S.; Dave, B. C. *Adv. Mater.* 2001, *13*, 274-276.
[290] Hench, L. L.; West, J. K. *Chem. Rev.* 1990, *90*, 33-72.
[291] Rao, M. S.; Dave, B. C. *Adv. Mater.* 2002, *14*, 443-447.
[292] Rao, G. V. R.; Balamurugan, S.; Meyer, D, E.; Chilkoti, A.; Lopez, G. P. *Langmuir* 2002, *18*, 1819-1824.

In: Research in Hybrid Materials
Editors: Simon J. Brunner and Julian W. Egger

ISBN 978-1-60456-718-2

Chapter 2

NOVEL BONE-REPAIRING MATERIALS: BIOACTIVE ORGANIC-INORGANIC HYBRIDS

Masanobu Kamitakahara*
Graduate School of Materials Science, Nara Institute of Science and Technology, 8916-5, Takayama-cho, Ikoma-shi, Nara 630-0192, Japan

ABSTRACT

In general, artificial materials implanted into bone defects are encapsulated by a fibrous tissue. Some ceramics, such as Bioglass®, sintered hydroxyapatite and glass-ceramic A-W, however, form a bone-like apatite layer on their surfaces in the living body and bond to living bone through this apatite layer, i.e. they show bioactivity. Although these bioactive ceramics are used clinically as important bone-repairing materials, they are essentially brittle and hence limited in their applications. It is desirable to develop new types of deformable bioactive materials. Organic–inorganic hybrids prepared by a sol-gel method are expected to exhibit bioactivity as well as deformability. Polydimethylsiloxane (PDMS)-CaO-SiO_2-based hybrids were prepared. To evaluate their potential of bone-bonding property, their ability of apatite formation was examined in a simulated body fluid (SBF). Since the apatite formation on bioactive materials in the body can be reproduced even in SBF, we can estimate the bioactivity of the material by using SBF. Mechanical properties were also examined. Some of them showed apatite-forming ability and mechanical properties analogous to those of human cancellous bone. These CaO-containing hybrids, however, showed a decrease in mechanical strength in SBF. These hybrids containing no CaO do not form apatite on their surfaces in SBF. We then prepared CaO-free PDMS-TiO_2 and poly(tetramethylene oxide) (PTMO)-TiO_2 hybrids in which anatase was precipitated by hot-water treatments. These hybrids also showed apatite-forming ability on their surfaces in SBF and flexibility, and they showed little decrease in mechanical strength in SBF. In conclusion, it was revealed that we can design new bioactive materials for bone-repair by controlling the compositions and structures in organic-inorganic hybrid systems.

* E-mail address: kamitaka@ms.naist.jp. Tel: +81-743-72-6122, Fax: +81-743-72-6129

1. INTRODUCTION

Bone supports our body, protects our organs, such as brain and heart against external force, and enables us to do various kinds of motion. Once even one of the bones is damaged by a disease or an accident, quality of our lives is much decreased. Defects of bones have been repaired by autogenous or allogenic bones. In the case of autogenous bones, however, a healthy part of the patient is damaged and available amount of bone is limited. In the case of allogenic bones, infections are sometimes caused through viruses, bacteria, *etc*. Therefore, artificial materials, which are free from virus and bacteria and not limited in its available amount, are needed to repair the defects of bones. Artificial materials implanted into bone defects, however, are generally encapsulated by a non-calcified fibrous tissue to be isolated from the surrounding bone [1].

In early 1970s, Hench *et al.* showed that some glasses in the system Na_2O-CaO-SiO_2-P_2O_5 spontaneously bond to living bone without forming the fibrous tissue around them and named them Bioglass® [2-4]. Since the discovery of Bioglass®, various kinds of ceramics, such as glass-ceramic Ceravital® containing crystalline apatite [5, 6], sintered hydroxyapatite ($Ca_{10}(PO_4)_6(OH)_2$) [7, 8], glass-ceramic A-W containing crystalline apatite and wollastonite [9, 10], glass-ceramic Bioverit® containing crystalline apatite and fluorophlogopite ($(Na, K)Mg_3(AlSi_3O_{10})F_2$) [11, 12], MgO-CaO-SiO_2-P_2O_5 glasses [13] and CaO-SiO_2 glasses [14], have been found to bond to living bone. These materials are called bioactive materials, and some of them are already clinically used as important bone substitutes, such as artificial iliac crests, artificial vertebrae, artificial intervertebral discs, bone fillers, *etc* [15-24].

They, however, occupy only 10 to 30 % of whole the bone grafts needed, because they are brittle, poor in fracture toughness and deformability, and too high in elastic modulus. Among the problems described above, the low fracture toughness has been solved by development of bioactive metals with high fracture toughness. For example, Kim *et al.* showed that titanium metal and its alloys bond to living bone, if they are subjected to prior NaOH and heat treatments to form a sodium titanate layer on their surfaces [25-34]. Miyazaki *et al.* showed that tantalum metal also bonds to living bone, if it is subjected to prior NaOH and heat treatments to form a sodium tantalate layer on its surface [35-37]. These metallic materials, however, have higher elastic moduli than that of human cortical bones, and hence are liable to induce resorption of the surrounding bone due to stress shielding by them. Bioactive materials with low elastic moduli and high deformability are desired to be developed.

Living bone is a composite of 70 wt% inorganic component, apatite, and 30 wt% organic component, collagen, as shown in Fig. 1 [38, 39]. Due to its unique composition and structure, natural bone can show not only high strength and high fracture toughness, but also deformability and low elastic modulus. Therefore, it is expected that a bioactive material with lower elastic modulus and deformability could be obtained if a bioactive inorganic substance is combined with some organic substance. Bonfield *et al.* early developed a composite of hydroxyapatite granules with polyethylene [40, 41]. These composites, however, lose deformability when the hydroxyapatite content exceeds 40 vol%. When the hydroxyapatite content is limited to less than 40 vol%, the composite can not show high bioactivity [3].

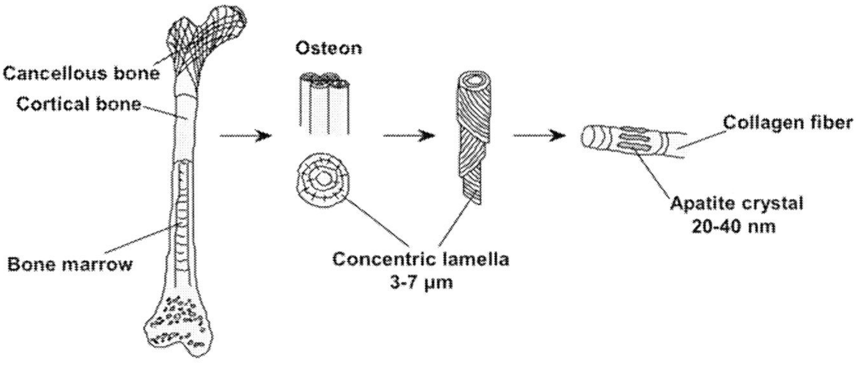

Fig. 1 Structure of bone [38].

2. Bone-Bonding Mechanism

It has been shown that the bioactive materials bond to living bone through an apatite layer which is formed on their surfaces in the body [3, 31, 42-49]. The schematic representation is shown in Fig. 2. Therefore, it is believed that the essential requirement for an artificial material to bond to living bone is formation of an apatite layer on their surfaces in the living body. This apatite layer consists of calcium-deficient and carbonate-containing nano-sized hydroxyapatite with low crystallinity and defect structure [50, 51]. This apatite is similar to the bone mineral in composition and structure, and hence bone-producing cell called osteoblast can proliferate preferentially on this surface apatite layer, and differentiate to produce bone matrix composed of apatite and collagen [4, 52, 53]. As a result, the surrounding bone can come into direct contact with the surface apatite layer. When this occurs, a tight chemical integration is formed between the bone apatite and the surface apatite on bioactive materials in order to decrease interfacial energy between them.

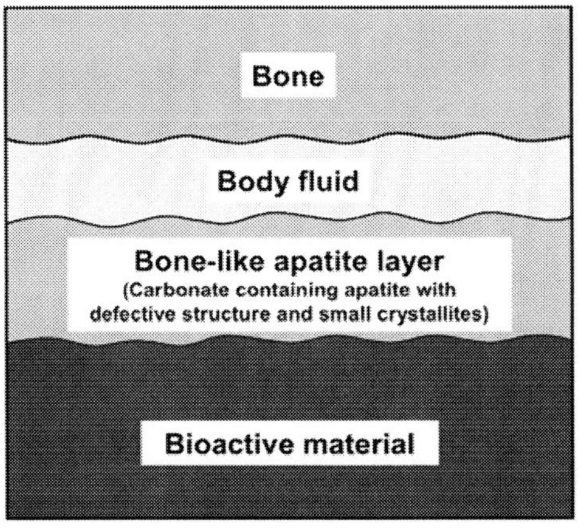

Fig. 2 Schematic representation of bonding mechanism of bioactive material to living bone.

Kokubo *et al.* [54] previously showed that the apatite formation on the surfaces of bioactive materials in the living body can be reproduced even in an acellular protein-free simulated body fluid (SBF) with ion concentrations nearly equal to those of human blood plasma. Ion concentrations of human blood plasma and SBF are shown in Fig. 3. This indicates that bioactivity of a material can be evaluated even *in vitro* by examining the apatite-formation on its surface in SBF.

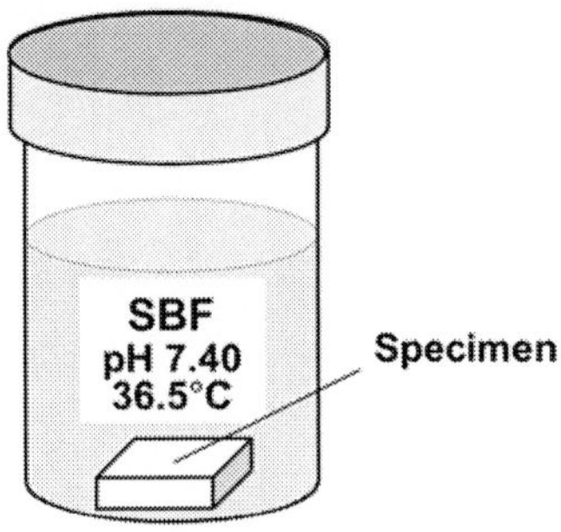

Ion concentrations of human blood plasma and Simulated Body Fluid (SBF)

Ion	Concentration / m*M*	
	Blood plasma	SBF
Na^+	142.0	142.0
K^+	5.0	5.0
Mg^{2+}	1.5	1.5
Ca^{2+}	2.5	2.5
Cl^-	103.0	148.8
HCO_3^-	27.0	4.2
HPO_4^{2-}	1.0	1.0
SO_4^{2-}	0.5	0.5

SBF was buffered with 50 m*M* tris(hydroxymethyl)aminomethane and appropriate amount of hydrochloric acid

Fig. 3 Estimation of bioactivity by using SBF [54].

3. Design of Bioactive Hybrids

Currently, organically modified sol-gel-derived metal oxides are paid attention because of their unique properties, such as low elastic modulus and deformability [55-59]. In these hybrid materials, organic components are chemically incorporated into inorganic networks at nanometer level. It has been revealed that some sol-gel-derived metal oxide gels, such as silicon oxide [60-63], titanium oxide [63-65], tantalum oxide [66], zirconium oxide [67, 68] and niobium oxide [69], form a bone-like apatite layer on their surfaces in SBF. Therefore, it is expected that if some organic component is incorporated into a network of these metal oxide gels by a sol-gel method, the obtained hybrids could show not only apatite-forming ability but also deformability. Moreover, if calcium ions are incorporated into these hybrids, the obtained hybrid could show higher apatite-forming ability, because the incorporated calcium ions are released into the surrounding body fluid to accelerate apatite nucleation by increasing the ionic activity product of apatite. [70].

4. Polydimethylsiloxane (PDMS)-CaO-SiO$_2$-Based Hybrids

Modification of CaO-SiO$_2$ system with organic polymer may provide bioactive organic-inorganic hybrids. Tsuru *et al.* [71] prepared polydimethylsiloxane (PDMS)-CaO-SiO$_2$ hybrids by a sol-gel method. Tetraethoxysilane (TEOS) was hydrolyzed and reacted with

PDMS, and calcium nitrate ($Ca(NO_3)_2$) was incorporated. The model structure of the PDMS-CaO-SiO_2 hybrid is shown in Fig. 4. Some of the resultant hybrids formed an apatite layer on their surfaces in SBF. This indicates that these hybrids are expected to form a bone-like apatite layer on their surfaces in the living body and bond to living bone through this apatite layer [72]. However, they did not show their mechanical properties because they could not prepare large enough specimens for mechanical test. Recently, Chen *et al.* [73-77] and we [78, 79] successfully synthesized large samples enough for mechanical test in PDMS-CaO-SiO_2-TiO_2 and PDMS-CaO-SiO_2 hybrids and revealed that some of the hybrids show both apatite-forming ability on their surfaces in SBF and deformability.

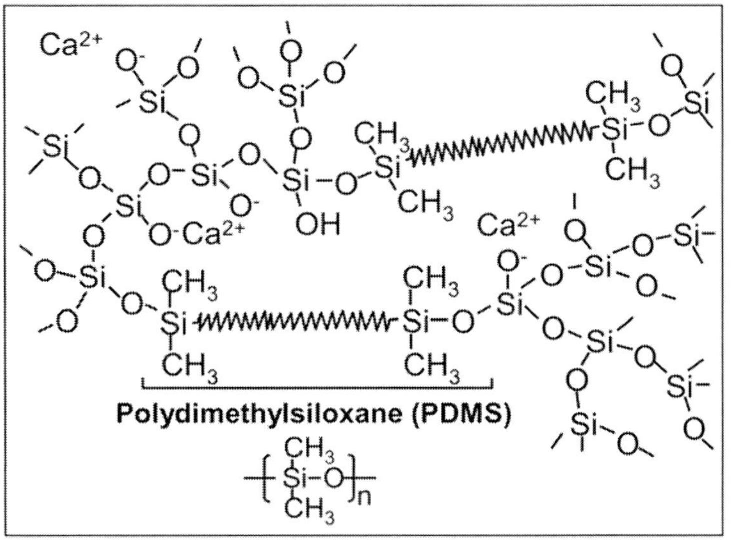

Fig. 4 Structural model of PDMS-CaO-SiO_2 hybrid.

We prepared the PDMS-CaO-SiO_2 hybrids whose starting compositions are given in Table 1 [78, 79]. The surfaces of the hybrids before and after soaking in SBF were observed by scanning electron microscopy (SEM). Figure 5 shows SEM photographs of the surfaces of the PDMS-CaO-SiO_2 hybrids before and after soaking in SBF for 7 days. Apatite-forming ability of the PDMS-CaO-SiO_2 hybrids increased with decreasing PDMS content. When PDMS content increases, the amount of silanol groups which induce apatite nucleation decreases, accordingly the apatite-forming ability decreases. The CaO-free hybrid (Ca0) did not form apatite on its surface. The calcium-containing hybrids release calcium ions and the released calcium ions accelerated apatite nucleation of apatite [70]. Once the apatite nuclei are formed, they can grow spontaneously by consuming calcium and phosphate ions from the surrounding SBF, since SBF is highly supersaturated with respect to apatite. Figure 6 shows stress-strain curve of one of the PDMS-CaO-SiO_2 hybrids, in comparison with that reported for human cancellous bone [80]. Unlike usual brittle ceramics, the hybrid was deformable and showed mechanical properties analogous to those of human cancellous bone.

Modification of inorganic component $CaO-SiO_2$ with organic component PDMS gives a hybrid material with apatite-forming ability and mechanical properties analogous to those of human cancellous bone.

Table 1 Starting compositions of the $PDMS-CaO-SiO_2$ hybrids with different PDMS and CaO contents [78, 79]

Notation	Composition					
	PDMS/TEOS (weight ratio)	$Ca(NO_3)_2$/TEOS (molar ratio)	H_2O/TEOS (molar ratio)	HCl/TEOS (molar ratio)	IPA/TEOS (molar ratio)	THF/TEOS (molar ratio)
PD20	20/80	0.15	2	0.045	2.47	0.49
PD25	25/75	0.15	2	0.045	2.47	0.49
PD30	30/70	0.15	2	0.045	2.47	0.49
PD35	35/65	0.15	2	0.045	2.47	0.49
Ca0	30/70	0	2	0.045	2.47	0.49

PDMS: Polydimethylsiloxane, M=1100
TEOS: Tetraethoxysilane
IPA: Isopropylalcohol
THF: Tetrahydrofuran

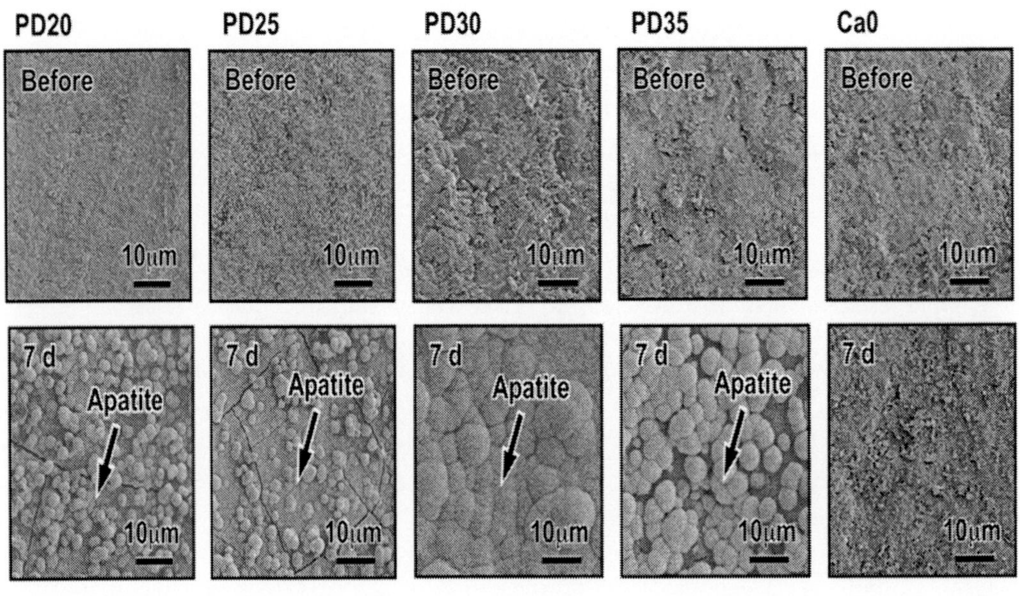

Fig. 5 SEM photographs of surfaces of hybrids PD20, PD25, PD30, PD35 and Ca0 before and after soaking in SBF for 7 days [78, 79].

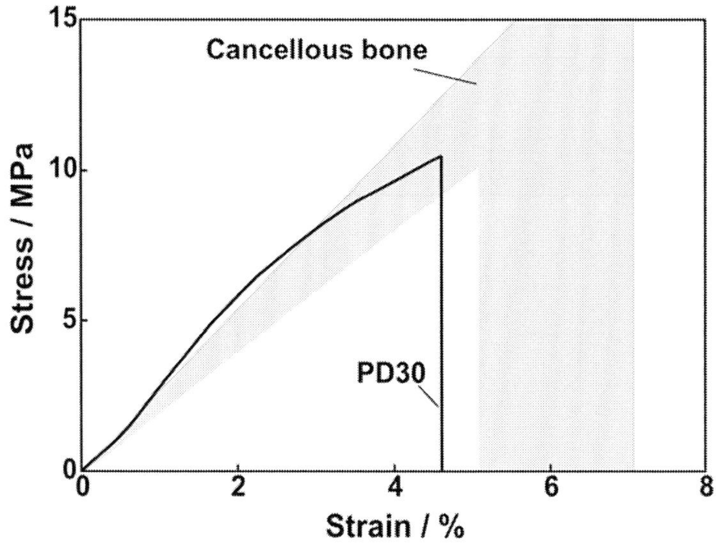

Fig. 6 Representative stress-strain curves of hybrid PD30 [78], in comparison with that reported for human cancellous bone [80].

5. Degradation of PDMS-CaO-SiO_2-Based Hybrids in SBF

For the clinical application of these hybrids, to know their changes in mechanical properties in the body environment is important. The mechanical properties of PDMS-CaO-SiO_2-TiO_2 hybrids were examined before and after soaking in SBF [81]. The starting compositions of PDMS-CaO-SiO_2-TiO_2 hybrids are given in Table 2. The changes in mechanical strength and strain at failure of the PDMS-CaO-SiO_2-TiO_2 hybrids before and after soaking in SBF are shown in Fig. 7. After soaking in SBF, the bending strength and strain to failure of hybrids Ti0Ca0 and Ti10Ca0 only slightly decreased, whereas those of hybrids Ti0Ca15, Ti10Ca15, Ti30Ca0, and Ti30Ca15 significantly decreased after soaking in SBF. The decreases in both the bending strength and the strain to failure of the calcium-containing hybrids due to soaking in SBF were larger than those of the corresponding calcium-free hybrids. It has been reported that calcium ions inhibit the polycondensation of TEOS in the preparation of a CaO-SiO_2 glass by a sol-gel method [82]. Therefore, it is considered that calcium ions inhibit the formation of a continuous three-dimensional network of the hybrids in the calcium-containing hybrids. In such cases, calcium ions are easily released via exchange with hydronium ions in SBF and water molecules also easily enter into the network of calcium-containing hybrids to break the Si-O-Si bonds of the hybrids. These may be the reasons why the deterioration of the calcium-containing hybrids was larger than that of the calcium-free hybrids in SBF.

It can be concluded that incorporation of a large amount of CaO into the PDMS-CaO-SiO_2-based hybrid system results in deterioration of the hybrids in the body environment.

Table 2 Starting compositions of PDMS-CaO-SiO_2-TiO_2 hybrids [81]

Notation	Composition in molar ratio						
	PDMS / Alkoxide	TiPT / Alkoxide	$Ca(NO_3)_2$ / Alkoxide	H_2O / Alkoxide	HCl / Alkoxide	IPA / Alkoxide	THF / Alkoxide
Ti0Ca0	0.11	0	0	2	0.045	2.47	0.49
Ti0Ca15	0.11	0	0.15	2	0.045	2.47	0.49
Ti10Ca0	0.11	0.1	0	2	0.045	2.47	0.49
Ti10Ca15	0.11	0.1	0.15	2	0.045	2.47	0.49
Ti30Ca0	0.11	0.3	0	2	0.045	2.47	0.49
Ti30Ca15	0.11	0.3	0.15	2	0.045	2.47	0.49

Alkoxide: (Tetraethoxysilane + Tetraisopropyltitanate)
PDMS: Polydimethylsiloxane, M=1100
TiPT: Tetraisopropyltitanate
IPA: Isopropylalcohol
THF: Tetrahydrofuran

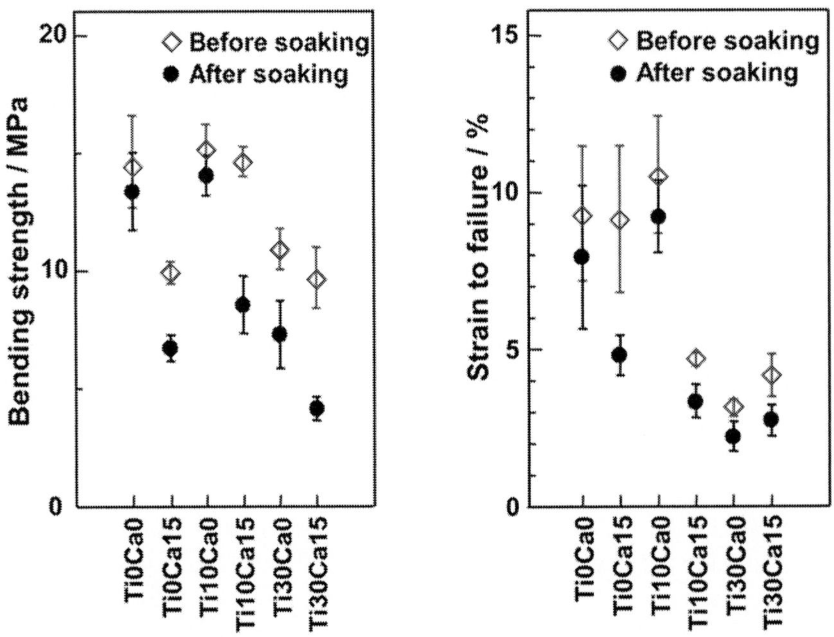

Fig. 7 Bending strength and strain to failure of PDMS-CaO-SiO_2-TiO_2 hybrids before and after soaking in SBF for 28 days [81].

6. PDMS-TiO_2 HYBRIDS

Preparation of bioactive CaO-free PDMS-modified hybrids that can form apatite on their surfaces was attempted. Recently, it has been shown that apatite formation on titania gel significantly depends on its structure. The Ti-OH groups in the anatase structure are most effectively induce apatite nucleation, whereas Ti-OH groups in the amorphous and rutile structure are not so effective [65]. It is expected that even a CaO-free hybrid can show high apatite-forming ability if the structure of a TiO_2-based hybrid is controlled.

A PDMS-TiO_2 hybrid was synthesized and its structure was controlled by hot-water treatment [83]. A polydimethylsiloxane (PDMS)-TiO_2 hybrid was prepared by a sol-gel method from PDMS (M=550) and tetraethylorthotitanate with PDMS/tetraethylorthotitanate molar ratios at 1.35. The as-prepared hybrid was amorphous and did not form apatite on its surface after soaking in SBF for 7 days. The PDMS-TiO_2 hybrid was then soaked in hot water at 60°C or 80°C for various periods. The structural change was examined by thin film X-ray diffraction (TF-XRD). Figure 8 shows TF-XRD patterns of the surfaces of the PDMS-TiO_2 hybrid subjected to the hot-water treatment at 60°C or 80°C for various periods. Anatase peaks were observed for the PDMS-TiO_2 hybrid treated with hot water at 60°C within 7 d and at 80 °C within 1 d. The peak intensities of anatase increased with increasing period and temperature of the hot-water treatment. Transmission electron microscopic (TEM) observation showed that the hybrid took a homogeneous amorphous structure before the hot-water treatment, and precipitated anatase particles 10-20 nm in size after the hot-water treatment at 80°C for 7 days (Fig. 9). The hot-water-treated hybrids were soaked in SBF. The SEM photographs of surfaces of the hot-water-treated PDMS-TiO_2 hybrids after soaking in SBF are shown in Fig. 10. The apatite was formed on the hybrid treated with hot water at 60°C for the periods longer than 3 days and at 80°C for the periods longer than 1 day. The amount of apatite formed on them increased with increasing period and temperature of the hot-water treatment. This indicates that the anatase precipitated in the hybrid induces apatite formation on the surface of the hybrid. It has been shown that apatite formation on titania depends mainly on its crystalline phase, and anatase is more effective for apatite formation than amorphous and rutile-type titania [65].

The obtained hybrid showed rubber elasticity even after the hot water treatment, as shown in Fig. 11. The representative stress-strain curves of the PDMS-TiO_2 hybrid before and after the hot-water treatment, and after both hot-water treatment and SBF treatment are shown in Fig. 12. Both tensile strength and Young's modulus of the hybrid decreased, and its strain to failure increased after the hot-water treatment. It is considered that the matrix of the hybrid was enriched with PDMS due to the segregation of TiO_2 by the hot-water treatment and hence the mechanical properties of the hybrid subjected to the hot-water treatment became close to the matrix rich in PDMS, since the continuous phase predominantly governs the overall mechanical behavior of composite materials. The observation that the strain energies stored until failure were not significantly changed by the hot-water treatment might indicate that rearrangement of the hybrid by the hot-water treatment does not cause deterioration of the hybrid. After soaking in SBF, its strain to failure decreased, but its tensile strength did not change and its Young's modulus a little increased. It is considered that an increase in its Young's modulus after soaking in SBF is due to apatite formation on its surface, because apatite shows much higher Young's modulus than the hybrid. In this case, its strain to failure decreases when its tensile strength does not change.

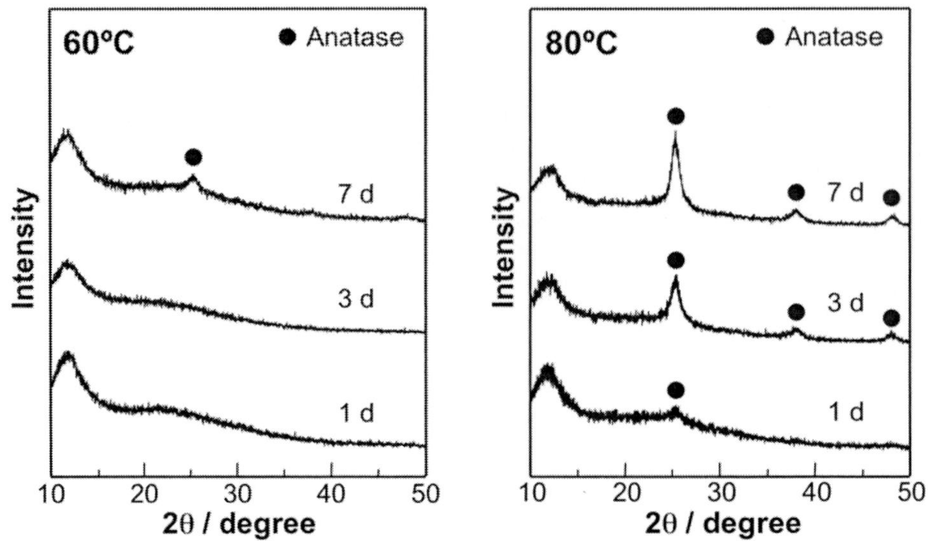

Fig. 8 TF-XRD patterns of the surfaces of the PDMS-TiO_2 hybrid subjected to hot-water treatment at 60°C or 80°C for various periods [83].

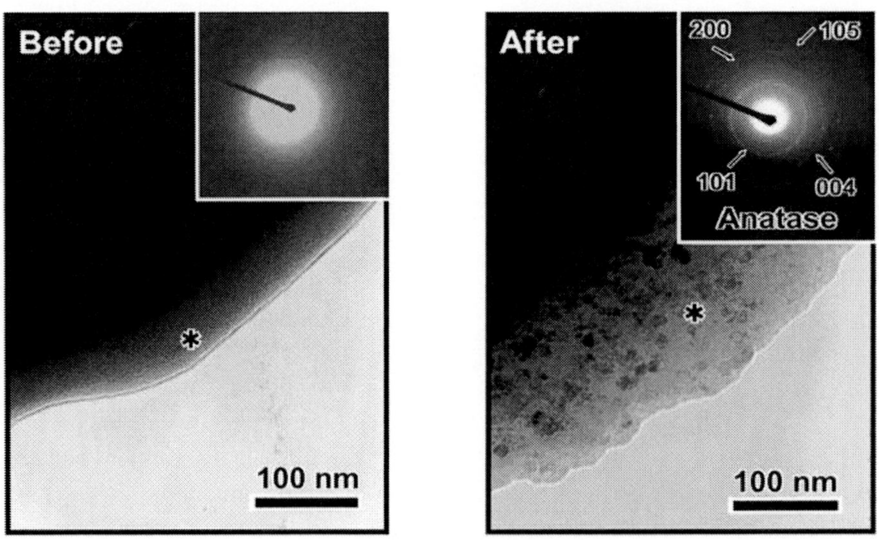

Fig. 9 TEM photographs of the PDMS-TiO_2 hybrid before and after hot-water treatment at 80°C for 7 days [83].
(* : center of electron diffraction)

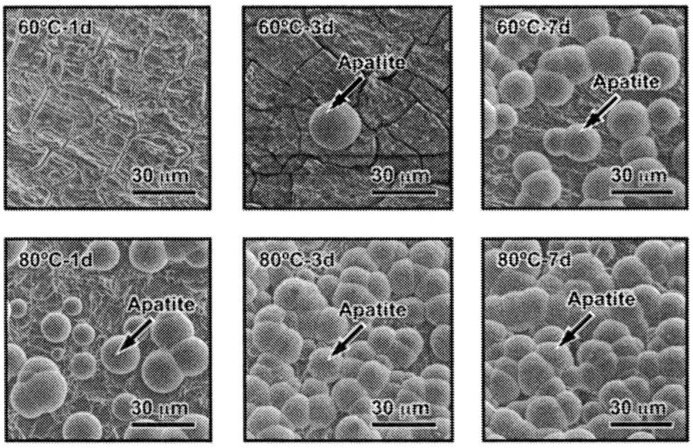

Fig. 10 SEM photographs of surfaces of the PDMS-TiO_2 hybrid treated with hot water at 60°C or 80°C for various periods and soaked in SBF for 7 days [83].

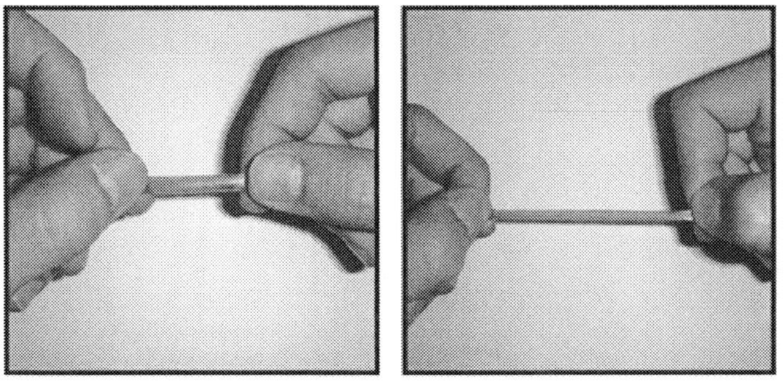

Fig. 11 Photographs of the PDMS-TiO_2 hybrid after hot-water treatment at 80°C for 7 days.

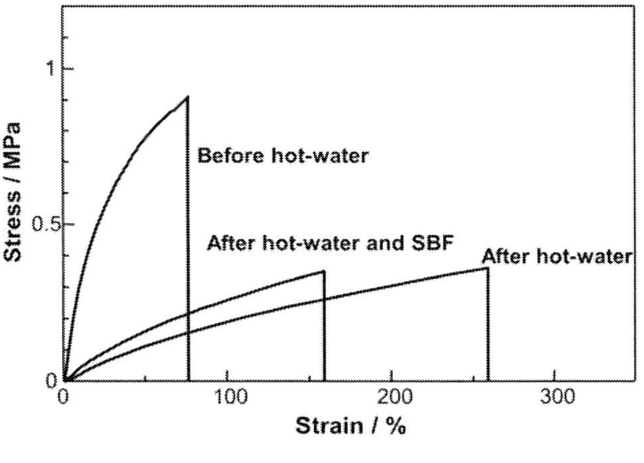

Fig. 12 Representative stress-strain curves of the PDMS-TiO_2 hybrid before and after the hot-water treatment at 80°C for 7 days, and that after both the hot-water treatment and SBF treatment for 28 days [83].

Since the environment in hot water is more severe than in the body environment, the unreacted reagents or solvents remaining in the hybrid might be released into hot water, and therefore the resultant hybrid is expected to be non-toxic and stable in the living body at around 36.5 °C.

A highly deformable PDMS-TiO_2 hybrid with apatite-forming ability and durability in SBF was obtained by the present method. Besides deformability, rubbery elasticity of the hybrid would also be an advantage, because the hybrid can return to the original shape, even after a large deformation.

7. Poly(Tetramethylene Oxide) (PTMO)-TiO_2 Hybrids

The PDMS-TiO_2 hybrid treated with hot water showed apatite-forming ability and high deformability. The mechanical strength (≈0.4 MPa) and Young's modulus (≈0.2 MPa) of the obtained hybrid, however, were much lower than those of human bones. It is expected that if PDMS is replaced with poly(tetramethylene oxide) PTMO, a deformable bioactive material with higher mechanical strength might be obtained [84].

Poly(tetramethylene oxide) (PTMO)-TiO_2 hybrids were prepared by a sol-gel method from triethoxysilane functionalized PTMO (Si-PTMO) and tetraisopropyltitanate with weight ratios of 30/70, 40/60 and 50/50 (hybrids PT30, PT40 and PT50, respectively), and subsequently subjected to a hot-water treatment at 95 °C for 2 d [85]. The structural model of the PTMO-TiO_2 hybrid is shown in Fig. 13. All of the obtained hybrids were amorphous before the hot-water treatment, and precipitated nanosized anatase after the hot-water treatment. The amount of precipitated anatase increased with decreasing PTMO content. Apatite was not formed on the surfaces of the hybrids in SBF before the hot-water treatment, but was formed after the hot-water treatment, and its amount increased with decreasing PTMO content, as shown in Fig. 14. This indicates that apatite-formation was induced by the anatase precipitated by hot-water treatment, as the case of the PDMS-TiO_2 hybrid. Figure 15 shows representative stress-strain curves of hybrid PT40 before and the hot-water treatment, and that after both the hot-water treatment and SBF treatment for 28 days, in comparison with that reported for human cancellous bone. Hybrid PT40 showed strength and Young's modulus analogous to those of human cancellous bones, and high deformability after the hot-water treatment. The PDMS-TiO_2 hybrid treated with hot water showed strength of only 0.4 MPa and Young's modulus of 0.2 MPa. The mechanical properties of the hybrid were improved by changing the organic component from PDMS to PTMO, from the point of view of strength and Young's modulus. Although the PDMS-TiO_2 hybrid treated with hot water can deform up to 200%, such extremely high deformability is not needed and 10% deformability of the PTMO-TiO_2 hybrid treated with hot water is considered to be enough for a bone-repairing material. After soaking in SBF, its Young's modulus increased, and its bending strength and strain to failure decreased. The increase in its Young's modulus after soaking in SBF might be due to the apatite-formation on its surface because apatite shows much higher Young's modulus than the hybrid. The decreases in its bending strength and strain to failure after soaking in SBF might be due to the degradation of the hybrid. These decreases were, however, much lower than those of a CaO-containing PTMO-CaO-TiO_2 hybrid [81].

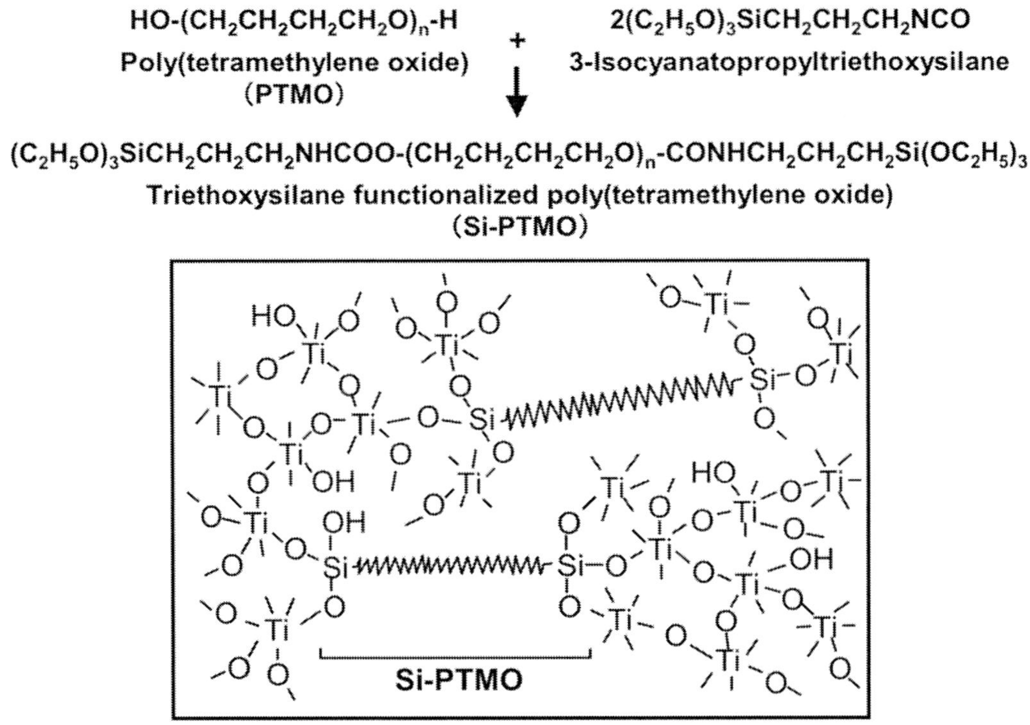

Fig. 13 Structural model of PTMO-TiO_2 hybrid.

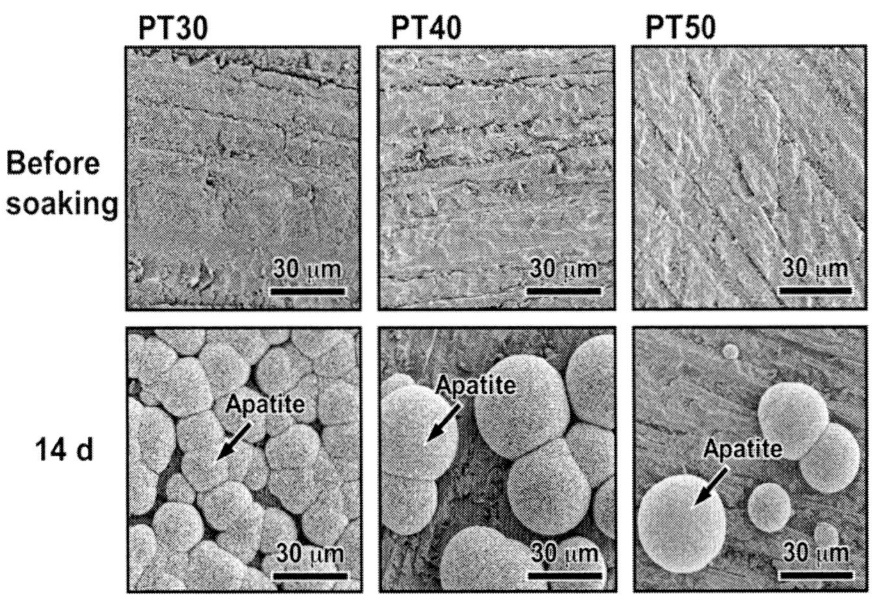

Fig. 14 SEM photographs of surfaces of hybrids PT30, PT40 and PT50 treated with hot water at 95 °C for 2 days before and after soaking in SBF for 14 days [85].

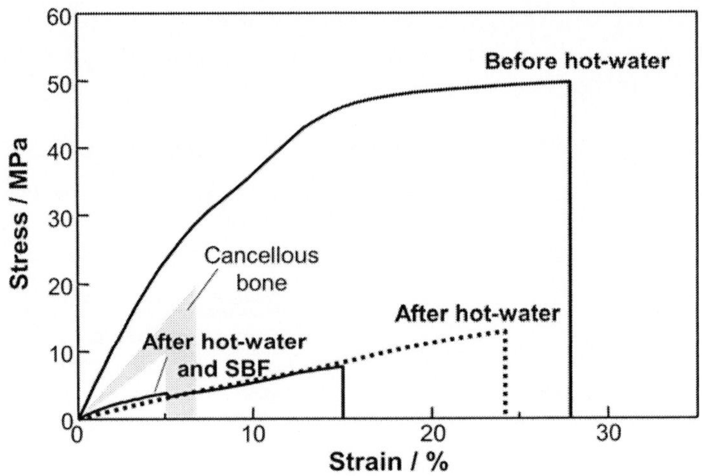

Fig. 15 Representative stress-strain curves of hybrid PT40 before and after the hot-water treatment, and that after both the hot-water treatment and SBF treatment for 28 days [85], in comparison with that reported for human cancellous bone [80].

8. Summary

Development of new types of deformable bioactive materials is desired for bone repair. A bioactive material which shows mechanical properties analogous to those of human bones and high durability in the body environment can be obtained by controlling not only the compositions but also the structures of sol-gel-derived organic-inorganic hybrids. The hybrid material can be a candidate for a novel bone-repairing material.

References

[1] Hulbert, SF. The use of alumina and zirconia in surgical implants. In: Hench LL, Wilson J. editors. *An Introduction to Bioceramics*. Singapore: World Scientific; 1993; 25-40.

[2] Hench, LL; Splinter, RJ; Allen, WC; Greenlee, TK. Bonding mechanism at the interface of ceramics prosthetic materials. *J. Biomed. Mater. Res. Symp.*, 1971, 2, 117-141.

[3] Hench, LL. Bioceramics: From concept to clinic. *J. Am. Ceram. Soc.*, 1991, 74, 1487-1510.

[4] Hench, LL; Andersson, Ö. Bioactive glasses, In: Hench LL, Wilson J. editors. *An Introduction to Bioceramics*. Singapore: World Scientific; 1993; 41-62.

[5] Brömer, H; Pfeil, E; Käs, HH. Glass-ceramic material. *German Patent* No. 2, 1973, 326, 100.

[6] Gross, UM; Müller-Mai, C; Voigt, C. Ceravital® bioactive ceramics. In: Hench LL, Wilson J. editors. *An Introduction to Bioceramics*. Singapore: World Scientific; 1993; 105-124.

[7] Jarcho, M; Kay, JL; Gumaer, RH; Drobeck, HP. Tissue, cellular and subcellular events at bone-ceramic hydroxyapatite interface. *J. Bioeng.*, 1997, 1, 79-92.

[8] LeGeros, RZ; LeGeros, JP. Dense hydroxyapatite. In: Hench LL, Wilson J. editors. *An Introduction to Bioceramics*. Singapore: World Scientific; 1993; 139-180.

[9] Kokubo, T; Shigematsu, M; Nagashima, Y; Tashiro, M; Nakamura, T; Yamamuro, T; Higashi, S. Apatite- and wollastonite-containing glass-ceramics for prosthetic application. *Bull. Inst. Chem. Res., Kyoto Univ.*, 1982, 60, 260-268.

[10] Kokubo, T. A/W glass-ceramic: Processing and properties. In: Hench LL, Wilson J. editors. *An Introduction to Bioceramics*. Singapore: World Scientific; 1993; 75-88.

[11] Höland, W; Nawmann, K; Vogel, W; Gummel, J. Machinable bioactive glass ceramic. *Wiss. Ztschr. Friedrich-Schiller-Univ. Jena, Math.-Naturwiss. Reihe.*, 1983, 32, 571-580.

[12] Höland, W; Vogel, W. Machinable and phosphate glass-ceramics. In: Hench LL, Wilson J. editors. *An Introduction to Bioceramics*. Singapore: World Scientific; 1993; 125-137.

[13] Kitsugi, T; Yamamuro, T; Nakamura, T; Kokubo, T. Bone bonding behavior of MgO-CaO-SiO_2-P_2O_5-CaF_2 glass (mother glass of A·W glass-ceramics). *J. Biomed. Mater. Res.*, 1989, 23, 631-648.

[14] Ohura, K; Nakamura, T; Yamamuro, T; Kokubo, T; Ebisawa, T; Kotoura, Y; Oka, M. Bone-bonding ability of P_2O_5-free CaO·SiO_2 glasses. *J. Biomed. Mater. Res.*, 1991, 25, 357-365.

[15] Reck, R; Störkel, S; Meyer, A. Bioactive glass-ceramics in middle ear surgery: An 8-year review. In: Ducheyne P, Lemons JE editors. *Bioceramics: Material characteristics versus in vivo behavior* Vol. 523. New York: The New York Academy of Science; 1988; 100-106.

[16] Yamamuro, T. Replacement of the spine with bioactive glass-ceramic prostheses. In: Yamamuro T, Hench LL, Wilson J. editors. *Handbook of bioactive ceramics* Vol. 1*: Bioactive glasses and glass-ceramics.* Boca Raton; CRC Press Inc.; 1990; 343-351.

[17] Ono, K; Yamamuro, T; Nakamura T; Kokubo, T. Apatite-wollastonite containing glass ceramic granule-fibrin mixture as a bone graft filler: Use with low granular density. *J. Biomed. Mater. Res.*, 1990, 24, 11-20.

[18] Gummel, J; Zippel, H; Hahnel, H. Replacement of the lumbar vertebral bodies with machinable bioactive glass ceramic. *Z. Klin. Med.*, 1988, 43, 1791-1793.

[19] Merwin, GE. Review of bioactive materials for otologic and maxillofacial applications. In: Yamamuro T, Hench LL, Wilson J. editors. *Handbook of bioactive ceramics* Vol. 1*: Bioactive glasses and glass-ceramics.* Boca Raton; CRC Press Inc.; 1990; 323-328.

[20] Yamamuro, T; Shikata, J; Okumura, H; Kitsugi, T; Kakutani, Y; Matsui, T; Kokubo, T. Replacement of the lumber vertebrae of sheep with ceramic prostheses. *J. Bone and Joint Surg.*, 1990, 72-B, 889-893.

[21] Hench, LL; Stanley, HR; Clark, AE; Hall, M; Wilson, J. Dental applications of bioglass® implant. In Bonfield W, Hastings GW, Tanner KE editors. *Bioceramics* Vol. 4. Oxford: Butterworth-Heinemann Ltd; 1991; 231-238.

[22] Wilson, J; Yli-Urpo A; Happonen, RP. Bioactive glasses: Clinical applications. In: Hench LL, Wilson J. editors. *An Introduction to Bioceramics*. Singapore: World Scientific; 1993; 63-73

[23] Yamamuro, T. A/W glass-ceramic: Clinical applications. In: Hench LL, Wilson J. editors. *An Introduction to Bioceramics*. Singapore: World Scientific; 1993; 89-103.

[24] Shors, EC; Holmes, RE. Porous hydroxyapatite. In: Hench LL, Wilson J. editors. *An Introduction to Bioceramics*. Singapore: World Scientific; 1993; 181-198.

[25] Kim, HM; Miyaji, F; Kokubo, T; Nakamura, T. Preparation of bioactive Ti and its alloys via simple chemical treatment. *J. Biomed. Mater. Res.*, 1996, 32, 409-417.

[26] Kokubo, T; Miyaji, F; Kim HM; Nakamura, T. Spontaneous formation of bonelike apatite layer on chemically treated titanium metals. *J. Am. Ceram. Soc.*, 1996, 79, 1127-1129.

[27] Kim, HM; Miyaji, F; Kokubo, T; Nakamura, T. Apatite-forming ability of alkali-treated Ti metal in body environment. *J. Ceram. Soc. Japan*, 1997, 105, 111-116.

[28] Kim, HM; Miyaji, F; Kokubo, T; Nakamura, T. Effect of heat treatment on apatite-forming ability of Ti metal induced by alkali treatment. *J. Mater. Sci.: Mater. Med.*, 1997, 8, 341-347.

[29] Kim, HM; Miyaji, F; Kokubo, T; Nakamura, T. Bonding strength of bonelike apatite layer to Ti metal substrate. *J. Biomed. Mater. Res.: Appl. Biomater.*, 1997, 38, 121-127.

[30] Kim, HM; Miyaji, F; Kokubo, T; Nishiguchi, S; Nakamura, T. Graded surface structure of bioactive titanium prepared by chemical treatment. *J. Biomed. Mater. Res.*, 1999, 45, 100-107.

[31] Yan, WQ; Nakamura, T; Kobayashi, M; Kim, HM; Miyaji, F; Kokubo, T. Bonding of chemically treated titanium implants to bone. *J. Biomed. Mater. Res.*, 1997, 37, 267-275.

[32] Yan, WQ; Nakamura, T; Kawanabe, K; Nishiguchi, S; Oka, M; Kokubo, T. Apatite layer-coated titanium for use as bone bonding implants. *Biomaterials*, 1997, 18, 1185-1190.

[33] Nishiguchi, S; Nakamura, T; Kobayashi, M; Kim, HM; Miyaji, F; Kokubo, T. The effect of heat treatment on bone-bonding ability of alkali-treated titanium. *Biomaterials*, 1999, 20, 491-500.

[34] Nishiguchi, S; Kato, H; Fujita, H; Kim, HM; Miyaji, F; Kokubo, T; Nakamura, T. Enhancement of bone-bonding strengths of titanium alloy implants by alkali and heat treatments. *J. Biomed. Mater. Res.: Appl. Biomater.*, 1999, 48, 689-696.

[35] Miyazaki, T; Kim, HM; Miyaji, F; Kokubo, T; Kato, H; Nakamura, T. Bioactive tantalum metal prepared by NaOH treatment. *J. Biomed. Mater. Res.*, 2000, 50, 35-42.

[36] Miyazaki, T; Kim, HM; Kokubo, T; Miyaji, F; Kato, H; Nakamura, T. Effect of thermal treatment on apatite-forming ability of NaOH-treated tantalum metal. *J. Mater. Sci.: Mater. Med.*, 2001, 12, 683-687.

[37] Kato, H; Nakamura, T; Nishiguchi, S; Matsusue, Y; Kobayashi, M; Miyazaki, T; Miyaji, F; Kim, HM; Kokubo, T. Bonding of alkali- and heat-treated tantalum implants to bone. *J. Biomed. Mater. Res.: Appl. Biomater.*, 2000, 53, 28-35.

[38] Park, JB; Lakes, RS. *Biomaterials 2nd Ed.* New York: Plenum Press; 1992; 185-222.

[39] Park, JB. *Biomaterials*. New York: Plenum Press; 1979; 105.

[40] Bonfield, W; Grynpas, MD; Tully, AE; Bowman, J; Abram, J. Hydroxyapatite reinforced polyethylene – a mechanically compatible implant. *Biomaterials,* 1981, 2, 185-186.

[41] Bonfield, W. Design of bioactive ceramic-polymer composites. In: Hench LL, Wilson J. editor. *An Introduction to Bioceramics*. Singapore: World Scientific; 1993; 299-303.

[42] Hench, LL. Stability of ceramics in the physiological environment. In Williams DF editor. *Fundamental Aspects of Biocompatibility* Vol. 1. Boca Raton: CRC Press Inc., 1981; 67-85.

[43] Hench, LL; Clark, AE. Adhesion to bone. In Williams DF editor. *Biocompatibility of Orthopedic Implants* Vol. 2. Boca Raton: CRC Press Inc., 1982; 129-170.

[44] Höland, W; Vogel, W; Nawmann, K; Gummel, J. Interface reaction between machinable bioactive glass-ceramics and bone. *J. Biomed. Mater. Res.*, 1985, 19, 303-312.

[45] Kitsugi, T; Nakamura, T; Yamamuro, T; Kokubo, T; Shibuya, T; Takagi, M. SEM-EPMA observation of three types of apatite-containing glass-ceramics implanted in bone: The variance of a Ca-P-rich layer. *J. Biomed. Mater. Res.*, 1987, 21, 1255-1271.

[46] Kokubo, T; Ohtsuki, C; Kotani, S; Kitsugi T; Yamamuro, T. Surface structure of bioactive glass-ceramics A-W implanted into sheep and human vertebra. In Heimke G editor. *Bioceramics* Vol. 2. Cologne: German Ceramic Society; 1990; 113-120.

[47] Kokubo, T. Surface chemistry of bioactive glass-ceramics. *J. Non-Cryst. Solids*, 1990, 120, 138-151.

[48] Ohtsuki, C; Kushitani, H; Kokubo, T; Kotani, S; Yamamuro, T. Apatite formation on the surface of Ceravital-type glass-ceramic in the body. *J. Biomed. Mater. Res.*, 1991, 25, 1363-1370.

[49] Neo, M; Kotani, S; Yamamuro, T; Ohtsuki, C; Kokubo T; Bando, Y. A comparasive study of ultrastructures of the interfaces between four kinds of surface-active ceramic and bone. *J. Biomed. Mater. Res.*, 1992, 26, 1419-1432.

[50] Kokubo, T; Hayashi, T; Sakka, S; Kitsugi, T; Yamamuro, T; Takagi, M; Shibuya, T. Surface structure of a load-bearable bioactive glass-ceramic A-W. In Vincenzini P editor. *Ceramics in Clinical Applications*. Amsterdam, Elsevier Science Publishing; 1987; 175-184.

[51] Kokubo, T; Ito, S; Huang, ZT; Hayashi, T; Sakka, S; Kitsugi, T; Yamamuro, T. Ca, P-rich layer formed on high-strength bioactive glass-ceramic A-W. *J. Biomed. Mater. Res.*, 1990, 24, 331-343.

[52] Neo, M; Nakamura, T; Yamamuro, T; Ohtsuki, C; Kokubo, T. Apatite formation on three kinds of bioactive material at an early stage in vivo: A comparative study by transmission electron microscopy. *J. Biomed. Mater. Res.*, 1993, 27, 999-1006.

[53] Loty, C; Sautier, JM; Boulekbache, H; Kokubo, T; Kim, HM; Forest, N. *In vitro* bone formation on a bone-like apatite layer prepared by a biomimetic process on a bioactive glass-ceramic. *J. Biomed. Mater. Res.*, 2000, 49, 423-434.

[54] Kokubo, T; Kushitani, H; Sakka, S; Kitsugi, T; Yamamuro, T. Solutions able to reproduce *in vivo* surface-structure changes in bioactive glass-ceramic A-W. *J. Biomed. Mater. Res.*, 1990, 24, 721-734.

[55] Schmidt, H. New type of non-crystalline solids between inorganic and organic materials. *J. Non-Cryst. Solids,* 1985, 73, 681-91.

[56] Wilkes, GL; Orler, B; Huang, HH. 'Ceramers': Hybrid materials incorporating polymeric/oligomeric species into inorganic glasses unitizing a sol-gel approach. *Polym. Prepr.,* 1985, 26, 300-302.
[57] Huang, HH; Orler, B; Wilkes, GL. Ceramers: Hybrid materials incorporating polymeric/oligomeric species with inorganic glasses by a sol-gel process. 2. Effect of acid content on the final properties. *Polymer Bulletin*, 1985, 14, 557-564.
[58] Mackenzie, JD; Chung, YJ; Hu, Y. Rubbery ormosils and their applications. *J. Non-Cryst. Solids*, 1992, 147&148, 271-279.
[59] Mackenzie, JD. Structures and properties of ormosils. *J. Sol-Gel Sci. and Tech.*, 1994, 2, 81-86.
[60] Li, P; Ohtsuki, C; Kokubo, T; Nakanishi, K; Soga, N; Nakamura, T; Yamamuro, T. Apatite formation induced by silica gel in a simulated body fluid. *J. Am. Ceram. Soc.*, 1992, 75, 2094-2097.
[61] Li, P; Ohtsuki, C; Kokubo, T; Nakanishi, K; Soga, N; Nakamura T; Yamamuro, T. A role of hydrated silica, titania and alumina in forming biologically active apatite on implant. *Trans. 4th World Biomater. Congress*. Berlin: Society for Biomaterials; 1992; 4.
[62] Li, P; Ohtsuki, C; Kokubo, T; Nakanishi, K; Soga, N; Nakamura, T; Yamamuro, T. Effects of ion in aqueous media on hydroxyapatite induction by silica gel and its relevance to bioactivity of bioactive glasses and glass-ceramics. *J. Appl. Biomater.*, 1993, 4, 221-229.
[63] Li, P; Ohtsuki, C; Kokubo, T; Nakanishi, K; Soga, N; de Groot, K. The role of hydrated silica, titania and alumina in inducing apatite on implants. *J. Biomed. Mater. Res.*, 1994, 28, 7-15.
[64] Li, P; Kangasniemi, I; de Groot, K; Kokubo, T. Bonelike hydroxyapatite induction by a gel-derived titania on a titanium substrate. *J. Am. Ceram. Soc.*, 1994, 77, 1307-1312.
[65] Uchida, M; Kim, HM; Kokubo, T; Fujibayashi, S; Nakamura, T. Structural dependence of apatite formation on titania gels in a simulated body fluid. *J. Biomed. Mater. Res.,* 2003, 64A, 164-170.
[66] Miyazaki, T; Kim, HM; Kokubo, T; Kato, H; Nakamura, T. Induction and acceleration of bonelike apatite formation on tantalum oxide gel in simulated body fluid. *J. Sol-Gel. Sci. Tech.,* 2001, 21, 83-88.
[67] Uchida, M; Kim, HM; Miyaji, F; Kokubo, T; Nakamura, T. Bonelike apatite formation induced on zirconia gel in a simulated body fluid and its modified solutions. *J. Am. Ceram. Soc.,* 2001, 84, 2041-44.
[68] Uchida, M; Kim, HM; Kokubo, T; Tanaka, K; Nakamura, T. Structural dependence of apatite formation on zirconia gels in a simulated body fluid. *J. Ceram. Soc. Japan,* 2002, 110, 710-715.
[69] Miyazaki, T; Kim, HM; Kokubo, T; Ohtsuki, C; Nakamura, T. Apatite-forming ability of niobium oxide gels in a simulated body fluid. *J. Ceram. Soc. Japan*, 2001, 109, 929-933.
[70] Ohtsuki, C; Kokubo T; Yamamuro, T. Mechanism of apatite formation on $CaO-SiO_2-P_2O_5$ glasses in a simulated body fluid. *J. Non-Cryst. Solids,* 1992, 143, 84-92.
[71] Tsuru, K; Ohtsuki, C; Osaka, A; Iwamoto T; Mackenzie, JD. Bioactivity of sol-gel derived organically modified silicates, part I: in vitro examination. *J. Mater. Sci.: Mater. Med.,* 1997, 8, 157-161.

[72] Kokubo, T. Bioactive glass ceramics; Properties and applications. *Biomaterials*, 1991, 12, 155-163.

[73] Chen, Q; Miyaji, F; Kokubo, T; Nakamura, T. Apatite formation on PDMS-modified CaO-SiO_2-TiO_2 hybrids prepared by sol-gel method. *Biomaterials,* 1999, 20, 1127-1132.

[74] Chen, Q; Miyata, N; Kokubo, T; Nakamura, T. Bioactivity and mechanical properties of PDMS-modified CaO-SiO_2-TiO_2 hybrids prepared by sol-gel process. *J. Biomed. Mater. Res.,* 2000, 51, 605-611.

[75] Chen, Q; Miyata, N; Kokubo, T; Nakamura, T. Bioactivity and mechanical properties of PDMS-modified calcia-silica hybrids with added titania. *J. Am. Ceram. Soc.*, 2003, 86, 806-810.

[76] Chen, Q; Kamitakahara, M; Miyata, N; Kokubo, T; Nakamura, T. Preparation of bioactive PDMS-modified CaO-SiO_2-TiO_2 hybrids by the sol-gel method. *J. Sol-Gel Sci. Tech.*, 2000, 19, 101-105.

[77] Chen, Q; Miyata, N; Kokubo, T; Nakamura, T. Effect of heat treatment on bioactivity and mechanical properties of PDMS-modified CaO-SiO_2-TiO_2 hybrids via sol-gel process. *J. Mater. Sci.: Mater. Med.,* 2001, 12, 515-522.

[78] Kamitakahara, M; Kawashita, M; Miyata, N; Kokubo, T; Nakamura, T. Bioactivity and mechanical properties of polydimethylsiloxane(PDMS)-CaO-SiO_2 hybrids with different PDMS contents. *J. Sol-Gel Sci. Tech.*, 2001, 21, 75-81.

[79] Kamitakahara, M; Kawashita, M; Miyata, N; Kokubo, T; Nakamura, T. Bioactivity and mechanical properties of polydimethylsiloxane(PDMS)-CaO-SiO_2 hybrids with different calcium contents. *J. Mater. Sci.: Mater. Med.,* 2002, 13, 1015-1020.

[80] Hench, LL; Wilson, J. In: Hench LL, Wilson J. editors. *An Introduction to Bioceramics*. Singapore: World Scientific; 1993; 12.

[81] Kamitakahara, M; Kawashita, M; Miyata, N; Kokubo, T; Nakamura, T. Degradation of bioactive polydimethylxiloxane-CaO-SiO_2-TiO_2 and poly(tetramethylene oxide)-CaO-TiO_2 hybrids in a simulated body fluid", *J. Am. Ceram. Soc.*, 2004, 87, 235-239.

[82] Hayashi, T; Saito, H. Preparation of CaO-SiO_2 glasses by the gel method. *J. Mater. Sci.*, 1980, 15, 1971-1977.

[83] Kamitakahara, M; Kawashita, M; Miyata, N; Kokubo, T; Nakamura, T. Apatite formation on CaO-free polydimethylsiloxane(PDMS)-TiO_2 hybrids", *J. Mater. Sci.: Mater. Med.*, 2003, 14, 1067-1072.

[84] Wen, J; Wilkes, GL. Organic/inorganic hybrid network materials by the sol-gel approach. *Chem. Mater.*, 1996, 8, 1667-1681.

[85] Kamitakahara, M; Kawashita, M; Miyata, N; Kokubo, T; Nakamura, T. Apatite-forming ability and mechanical properties of CaO-free poly(tetramethylene oxide) (PTMO)-TiO_2 hybrids treated with hot water. *Biomaterials*, 2003, 24, 1357-1363.

In: Research in Hybrid Materials
Editors: Simon J. Brunner and Julian W. Egger
ISBN 978-1-60456-718-2

Chapter 3

ORGANIC-INORGANIC LAYERED HYBRID MATERIALS

***B. Ingham*[1], *S.V. Chong*[2] *and J.L. Tallon*[1,2]**

[1] The MacDiarmid Institute for Advanced Materials and Nanotechnology, Victoria University of Wellington, Wellington, New Zealand

[2] Industrial Research Limited, Lower Hutt, New Zealand

ABSTRACT

Organic-inorganic hybrid materials based on simple two-dimensional layers of tungsten oxide separated by organic molecules are merely one example of the huge variety of systems classed as "organic-inorganic hybrids". The materials in question have the potential to combine the high-electron mobility of the metal oxide layers with the flexibility of the organic layers. More importantly they identify the primary elements that might enable the formation of more complex nanostructures through self-assembling processes and the templating effect of the organic component. One such example that will be discussed here is the possibility of including magnetic ions within the inorganic layers, which is observed to have profound effects upon the magnetic behaviour of the resulting hybrids.

Tungsten oxide-organic amine hybrid materials have been synthesised via a range of methods and characterised using conventional materials characterisation techniques, such as X-ray diffraction, infrared and Raman spectroscopy, UV-visible spectroscopy and electron microscopy. The work is also complemented by *ab initio* calculations of the physical and electronic structure of the tungsten oxide hybrids. For compounds incorporating transition metal elements, the magnetisation was also measured. The effect of the low-dimensionality of the inorganic component is clearly seen in materials that include certain transition metals, in the form of magnetic transitions that indicate a crossover from 2-dimensional to 3-dimensional behaviour.

INTRODUCTION

There has been much interest in recent years in the field of organic-inorganic hybrid materials. In the literature this title is used quite broadly, ranging from metal-organic complexes to macrostructural inorganic particles in an organic matrix. Between these two extremes, organic-inorganic compounds can be found which are comprised of molecular units in 3-dimensional networks, 2-dimensional layered structures, 1-dimensional chains or even 0-dimensional clusters. Of these, one can find a variety of systems from examples where the organic component is used merely as a templating agent for the inorganic layer, to cases where several different organic species are used in a structural sense via co-ordination to the inorganic species. The organic molecule can be chosen to have a broad range of functionality, provided it can be incorporated within the inorganic structure. The lower the dimensionality of the inorganic framework, the greater the variety of organic molecules that can be included.

Some of the major categories of organic-inorganic hybrid systems are as follows:

a) Phosphate-based organic-inorganic hybrid materials. These are normally synthesised using hydro- or solvo-thermal methods and typically form as networks (1-, 2- or 3-D) of metal atoms linked via the oxygen atoms of the PO_4^{3-} tetrahedra linkages [1-2]. Aminoalkanes are the most common organic intercalate [3]. The size of the organic molecules and the synthesis conditions (pH, temperature, pressure, concentrations) have a strong influence on the structure of the inorganic layer.
b) Expanded perovskite-based organic/inorganic hybrid materials. Many different layered perovskites have been used as the inorganic layers in organic-inorganic hybrids formed by *chimie douce* methods. These include the perovskite $Ca_2Nb_3O_{10}$ and its derivatives [4], the Aurivillius phases $Bi_2W_nO_{3n+3}$ [5], the MnO_2 'slab' structures birnessite and buserite [6-7], the layered double hydroxide (LDH) system, based on alumino-silicate clays [8], and even the Bi-2212 ($Bi_2Sr_2CaCu_2O_8$) superconductor [9-13]. The organic molecules act primarily to separate the inorganic layers, and they may be present as neutral species or positively charged to maintain charge balance.
c) Cluster ions. Zero-dimensional hybrids of organic molecules and Keggin ions or Wells-Dawson structures consisting of clusters of bonded MO_6 octahedra (M = Mo or W) may be produced via a range of techniques, including hydrothermal synthesis [14], precipitation from solution [15], sol-gel method [16], or successive layers built up by dip-coating [17].
d) Other miscellaneous systems. There are of course many other possible combinations for forming organic-inorganic hybrid materials, however in literature these appear to be for the most part unrelated. One system worthy of mention however is the vanadate hybrid structures, which are able to form a variety of structures, most of them quite complicated [18].
e) Metal halide layer-based organic-inorganic hybrid materials. There are a wide variety of these materials, which are comprised of MX_6 octahedra which are corner-shared to form simple regular 2-dimensional sheets, MX_4^{2-} (M = M^{2+} metal ion, X = Cl, Br, I). The layers are then separated by amine-terminated organic molecules. Ref [19] is an excellent review of these materials, displaying the wide variety of inorganic species and organic amine molecules that together can form this relatively simple structure. These systems

have been studied for over 70 years [20] and have been shown to display remarkable photoluminescence [21-22] and have been applied as materials in thin film transistors [23]. Their magnetic properties have also been extensively studied and represent a near-perfect 2-dimensional magnetic system [24-27].

One major detrimental factor of the metal halide-organic amine system is that the compounds are easily oxidised in air and therefore must be synthesised, stored and handled under an inert atmosphere. If one were able to form an identical structure with an appropriate metal oxide, this issue could be overcome. Such an oxide structure requires a hexavalent metal ion, which nominally limits such structures to molybdenum and tungsten oxides. Both of these have been used as the inorganic component of hybrid materials before [28-29], but not extensively. This may be due in part to the formation of heteropolytungstate and -molybdate ions at low pH [30]. Several other molybdenum oxide-organic hybrids have been reported [31-34] but the inorganic frameworks are not simple corner-shared arrangements of the MoO_6 octahedra. For this reason the scope of this chapter will concentrate on tungsten oxide hybrids that do exhibit this structure.

While tungsten trioxide WO_3 forms a relatively simple, almost cubic structure of corner-shared WO_6 octahedra [35], molybdenum trioxide forms into zig-zagged layers, known as the MoO_3 L-structure [35-36]. Both molybdenum and tungsten oxide hydrates exist, with the general formula $MO_3{\cdot}nH_2O$ (n = 1, 2, ...). They are isostructural and form layers of corner shared MO_6 octahedra separated by water molecules [37]. For n=1, the oxygen atom of the water molecule is co-ordinated to the metal ion as one of the apical oxygen atoms [38]. Subsequent water molecules (n > 1) fill the space between the layers [39]. If the layered structure is preserved through moderate synthesis conditions, one ought to be able to form tungsten and/or molybdenum oxide hybrids with layered structures similar to those of the metal halide hybrids (Figure 1).

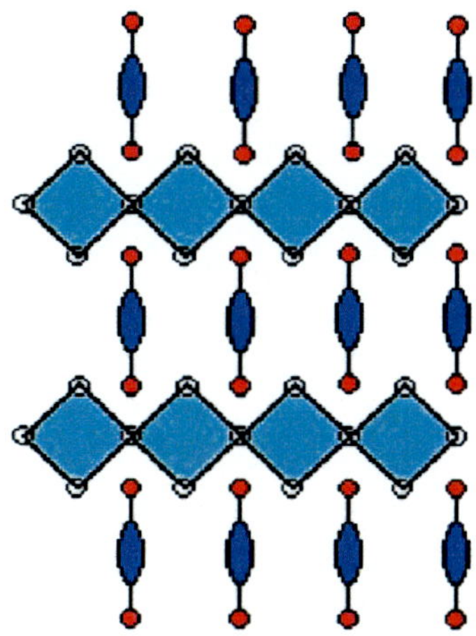

Figure 1. General schematic diagram of the layered tungsten oxide-organic hybrid materials. Blue diamonds represent corner-shared tungsten oxide octahedra, red circles are ammonium terminal groups, and blue ovals the organic R group (e.g. alkane).

Tungsten trioxide also exhibits interesting electrical and optical properties upon doping. Coloured "tungsten bronzes" with metallic-like conductivity are formed when small metal ions (e.g. sodium) are inserted in the vacancies within the oxide structure [40-41]. Removal of oxygen also causes a stark colour change [42] and increased conductivity [43]. For this reason tungsten oxide is a common material employed in electrochromic applications [44]. If these aspects of the oxide can be incorporated within the hybrid also, this would prove to be a very interesting system.

EXPERIMENTAL

Tungsten oxide-organic amine hybrid materials have been formed using a variety of synthesis techniques [45]:

a) Evaporation of solvent. Tungstic acid (H_2WO_4) is dissolved in aqueous ammonia solution (30 wt. %) with heating. The desired organic amine is then dissolved in an appropriate solvent and added to the hot tungstate solution. The remaining solvent is evaporated and a white-cream coloured powder forms. The product is washed and filtered with ethanol and dried in air at 80°C. It is possible to fabricate thin films by spin- or dip-coating using the saturated solution before a precipitate forms.
b) *Chimie douce*. Tungstic acid is soaked at room temperature for a period of days-weeks in an appropriate solution of the organic amine. The intercalation of the amine molecules between the inorganic layers of the tungsten oxide hydrate is evidenced by the colour change from yellow to white-cream. The product is filtered, washed with ethanol, and dried.
c) 'Melt method'. Many simple aliphatic and aromatic amines melt cleanly at temperatures in the range of –50 to 200°C. To an excess of the liquid amine tungstic acid is added, followed by a few drops of aqueous ammonia solution, which appears to act as a templating agent. There is an immediate colour change from yellow to white-cream. The product is washed and filtered with ethanol and dried as per the other synthesis methods.

Transition metal-substituted tungstate hybrids have also been synthesised via the same three techniques. Manganese and iron tungstate hybrids can be formed using the evaporation of solvent technique, however cobalt, nickel and copper all react with the excess of ammonia solution needed to dissolve the tungstic acid. For this reason the *chimie douce* and melt methods are used to form cobalt, nickel and copper hybrids from their respective tungstate hydrates. Transition metal tungstate hydrates are easily formed as precipitates when aqueous solutions of the desired metal salt (e.g. chloride) and sodium tungstate are combined [46]. The transition metal hydrates and hybrids exhibit a wide variety of different colours.

The compounds are abbreviated herein as follows: the organic molecules are for the most part mono-ammonium-n-alkanes or di-ammonium-n-alkanes, labelled MAn (n = 1, 2, …) or DAn (n = 2, 4, …) respectively. Other organic intercalates include diaminobenzene (phenylenediamine, DAphen) and phenethylamine (phen, a mono-ammonium organic). The inorganic framework is identified by the metal ions present, i.e. W- or Mo- for tungsten and molybdenum oxide hybrids respectively, and (M,W)- for compounds including transition metal ions (M = Mn, Fe, Co, Ni, Cu). Thus (Cu,W)-DA6 corresponds to a copper tungstate hybrid with 1,6-diaminohexane as the organic intercalate.

X-ray diffraction (XRD) was performed on powder samples using a Philips PW1700 series powder diffractometer employing Co Kα radiation. Infrared spectra were recorded with a Bomem DA8 FT spectrometer using the KBr disc method over the range 400-4000 cm^{-1} with 2 cm^{-1} resolution. Raman spectra were collected using a Jobin-Yvon LabRam KR spectometer with an excitation wavelength of 633 nm. UV-visible spectroscopy of powder samples was measured using a GCA/McPherson 2051 monochromate spectrometer with a pre-disperser and diffuse reflectance integrating sphere set-up, with a spectral bandwidth of 1 nm. Scanning electron microscopy (SEM) and energy dispersive X-ray analysis were carried out using a Leo 440 microscope with up to 50,000 times magnification capability and an Oxford ISIS EDX system. Thermogravimetric analysis (TGA) was performed with a Pheometrics STA1500 TG-DTG/DSC thermogravimetric analyser with a heating rate of $5^{o}C$/min under flowing nitrogen gas. Elemental microanalyses for C, H and N were performed using the automatic analyser at the Campbell Microanalytical Laboratory at Otago University [47]. DC and AC magnetic susceptibilities were measured using a Quantum Design MPMS XL SQUID magnetometer. *Ab initio* calculations were performed using the VASP software [48-49], which employees a local-density approximation (LDA) to density functional theory (DFT) using ultra-soft plane-wave pseudopotentials.

RESULTS

I. Physical Properties of tungsten Oxide Hybrids

XRD spectra of various tungsten oxide hybrid materials with diaminoalkanes of carbon chain lengths from 2 to 12 carbons are shown in Figure 2a. They display a single intense peak at low 2-theta values which is characteristic of the spacing between inorganic layers. The progression of the position of this first peak with increasing organic length is illustrated in Figure 2b. From the slope and intercept of this line it is concluded that the organic molecules lie at an angle of 6.1° to the normal of the plane of the inorganic layers, the dimension of which corresponds to a single tungsten oxide corner-shared octahedral layer. The same phase is produced via each of the three different synthesis methods utilised.

A SEM micrograph of a W-DA6 (tungsten oxide with 1,6-diaminohexane as the organic intercalate) sample is shown in Figure 3. The platy morphology can clearly be seen. Due to the small size of the crystallites the structure of these materials could not be determined by single crystal X-ray studies and we have not yet succeeded in growing larger crystals. Therefore other techniques were used to determine information about the structure.

Infrared and Raman spectroscopy (Fig. 4 and 5, respectively) were used to examine the nature of the organic molecules, the inorganic layers and the interaction between them [50]. From these two methods we observe that the low-frequency peaks corresponding to the inorganic structure are unchanged as the organic molecule is lengthened from two carbons to twelve. However, the inorganic structure is different from that of the layered tungsten oxide hydrate ("tungstic acid"). There is also a difference in the organic molecule with two carbons

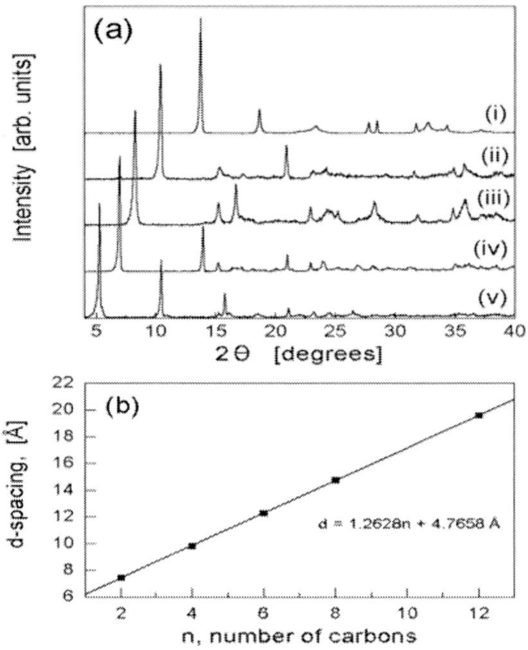

Figure 2. (a) XRD spectra of tungsten oxide hybrids of various length organic diamines, $H_2N(CH_2)_nNH_2$ (I) n=2, (ii) n=4, (iii) n=6, (iv) n=8, (v) n=12. (b) Progression of the interlayer spacing as determined from the series of 00l lines.

Figure 3. SEM image of W-DA6 sample.

as opposed to the longer alkyl chains. In the spectra of the longer alkyl chains a broad feature appears at ~2100 cm^{-1} which is believed to correspond to a combination of asymmetrical $-NH_3^+$ bending and torsional oscillation of the $-NH_3^+$ group as it interacts via hydrogen-bonding with the apical oxygen of the W-O framework [51]. Both $-NH_3^+$ bending and $-NH_2$ scissoring modes occur near 1600 cm^{-1}, the former displaying three bands (as seen in the longer alkyl chain hybrids) but the latter only two (as in W-DA2). These two features indicate that in W-DA2 the organic molecule exists as a neutral amine rather than being protonated, which is the case for the longer organic hybrids. This raises questions as to how charge balance is achieved in the shortest organic W-DA2 system. Possibly the protons required for charge balance are delocalised to provide shielding for the apical oxygen atoms of adjacent layers.

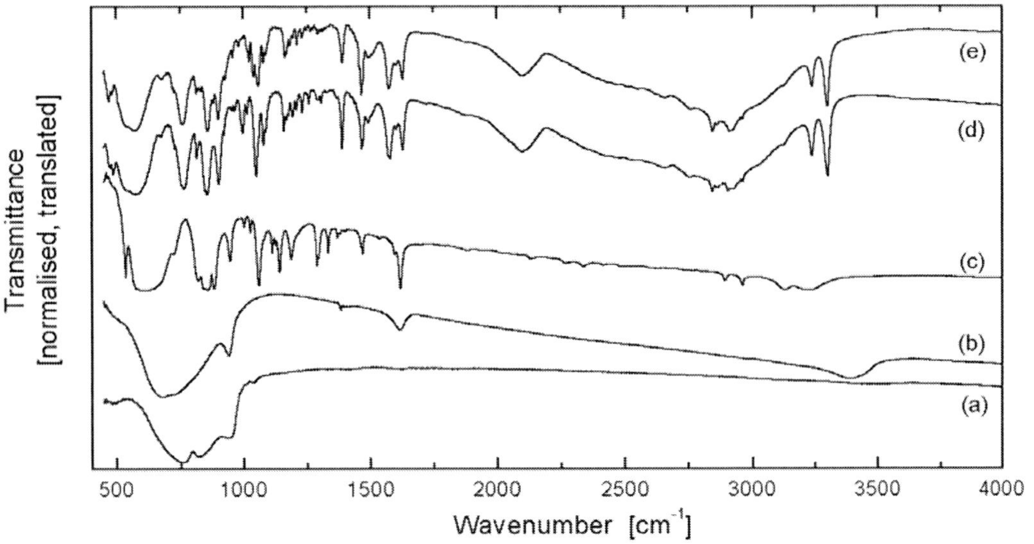

Figure 4. Infrared spectra of W-DAn hybrids and related materials, (a) WO_3, (b) H_2WO_4, (c) W-DA2, (d) W-DA6, (e) W-DA12.

The Raman spectra provide additional information about the inorganic layer. Comparing the hybrid spectra with that of H_2WO_4, the most dominant feature is a single intense peak near 900 cm^{-1}. This corresponds to an apical W=O bond [52]. The position of the peak shifts from 950 cm^{-1} in H_2WO_4 to 890-900 cm^{-1} in the W-DAn hybrids. This decrease in the frequency of the vibration indicates that the W=O bond is slightly weakened in the case of the hybrids, which is consistent with the change in environment of the apical oxygen atoms from H_2WO_4 to the hybrids. In H_2WO_4 one apical oxygen is contributed from a co-ordinated water molecule, leaving the other to bind to the tungsten atom relatively strongly. However in the W-DAn hybrids the influence by the intercalated organic with its amine/ammonium termination is more uniform towards each apical oxygen, which causes the W=O bond to become slightly weaker, and therefore probably longer. The Raman spectra also indicate the absence of any co-ordinated water in the hybrid samples (observed in H_2WO_4 as a peak at 377 cm^{-1}).

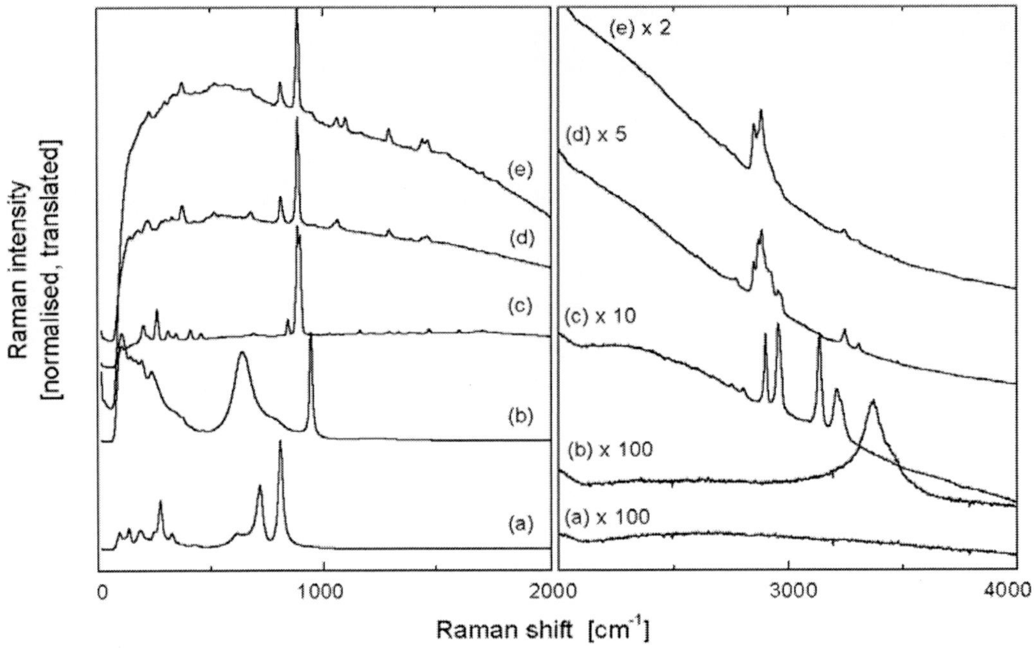

Figure 5. Raman spectra of tungsten oxide hybrids and related compounds: (a) WO_3, (b) H_2WO_4, (c) W-DA2, (d) W-DA6, (e) W-DA12.

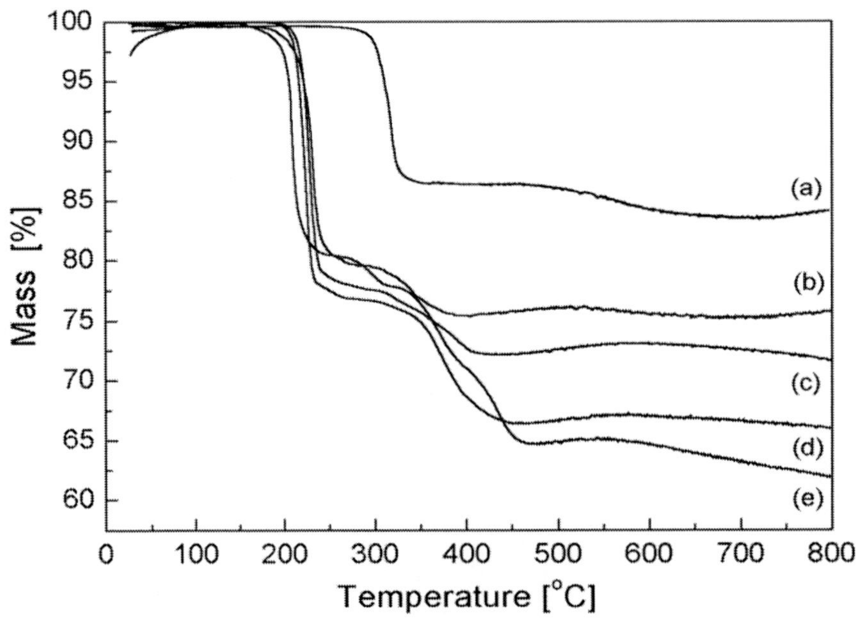

Figure 6. TGA curves of tungsten oxide hybrids W-DAn: (a) n=2, (b) n=4, (c) n=6, (d) n=8, (e) n=12.

TGA results (Figure 6) indicate two major features at around 230 and 370°C for the longer chain hybrids, and 315 and 500°C for the 2-carbon chain hybrid. These correspond respectively to the clean deintercalation of the organic molecules and decomposition to WO_3. As in the IR spectra, the shortest chain DA2 hybrid displays different behaviour from the other materials. The two results may both be related to a different bonding regime employed in the W-DA2 hybrid. The mass losses are consistent with the inorganic layer having the formula WO_4. This is also confirmed via elemental analysis, shown in Table I.

Table I. Elemental analysis results for selected tungsten oxide hybrid samples.

Sample	C (%)	H (%)	N (%)	Formula	Expect	Ratio to WO_4
W-DA2	8.38	2.69	9.63	$C_2H_{7.64}N_{1.97}$	$C_2H_{8\text{-}10}N_2$	0.917 : 1
W-DA6	20.19	4.53	7.74	$C_6H_{16.10}N_{1.98}$	$C_6H_{16\text{-}18}N_2$	0.977 : 1

The electronic structure of the tungsten oxide hybrids was investigated using UV-visible diffuse reflectance and complemented with *ab initio* computations. The diffuse reflectance spectra are given in Figure 7 and each display a single absorption edge at 300 nm (4.1 eV). The parent material, WO_3, also displays a single absorption edge, located at 450 nm (2.8 eV). This feature is thought to correspond to a band gap in each series. WO_3 is reported to have an indirect band gap of 2.6 eV [53], while the band gap of the hybrids is larger than the parent oxide. There is virtually no change in the size of the gap as one changes the length of the organic molecule in the hybrid materials.

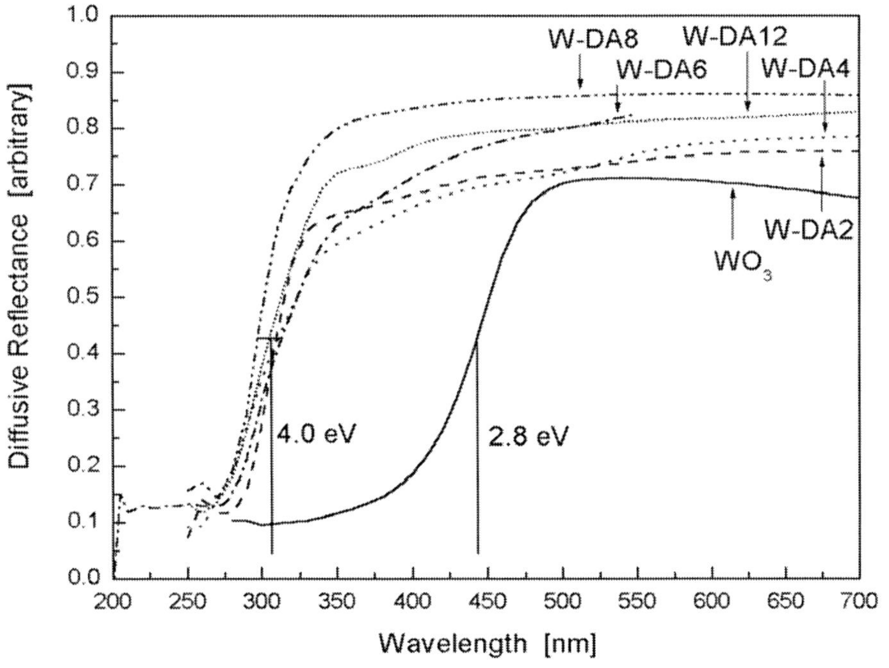

Figure 7. UV-visible diffuse reflectance spectra of WO_3 and W-DAn hybrids.

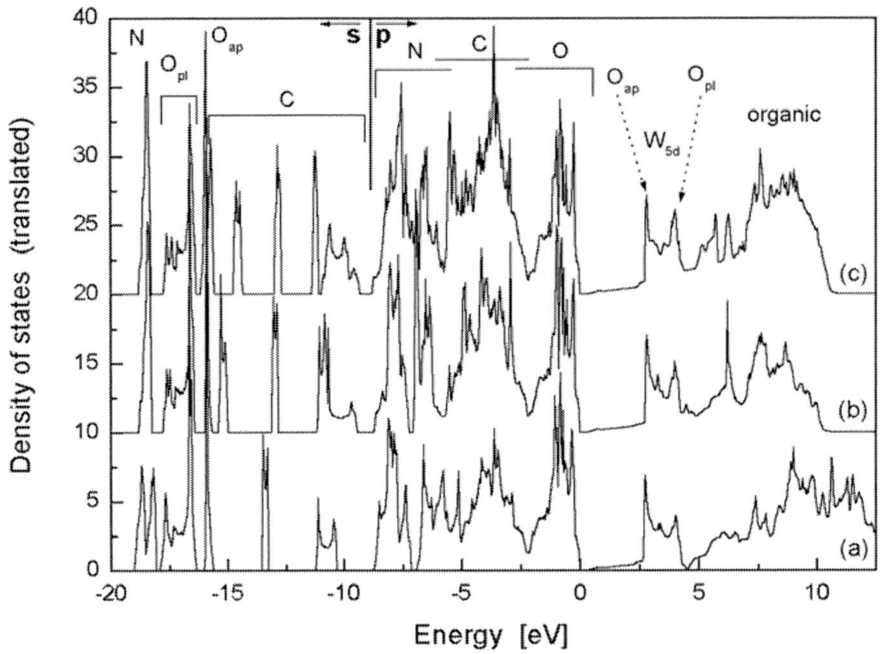

Figure 8. Calculated density of states for selected W-DAn hybrids: (a) n=2, (b) n=4, (c) n=6.

The calculated density of states for WO_3 and three hybrids with chain lengths of 2, 4 and 6 carbons are shown in Figure 8. Amongst the three hybrid systems studied, the only noticeable differences are well below the Fermi level and correspond to the addition of extra non-equivalent carbon bands. Near the Fermi level, the three hybrids display the same features: the valence band is comprised solely of oxygen *2p* states and the conduction band a mixture of tungsten *5d* and oxygen *2p* states. The Fermi level is located at the top of the valence band (scaled to E = 0). The gap is around 0.5 eV, which is much smaller than the 4.1 eV observed as the position of the absorption edge. However this difference is not too surprising as the density functional method utilised generally underestimates band gaps. The density of states is very low up to two peak features located at 2.8 and 4.0 eV, which may account for features observed in the optical spectra.

In the WO_3 density of states the gap is about 0.4 eV, but the density of states rises up to a strong peak at 3.8 eV. This may correspond to a direct band gap, which is observed in the calculated band structure [54-56]. A similar value (4.0 eV) is obtained from absorption coefficient analysis of transmittance and reflectance spectra of WO_3 thin films [57]. Analysis of the partial density of states reveals that the valence band is comprised solely of oxygen *2p* states, with the conduction band consisting only of tungsten *5d* states. The Fermi level is again located at the top of the valence band (scaled to E = 0), rendering both WO_3 and the hybrid system electrically insulating.

In summary, then, WO_3 is an insulator that exhibits an indirect band gap of 2.6-2.8 eV (which is observed in powder diffuse reflectance) and a direct band gap of 3.8-4.0 eV (which is observed in the absorption coefficient analysis). Extending this to the hybrid system indicates two peak features in the calculated density of states at 2.8 and 4.0 eV which may correspond to two indirect band gaps, of which the 4.0 eV gap is seen in the powder diffuse reflectance measurements. The length of the organic intercalated molecule does not affect the size of the gap.

II. Physical Properties of Transition Metal Tungstate Hybrids

Many compounds including transition metal ions exhibit bright colours, and the same is true of transition metal tungstate hybrids. These are indicated in Table II.

Table II. Colours of selected transition metal tungstate hybrid samples.

Intercalate	Transition metal ion included				
	Mn	Fe	Co	Ni	Cu
H_2O	Tan	Red-brown	Pink	Pale green	Green-blue
DA2	Tan	Red-brown	Orange	Pale purple	Blue-purple
DA6	Tan	Red-brown	Dark blue	Pale blue	Teal
phen	Tan	-	-	Pale green	Pale blue-green

For the remainder of this section we will focus primarily on the manganese tungstate hybrids as they have been studied the most extensively.

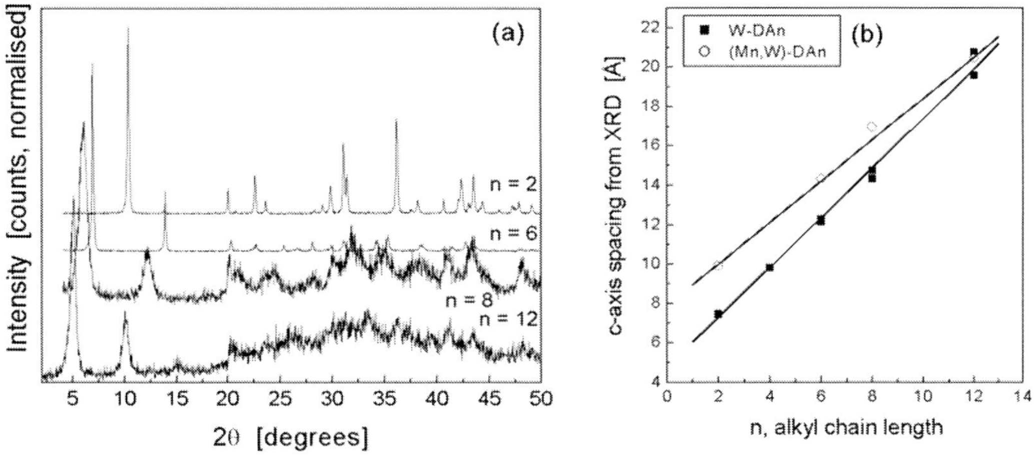

Figure 9. (a) XRD spectra of manganese tungstate (Mn,W)-DAn hybrids. (b) Progression of the interlayer spacing as determined from the series of 00l lines, compared with W-DAn.

The XRD spectra of the various transition metal tungstate hybrids exhibit different systematics with alkane length and included ion. Figure 9a shows spectra for manganese tungstate hybrid samples with different alkyl chain lengths, and Figure 9b shows how this compares with the 'pure' tungstate hybrid results. As can be seen from the slope, the change in interlayer spacing with increasing alkyl chain length is less than that for the tungstate system. The organics are calculated to lie at an angle of 34.2° to the normal, and the inorganic layer is thicker (associated with the incorporation of the manganese). Figure 10 shows XRD spectra of samples with different transition metals with DA6 as the organic spacer. All show a series of strong 00l reflections, which is indicative of layered structures. The five samples form at least two groups, based on the position of the first low-angle peak, which occurs at 12.2 Å in the case of tungsten only, nickel, and copper; and 14.5 Å for the manganese and cobalt samples.

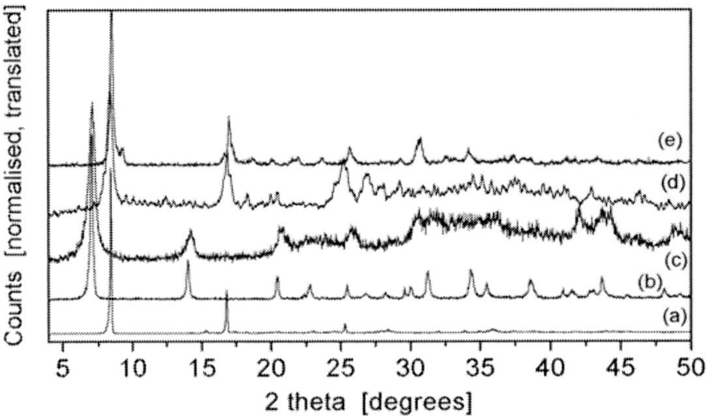

Figure 10. XRD spectra of various hybrid samples with DA6 as the organic intercalate: (a) W-DA6, (b) (Mn,W)-DA6, (c) (Co,W)-DA6, (d) (Ni,W)-DA6, (e) (Cu,W)-DA6.

SEM micrographs (Figure 11) show the formation of platelets of sizes up to 10 μm.

Figure 11. SEM micrograph of (Mn,W)-DA6.

Infrared and Raman spectra were obtained as in the 'plain' tungsten oxide hybrids case, and the spectra for a manganese tungstate hybrid series are shown in Figures 12-13 respectively.

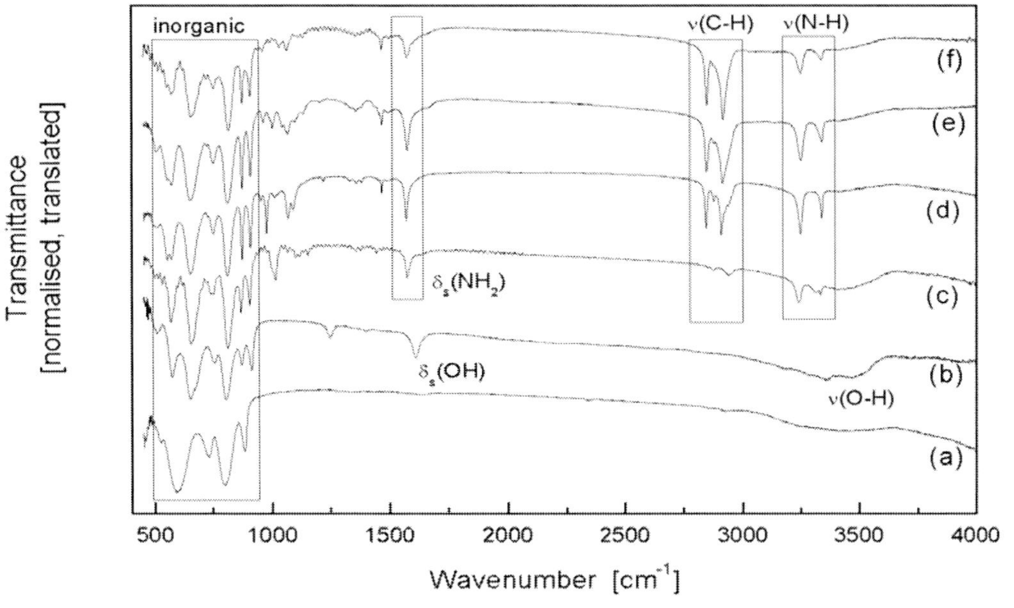

Figure 12. Infrared spectra of (Mn,W)-DAn hybrids and related materials, (a) $MnWO_4$, (b) $MnWO_4{\cdot}xH_2O$ (c) (Mn,W)-DA2, (d) (Mn,W)-DA6, (e) (Mn,W)-DA8, (f) (Mn,W)-DA12.

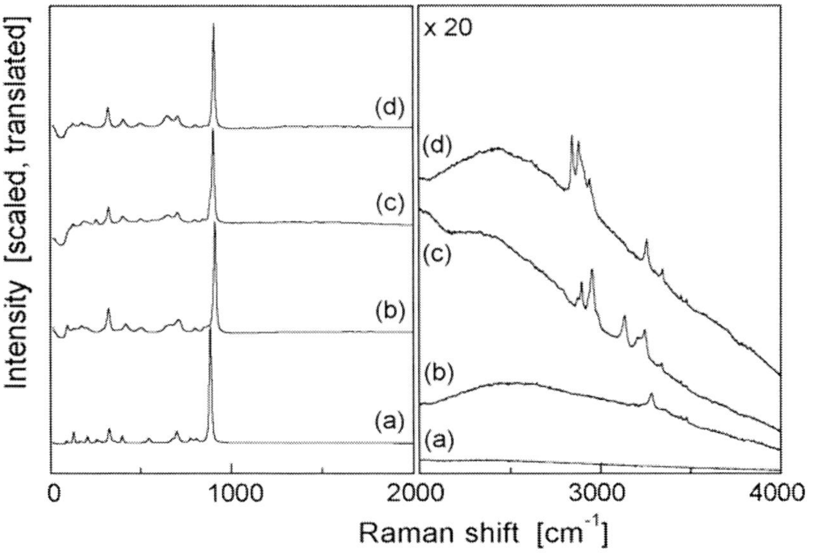

Figure 13. Raman spectra of manganese tungstate hybrids and related compounds: (a) $MnWO_4$, (b) $MnWO_4{\cdot}xH_2O$, (c) (Mn,W)-DA2, (d) (Mn,W)-DA6.

The first point to notice in both the infrared and Raman spectra is how similar the spectra are for the hybrids. In the infrared spectra the labelled regions show the major organic vibrational modes: N-H stretch at 3240-3340 cm^{-1}, C-H stretch 2840-2940 cm^{-1}, NH_2 bending 1570 cm^{-1}. Comparing these with the spectra of the tungsten oxide hybrids earlier, the N-H stretching lines are much sharper, there are only two NH_2 bending modes, and there are no broad features near 2100 cm^{-1}. These factors together indicate that the amine end-groups are neutral $-NH_2$ terminations rather than charged $-NH_3^+$ as in the tungstates. At the low-frequency end, the lines corresponding to inorganic vibrational modes are the same for the hybrids and the hydrate material. However they are different from the 'parent' compound, $MnWO_4$, which is also shown for reference. These lines have been reported previously for the hydrate compound, although no structural information was given [58].

The Raman spectra (Figure 13) also appear identical for the hybrids and hydrate system (with the exception, of course, of the organic modes at high Raman shifts). Again, there are differences between these and the reference compound $MnWO_4$. As in the tungsten oxide hybrids, a strong peak corresponding to W=O bonding is evident near 900 cm^{-1}. However, unlike the tungsten oxide hybrids, the frequency of this line increases from 886 cm^{-1} for $MnWO_4$ to 906 cm^{-1} for the hydrates and hybrids. The structure of $MnWO_4$ consists of an interlocking network of distorted MnO_6 and WO_6 octahedra which align in alternating planes of Mn and W cations [59]. It is therefore quite conceivable that when the layers form in the case of the hybrids and hydrate, some W=O bonds are formed which are then shorter and stronger than those in $MnWO_4$. The inorganic structures are identical for the hybrids and hydrates, and there is no direct interaction between the neutral organic or water molecules and the inorganic layer, as indicated from analysis of the infrared spectra.

III. Magnetic Properties of Transition Metal Tungstate Hybrids

A wide variety of different behaviours are exhibited by the various transition metal tungstate hybrids. Many samples display features corresponding to crossovers from 2D to 3D magnetic behaviour. In all cases the high-temperature part of the DC susceptibility can be approximated by the Curie-Weiss Law:

$$\chi = \frac{C}{T - \Theta} + TIP \qquad (1)$$

where C is the Curie constant, Θ is the Curie-Weiss temperature and TIP is any remnant temperature-independent paramagnetic component. A material with a negative Curie-Weiss temperature orders antiferromagnetically, while a material with a positive Curie-Weiss temperature orders ferromagnetically. Typical Curie-Weiss temperatures for some of the transition metal tungstate hybrids studied are given in Table III.

Table III. Curie-Weiss temperatures (in Kelvin) for transition metal tungstate hybrids.

	Mn	Fe	Co	Ni	Cu
MWO_4	-72.59		-98.07		
(M,W)-H_2O	-16.59	-23.4	-1.01	6.76	-2.83
(M,W)-DA2	-19.56	-9.46	-0.13	-0.15	-0.29
(M,W)-DA6	-19.55	-3.74	-17.58	6.33	0.28
(M,W)-DA12	-20.09	-4.42			
(M,W)-phen	-19.7			-0.87	3.31

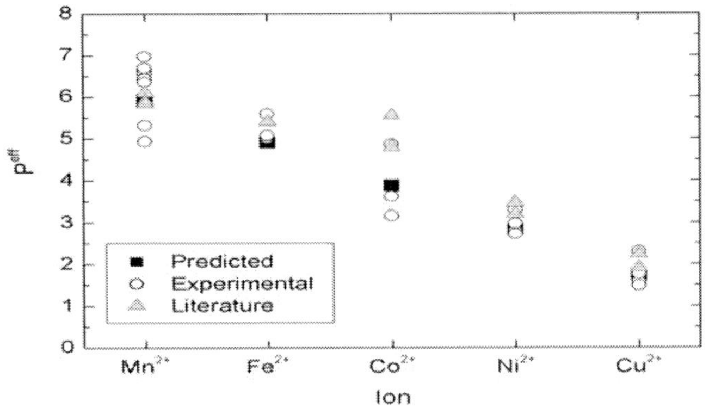

Figure 14. Effective moments per ion for various transition metal tungstate hybrids.

From the Curie constant C, one can determine the effective moment, P^{eff}, contributed by the transition metal ion present, from the formula

$$C = \frac{1}{3}\frac{N}{V}\frac{\mu_B{}^2 \mu_0 P^{eff\,2}}{k_B} \qquad \text{(SI units)} \quad (2)$$

where N is the number of magnetic ions per cell, V is the cell volume, and μ_B, μ_0 and k_B are the Bohr magneton, permeability of vacuum and Boltzmann constants respectively. This can be compared with the expected value calculated from the spin of the ion, $P^{eff} = g\sqrt{S(S+1)}$ [60]. Figure 14 illustrates the good agreement between the experimental values calculated from equation 2, literature values for transition metal tungstate systems [61], and the expected theoretical values [60]. The tungsten W^{6+} ions have a net zero spin (d^0) and therefore contribute nothing to the magnetisation.

We now examine some of the more interesting samples in greater detail.

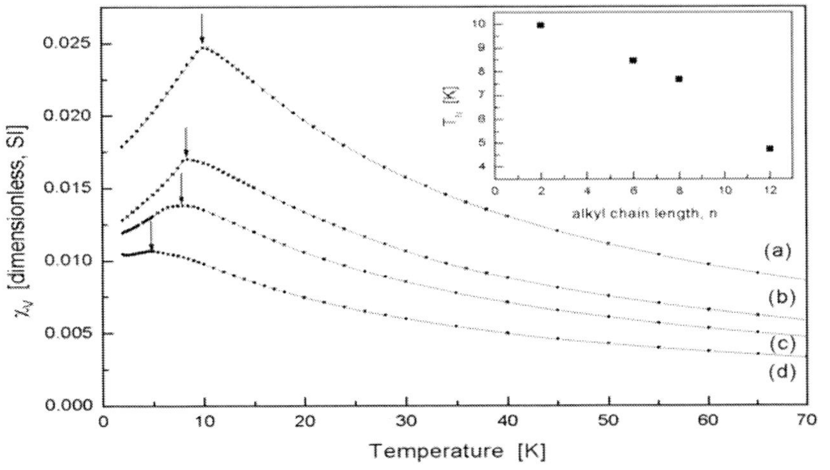

Figure 15. DC susceptibility of manganese tungstate hybrids with different length organic intercalates: (a) (Mn,W)-DA2, (b) (Mn,W)-DA6, (c) (Mn,W)-DA8, (d) (Mn,W)-DA12.

Figure 15 gives the DC susceptibility of a series of manganese tungstate hybrids with linear diaminoalkane molecules ranging in length from 2 to 12 carbons. In each case an antiferromagnetic (AF) transition is observed at low temperatures (indicated by the arrows). This transition temperature, T_N, is plotted versus the alkyl chain length (which varies linearly with the interlayer spacing, as evidenced by XRD results discussed in Section II) in the insert. It is expected that as the interlayer spacing is increased the interlayer coupling will decrease, which will cause the transition temperature to decrease. This is qualitatively what is observed.

From Table III we notice that the Curie-Weiss temperatures of the layered manganese tungstate hybrid materials are all similar, at about -19 to -20 K. This is in contrast to the 'parent' $MnWO_4$ compound which has a Curie-Weiss temperature of -72.6 K. This parent compound is not layered and hence exhibits strong AF ordering in 3 dimensions, which is weakened in the case of the layered hybrid materials.

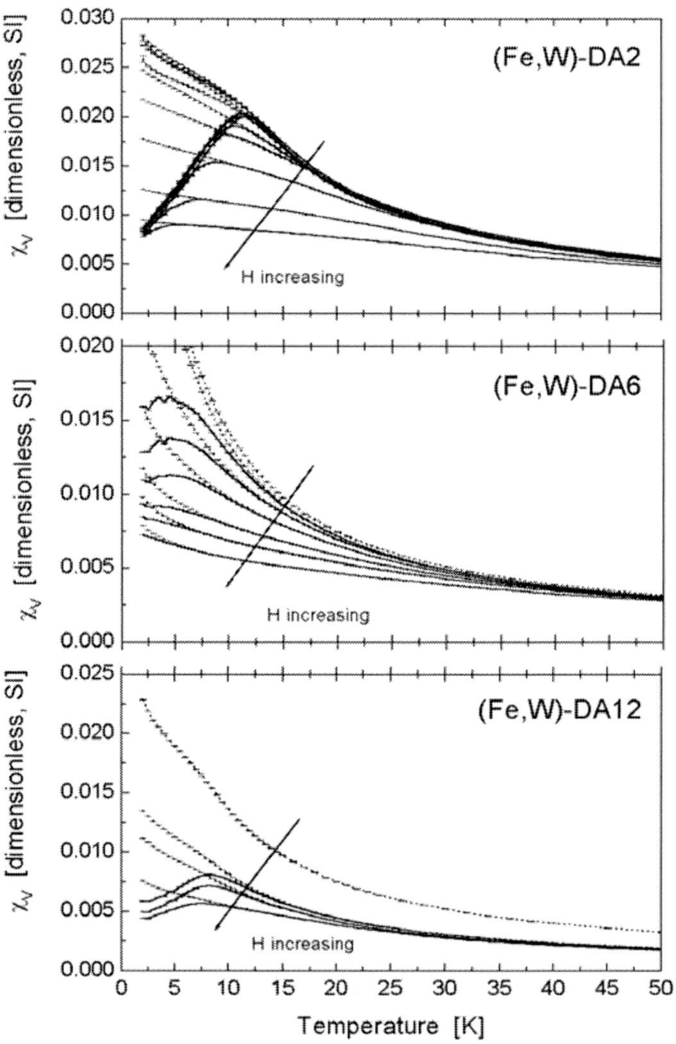

Figure 16. DC susceptibility of (Fe,W)-DAn samples taken at applied fields of 10 Oe to 1 T. Solid lines: zero-field-cooled, dotted lines: field-cooled.

The iron tungstate hybrids investigated also display antiferromagnetic correlations, as evidenced by the negative Curie-Weiss temperatures when fitted in the high-temperature part of the DC susceptibility plots (shown in Figure 16). However, unlike the manganese tungstate hybrids, there is irreversibility below a certain transition temperature, which varies strongly with the applied field. In the shortest 2-carbon chain case, the zero-field cooled data passes through an abrupt peak which is suppressed at higher fields, and then follows a uniform trend in what appears to be an antiferromagnetically-ordered 3D coupled state. A similar well-defined peak is observed in the 12-carbon chain sample (the longest chain length studied), but without the uniform trend behaviour. However for the sample with a 6-carbon chain, the zero-field-cooled and field-cooled curves separate very gradually and there is no well-defined transition temperature.

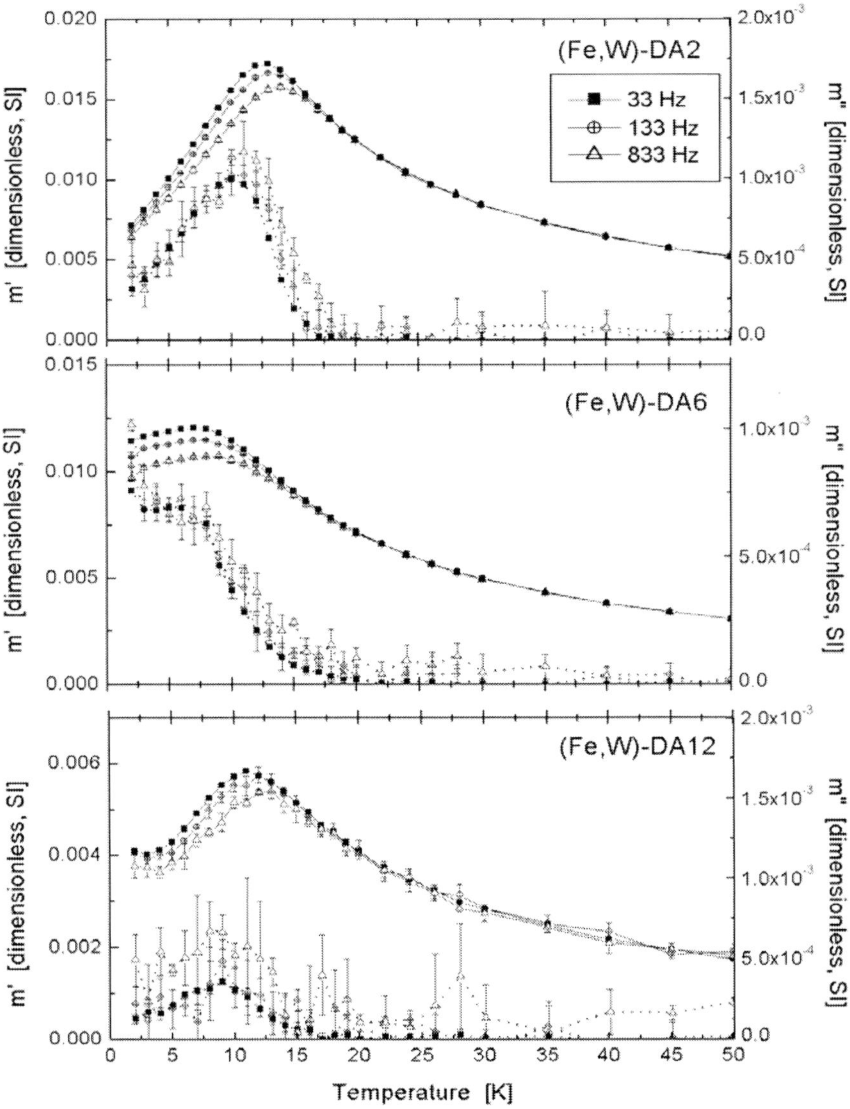

Figure 17. AC susceptibility of (Fe,W)-DAn samples at 3 Oe amplitude and various frequencies. Solid lines: real part, dotted lines: imaginary part.

The AC-susceptibility of these samples (given in Figure 17) is even more interesting. All three samples show frequency dependence over the range studied (33-833 Hz). This corresponds to quite long relaxation times, which may be indicative of a spin glass, or perhaps valence fluctuations.

Divalent copper has a spin of only ½, in contrast with manganese with spin **5/2**. This results in a much lower signal-to-noise ratio in the copper tungstate hybrids. A copper tungstate 1,6-diaminohexane ((Cu,W)-DA6) sample was measured and found to display broad features which can be fitted extremely well to a modified Bleaney-Bowers model for copper dimers [62]. This is shown in Figure 18.

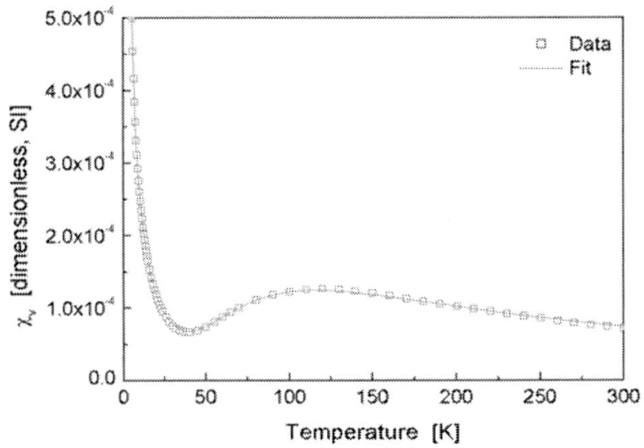

Figure 18. DC susceptibility of (Cu,W)-DA6 (ZFC, 0.1 T) and the results of a fit to a modified Bleaney-Bowers model.

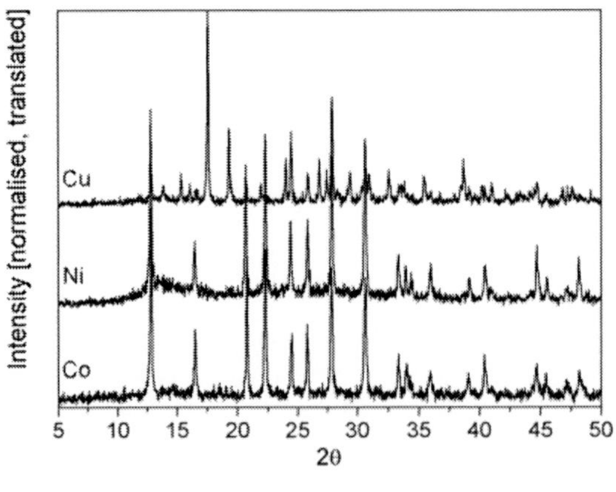

Figure 19. XRD of three (M,W)-DA2 samples (M = Co, Ni, and Cu).

Samples of copper, cobalt and nickel tungstates with 1,2-diaminoethane (DA2) as the organic molecule, have nearly identical structures as indicated by XRD (Figure 19). The copper sample displays several extra peaks, which are most likely due to symmetry breaking due to copper ions often displaying several non-equivalent bond lengths due to the Jahn-Teller effect [63-64]. The magnetisation of all three samples follows the Curie-Weiss law exactly, indicating that free-spin pure paramagnetism with no phase transitions. This is demonstrated in Figure 20, which shows the inverse DC susceptibility is linear and passes through the origin. The parameters from the Curie-Weiss fits are given in Table IV. The Curie-Weiss temperatures are all very similar, and the Curie constants vary in accordance with **3/2** the spin of the ions present (for Co, 1 for Ni, ½ for copper).

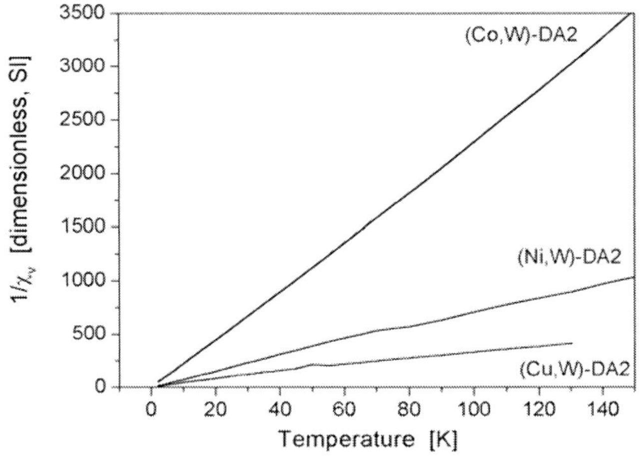

Figure 20. Inverse of the DC susceptibility of (M,W)-DA2 (M = Co, Ni, Cu).

Table IV. Curie-Weiss fit parameters for Co, Ni, and Cu (M,W)-DA2 samples.

	C [K^{-1}]	**Θ [K]**
(Co,W)-DA2	0.207	-0.23
(Ni,W)-DA2	0.132	-0.15
(Cu,W)-DA2	0.046	-0.28

As shown in Table III, the nickel tungstate 1,6-diaminohexane (DA6) sample is one of the few to have a positive Curie-Weiss temperature, corresponding to ferromagnetic correlations. This sample also displays a sharp ferromagnetic transition at around 12.5 K, shown in Figure 21. Despite the apparent ferromagnetic ordering, the absolute value of the saturation magnetisation of the sample is much less than that expected per Ni^{2+} ion, indicating that the sample is in fact a canted antiferromagnet.

Hysteresis loops were performed on this sample at temperatures from 2-15 K and fields from -6 to +6 T. At each temperature the saturation magnetisation was estimated by extrapolating the linear part of each curve to H=0. These values are plotted in Figure 22. Extrapolating this curve to T=0, we can obtain an estimate for the canting angle α of 0.46°, from the formula

$$\tan\alpha = \frac{M_{nc}(T=0)}{g\mu_B S}$$

(taking $S = 1$ for Ni^{2+}) [65].

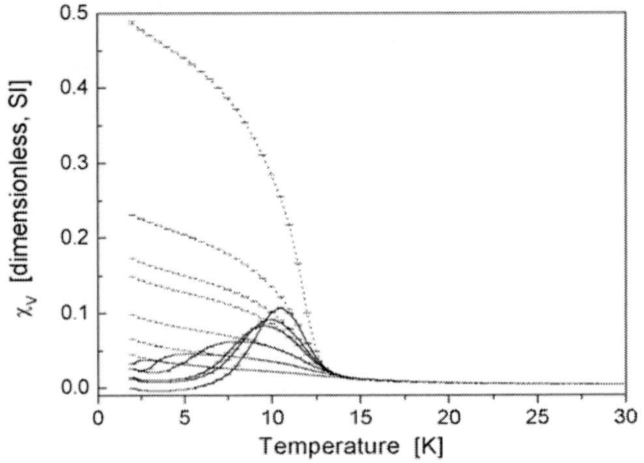

Figure 21. DC susceptibility of (Ni,W)-DA6 at applied fields of 10 Oe to 1 T. Solid lines: zero-field-cooled, dotted lines: field-cooled.

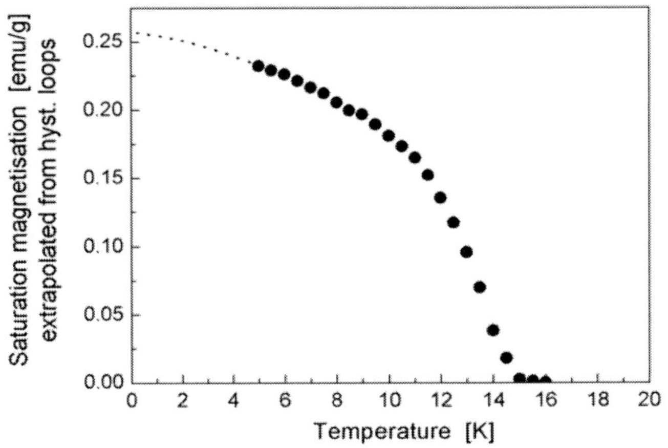

Figure 22. Saturation magnetisation analysis of (Ni,W)-DA6 versus temperature.

CONCLUSION

Layered organic-inorganic hybrid materials based on tungsten oxide and transition metal tungstates offer a wide range of interesting structural, optical, electronic and magnetic behaviour, due to the presence of inorganic layers which are separated by organic molecules of various lengths and functions. The properties of both the organic and inorganic components are observed to contribute to the attributes of the resulting hybrid material.

The electronic studies via optical spectroscopy and *ab initio* computations demonstrate the similarity between the tungsten oxide-organic hybrid materials and their parent compound WO_3. The hybrids are insulating and their band structures are virtually identical regardless of the length of organic diamine molecule inserted between the WO_4 layers. Like WO_3, the valence band consists of oxygen *2p* states and the conduction band of tungsten *5d* states.

Work is continuing to attempt to dope the hybrid systems in order to explore the conductivity, and compare this with the doped WO_3 and H_2WO_4 systems.

The addition of transition metal ions in the synthesis of the organic-inorganic hybrids results in a huge variety of different behaviours. In particular, the magnetic results of transition metal tungstate hybrids often display prominent features corresponding to crossovers from 2D to 3D magnetic ordering.

This is an exciting new field of soft condensed matter, which is only limited by the possible organic intercalates and the researcher's imagination.

REFERENCES

[1] Song, Y.; Zavalij, P.; Chernova, N.A.; Suzuki, M.; Whittingham, M.S. *J. Solid State Chem.* 2003, 185, 63-71.

[2] Kissick, J.L.; Cowley, A.R.; Chippindale, A.M. *J. Solid State Chem.* 2002, 167, 17-27.

[3] Fernandez, S.; Pizarro, J.L.; Mesa, J.L.; Lezama, L.; Arriortua, M.I.; Olazcuaga, R.; Rojo, T. *Inorg. Chem.* 2001, 40, 3476-3483.

[4] Jacobson, A.J.; Johnson, J.W.; Lewandowski, J.T. *Inorg. Chem.* 1985, 24, 3727-3729.

[5] Kudo, M.; Ohkawa, H.; Sugimoto, W.; Kumada, N.; Liu, Z.; Terasaki, O.; Sugahara, Y. *Inorg. Chem.* 2003, 42, 4479-4484.

[6] Gao, Q.; Giraldo, O.; Tong, W.; Suib, S.L. *Chem. Mater.* 2001, 13, 778-786.

[7] Liu, Z.; Ooi, K.; Kanoh, H.; Tang, W.; Tomida, T. *Langmuir* 2000, 16, 4154-4164.

[8] Kahn, A.I.; O'Hare, D. *J. Mater. Chem.* 2002, 12, 3191-3198.

[9] Choy, J.H.; Park, N.G.; Hwang, S.J.; Kim, D.H. *J. Am. Chem. Soc.* 1994, 116, 11564-11565.

[10] Choy, J.H.; Hwang, S.J.; Park, N.G. *J. Am. Chem. Soc.* 1997, 119, 1624-1633.

[11] Felner, I.; Asaf, U.; Weger, M.; Reich, S.; Leitus, G. *Physica B* 2002, 311, 191-194.

[12] Klauk, H. *Phys. World* 2000, 1, 18-19.

[13] Choy, J.H. *J. Phys. Chem. Solids* 2004, 65, 373-383.

[14] Rarig, R.S. Jr.; Zubieta, J. *J. Solid State Chem.* 2002, 167, 370-375.

[15] Shivaiah, V.; Nagaraju, M.; Das, S.K. *Inorg. Chem.* 2003, 42, 6604-6606.

[16] Yun, S.K.; Maier, J. *Chem. Mater.* 1999, 11, 1644-1647.

[17] Wang, K.Z.; Gao, L.H. *Mat. Res. Bull.* 2002, 37, 2447-2451.

[18] Gomez-Romero, P. *Adv. Mater.* 2001, 13, 163-174.

[19] Mitzi, D.B. *Prog. Inorg. Chem.* 1999, 48, 1-121.

[20] Remy, H.; Laves, G. *Ber. Deut. Ges.* 1933, 66, 401-407.

[21] Mitzi, D.B.; Liang, K. *Chem. Mater.* 1997, 9, 2990-2995.

[22] Era, M.; Moriomoto, S.; Tsutsui, T.; Saito, S. *Appl. Phys. Lett.* 1994, 65, 676-678.

[23] Kagan, C.R.; Mitzi, D.B.; Dimitrakopolous C.D. *Science* 1999, 286, 945-947.

[24] de Jongh, L.J.; Miedema, A.R. *Adv. Phys.* 1974, 23, 1-260.

[25] Mostafa, M.F.; Willett, R.D. *Phys. Rev. B* 1971, 4, 2213-2215.

[26] de Jongh, L.J.; van Amstel, W.D.; Miedema, A.R. *Physica* 1972, 58, 277-304.

[27] van Amstel, W.D.; de Jongh, L.J. *Solid State Comm.* 1972, 11, 1423-1429.

[28] Johnson, J.W.; Jacobson, A.J.; Rich, S.M.; Brody, J.F. *J. Am. Chem. Soc.* 1981, 103, 5246-5247.

[29] Ayyappan S.; Subbanna, G.N.; Rao, C.N.R. *Chem. Eur. J.* 1995, 1, 165-170.
[30] Kepert, D.L. *Prog. Inorg. Chem.* 1962, 4, 199-274.
[31] de Farias, R.F.; Airoldi, C. *J. Phys. Chem. Solids* 2003, 64, 2199-2204.
[32] de Farias, R.F. *Int. J. Inorg. Mater.* 2001, 3, 303-307.
[33] Xu, L.; Qin, C.; Wang, X.; Wei, Y.; Wang, E. *Inorg. Chem.* 2003, 42, 7342-7344.
[34] Lasocha, W.; Jansen, J.; Schenk, H. *J. Solid State Chem.* 1995, 117, 103-107.
[35] Figlarz, M. *Prog. Solid State Chem.* 1989, 19, 1-46.
[36] Kihlborg, L. *Ark. Kemi* 1963, 21, 357-364.
[37] Krebs, B. *Acta Cryst. B* 1972, 28, 2222-2231.
[38] Szymanski, J.T.; Roberts, A.C. *Can. Mineral.* 1984, 22, 681-688.
[39] Guenter, J. *J. Solid State Chem.* 1972, 5, 354-359.
[40] Brown, B.W.; Banks E. *J. Am. Chem. Soc.* 1954, 76, 963-966.
[41] Brown, B.W.; Banks E. *Phys. Rev.* 1951, 84, 609-610.
[42] Azens, A.; Hjelm, A.; Le Ballac, D.; Granqvist, C.G.; Barczynska, J.; Pentjuss, E.; Gabrusenoks, J.; Wills, J.M. *Solid State Ionics* 1996, 86, 943-948.
[43] Glemser, O.; Sauer, H. *Z. anorg. Chem.* 1943, 252, 144-159.
[44] Lampert, C.M. *Sol. En. Mater. Sol. Cells* 1998, 52, 207-221.
[45] Chong, S.V.; Ingham, B.; Tallon, J.L. *Mat. Res. Soc. Symp. Proc.* 2004, 847, EE13.14.
[46] Li, K.C.; Wang, C.Y. *Tungsten*; 2nd Edition; Reinhold Publishing Corporation: New York, NY, 1947; pp. 254-260.
[47] http://neon.otago.ac.nz/chemistry/chemsearch/micro.html
[48] Kresse, G.; Furthmueller, J. *Phys. Rev. B* 1996, 54, 11169-11186.
[49] Kresse, G.; Furthmueller, J. *Comput. Mater. Sci.* 1996, 6, 15-50.
[50] Ingham, B.; Chong, S.V.; Tallon, J.L. *J. Phys. Chem. B* 2005, 109, 4936-4940.
[51] Silverstein, R.M.; Basser, G.C.; Morrill, T.C. *Spectrometric Identification of Organic Compounds*, 5th ed.; Wiley: New York, 1991.
[52] Daniel, M.F.; Desbat, B.; Lassegues, J.C.; Gerand, B.; Figlarz, M. *J. Solid State Chem.* 1987, 67, 235-247.
[53] Likalter, A.A. *Physica B* 2002, 315, 252-260.
[54] Bullett, D.W. *J. Phys. C* 1983, 16, 2197-2207.
[55] Cora, F.; Stachiotti, M.G.; Catlow, C.R.A. *J. Phys. Chem. B* 1997, 101, 3945-3952.
[56] Ingham, B.; Hendy, S.C.; Chong, S.V.; Tallon, J.L. *Phys. Rev. B* 2005, 72, 075109.
[57] Ingham, B.; Chong, S.V.; Tallon, J.L. *Curr. Appl. Phys.* 2004, 4, 202-205.
[58] Krustev, S.; Ivanov, K.; Klissurski, D. *J. Alloys Comp.* 1992, 182, 189-193.
[59] Lautenschlaeger, G.; Weitzel, H.; Vogt, T.; Hock, R.; Boehm, A.; Bonnet, M.; Fuess, H. *Phys. Rev. B* 1993, 48, 6087-6098.
[60] Ashcroft, N.W.; Mermin, N.D. *Solid State Physics*, Saunders College Publishing: Philadelphia, PA, 1976; pp 656-658.
[61] van Uittert, L.G.; Sherwood, R.C.; Williams, H.J.; Rubin J.J.; Bonner W.A. *J. Phys. Chem. Solids* 1964, 25, 1447-1451.
[62] Rarig, R.S. Jr.; Lam, R.; Zavalij, P.Y.; Ngala, J.K.; LaDuca, R.L. Jr.; Greedan, J.E.; Zubieta, J. *Inorg. Chem.* 2002, 41, 2124-2133.
[63] Jahn, H.A.; Teller, E. Proc. Roy. Soc. London A, 1937, 161, 220-235.
[64] Halcrow, M.A. *Dalton Trans.* 2003, 4375-4384.
[65] Bellitto, C.; Federici, F.; Colapietro, M.; Portalone, G.; Caschera, D. *Inorg. Chem.* 2002, 41, 709-714.

In: Research in Hybrid Materials
Editors: Simon J. Brunner and Julian W. Egger
ISBN 978-1-60456-718-2

Chapter 4

THE CHEMISTRY OF ORGANOIMIDO DERIVATIVES OF POLYOXOMETALATES - ORGANIC CHEMISTRY AS A USEFUL TOOL FOR THE FUNCTIONALIZATION OF POMS AND THE PREPARATION OF NOVEL POM-BASED ORGANIC-INORGANIC HYBRID MOLECULAR MATERIALS

Yongge Wei[1, 2], Pingfan Wu[3], Yuan Wang[2] and Meicheng Shao[2]
[1] Department of Chemistry, Tsinghua University, Beijing 100080, China
[2] State Key Laboratory for Structural Chemistry of Unstable and Stable Species, College of Chemistry and Molecular Engineering, Peking University, Beijing 100871, China
[3] Department of Bioengineering, Hubei University of Technology, Wuhan 430068, China

ABSTRACT

The chemistry of organoimido derivatives of polyoxometalates (POMs), an innovative emerging organic chemistry of polyoxometalates, has received increasing interest in the supramolecular chemistry and chemistry of materials recently. Organoimido derivatives of POMs are valuable building blocks for constructing novel nanostructured organic-inorganic hybrid molecular materials with the so-called ''value-adding properties'' and possible synergistic effects, which could result in unique catalytic, photo and electronic properties dramatically different from the corresponding parent materials, since the π electrons in the organic component may extend their conjugation to the inorganic framework and dramatically modify the electronic structure and redox properties of the corresponding parent POMs. The surface modification of POMs, that is, the substitution of one or more terminal or bridged oxo groups with organoimido ligand, is the fundamental method to synthesize organoimido derivatives of POMs via common organic chemistry reactions. In this chapter of the book, the synthetic chemistry of organoimido derivatives of POMs using different organic reagents will be reviewed, of which the novel DCC-dehydrating protocol to prepare organoimido derivatives of POMs recently developed by us and co-authors will be concentrated. As it can be expected, the reaction chemistry of organoimido derivatives of POMs stands for the fascinating future of the chemistry of organoimido derivatives of POMs since it opens not only a new road to the chemical modification of POMs, but also an exciting research arena where a variety of hybrid materials containing covalently bonded POM clusters and organic conjugated segments can be prepared in a more controllable and rational manner. Hence, in the rest of this chapter, Pd-catalyzed coupling reactions of the functionalized organoimido derivatives of POMs will be introduced in detail for this purpose. Especially, different rationally assembled hybrid molecules and polymers where the organic molecules and inorganic clusters are bonded together by an imido nitrogen atom will then be highlighted.

1. INTRODUCTION

Polyoxometalates (POMs), formed by the early transition metal ions such as vanadium, molybdenum and tungsten in their high oxidization states with bridged oxide anions, are a unique class of well-defined inorganic nanocluster compounds with much diversity in size, composition, structure and function, which can be, in fact, considered as the "oligomeric" or molecular states and model compounds of the corresponding "polymeric" metal oxides with extended solid state structure [1]. Since the first salt of POM, ammonium 12-molybdophosphate, $(NH_4)_3[PMo_{12}O_{40}]$ was reported by Berzelius in 1826, as a yellow precipitate from the aqueous solution of phosphoric acid and ammonium molybdate [2], the chemistry of POMs has gained dramatic development and is remarkable for its significance in quite diverse disciplines including catalysis, medicine and materials science [3- 5]. Especially, during the last decade, a lot of nanoscale huge clusters of polyoxometalates and their assemblies have been prepared and structurally characterized by our research team at Peking University [6- 9], A. Müller's group at University of Bielefeld [10- 24] and others [25- 28]. As examples, in figures 1-3, there are shown three typical nanoscale polyoxomolybdates, *i. e.*, tyre-like $\{Mo_{176}\}$ (1) [7, 12], ball-like $\{Mo_{132}\}$ (2) [9, 15], and chain-like $\{Mo_{38}\}$ (3) self-assembled from $\{Mo_{36}\}$ [6], synthesized by A. Müller *et al.* and us, respectively. These high nuclearity huge clusters are not only ideal theoretical models for mesocopic physics but also provide us an exciting research arena for nanochemistry and materials science— The nanotyres 1 are potential building blocks for fabricating novel nanotubes if they are arranged to stack along their symmetrical axes. The nanoballs 2 have a spherical cavity of *ca* 8 nm^3 with accessible openings, indicating that they are a good candidate for nanocontainers and may be used in controllable drug-releasing. The nanochains 3 show significantly red-shif in its UV-vis spectra, which implies that they are potential applications in making unique nanoscale molecular quantum wells or one-dimensional conducting materials. Furthermore, T. B. Liu *et al.* have recently observed and confirmed, according to laser light scattering studies, that in their solution, such nanoclusters of polyoxometalates spontaneously assemble into almost monodisperse, well-defined vesicle-like nanoaggregates with spherical structures [29, 30], via an unusually slow self-association process [31]; and more interestingly, in the NaCl-containing aqueous solution, dissolution and precipitation, two opposite processes, of these hydrophilic nanoanions automatically and subsequently occur without changing external conditions [32].

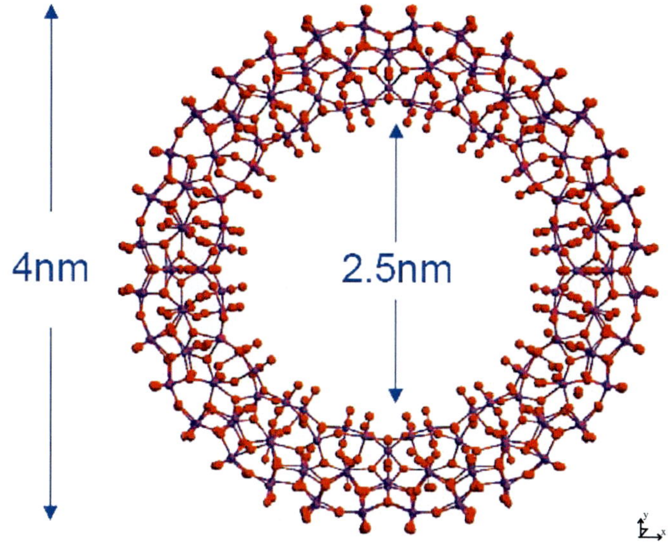

Figure 1. Structure of the nanotyre 1 with 176 Mo atoms in a cluster anion (view parallel to the C_8 axis of the cluster). Also see Ref. 7 and 12.

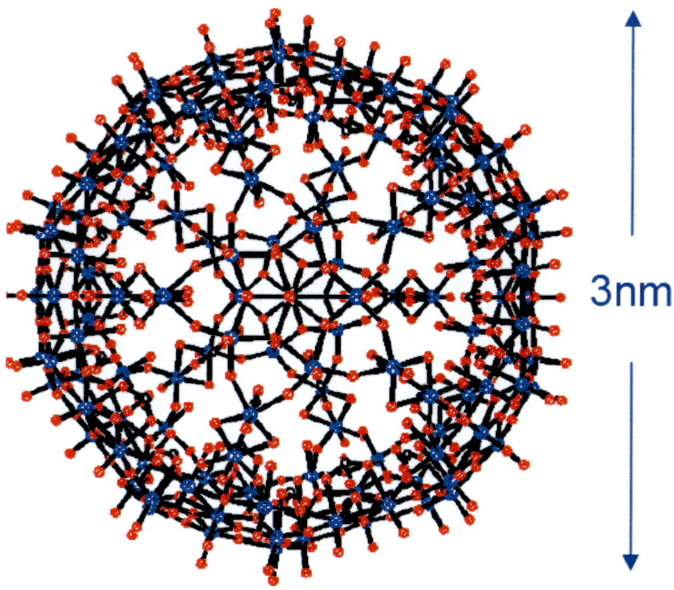

Figure 2. A view of the $\{Mo_{132}\}$ nanoball **2** along its C_5 axis. Also see Ref. 9 and 15.

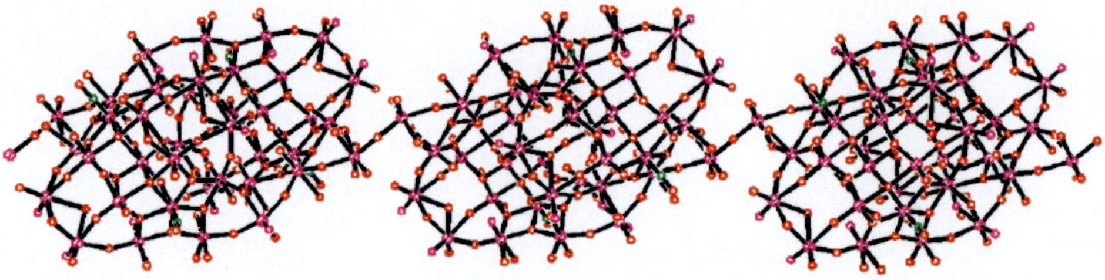

Figure 3. Drawing of the nanochain 3 extended along the 21 screw axis. Also see Ref. 6.

Parallel to the rapid and great progress made on these nanosized inorganic clusters of metal-oxide anions, studies on conjugated organic and polymeric materials have also exploded and flourished [33- 36], since such molecular materials have exciting technological applications in various photic and electronic devices, for instance, organic light emitting diodes (OLEDs) [37- 38], organic field-effect transistors (OFETs) or organic thin film transistors (OTFTs) [39- 42], and solid-state lasers [43, 44]. The fact that the 2000's Nobel Prize in chemistry was bestowed upon three preeminent scientists working on conjugated polymers is the ultimate testimony to the important developments in organic conjugated systems [45- 48].

Although they are vastly different in molecular structures, POMs and conjugated organic molecules/ polymers are both electrically active materials with similar electrical and optical properties such as photochromism, electrochromism, and conductivity [34, 49, 50]. The underlying mechanisms of these properties are, however, different for the two types of materials, with dπ electrons responsible for the inorganic POM clusters, and delocalized pπ electrons responsible for the organic counterpart. While both areas have been enjoying considerable success, there has been little success in bringing these two types of materials together through covalent bonds to make novel POM-based organic-inorganic hybrids [51] owing to the lack of reliable synthetical chemistry tools to functionalize POMs and prepare their organic derivatives in high throughput. As could be expected, however, such organic-inorganic hybrid materials will not only combine the advantages of organic materials, such as good processability and fine-regulable structure and electronic properties [34], with those of inorganic POM clusters, such as good chemical stability and strong electron acceptability [1], to produce so-called ''value-adding properties'' [52], but also may bring exciting synergistic effects due to the close interaction of delocalized organic pπ orbits with the inorganic POM cluster's dπ orbits. With their unique structures and potential novel properties, originating from the functional nanobuilding blocks of POMs [51], these sorts of hybrid materials are extremely interesting not only to synthetic chemists and materials scientists, but also to theoretical and experimental physicists.

Organically functionalized POMs, *i.e.*, species where one or some oxo or $\{Mo_x\}^{n+}$ groups have been replaced with organic functional groups, have been extensively studied and now form the largest subclass of POM derivatives [53], since the pioneering work at du pont, of which alkylated derivatives or esters of Keggin-structure phosphomolybdic and phosphotungstic acid, such as $(n\text{-}Bu_4N)_3[PMo_{12}O_{39}(OMe)]$ (4) and$(n\text{-}Bu_4N)_3[PW_{12}O_{39}(OMe)]$ (5), were prepared by Knoth et al in the early 1980s [54, 55]. There are a lot of reasons for the current interest in these derivatized POMs [53]: 1) Due to the perceived structural analogies of POMs to the surfaces of metal oxides, they can be viewed as soluble analogues of metal oxides and therefore are of special interest as model compounds for investigating the reactions and properties of metal oxides. Given the versatility of metal oxides in catalyzing organic conversions, and the difficulty in determining the intimate mechanism of these reactions, the study of the stoichiometric reactivity of well-defined surface POMs might provide an understanding of the elementary steps of heterogeneous reactions, particularly with respect to surface-bound intermediates. Therefore, the characterization of organic derivatives of POMs is relevant to the molecule-level modeling of catalytic reactions occurring at metal oxide surfaces. Furthermore, POM-supported transition-metal catalysts represent a new class of oxide-supported catalyst materials that can be fully studied at the atomic level, both structurally and mechanistically. 2) Derivatization may result in the activation of surface oxo groups of POMs, thus giving more chances to be modified. 3) Functionalization can conduce to the stabilization of otherwise labile POM frameworks in the derivatives, providing novel building blocks for the assembly of larger systems. 4) The construction of interconnected POM networks could be built up through the incorporation of organic groups with functionality and polysubstituted derivatives could be utilized in the context of the development of novel dendrimers. 5) Such derivatives can probably provide multifunctional oxidation or acidification catalysts that display selective recognition of substrates, thus higher selectivities, and might develop targeting of POMs in antiviral and anticancer chemotherapy.

Among many organic derivatives of POMs, organoimido derivatives have attracted increasing interest and recently, a lot of arylimido derivatives including polymers of the Lindqvist hexamolybdate cluster, $[Mo_6O_{19}]^{2-}$, have been synthesized continuously [56], since the π electrons in the organic component of such derivatives may extend their conjugation to the inorganic POM framework via the N atoms of the imido groups, thus result in strong d–pπ interactions and dramatically modify the electronic structure and redox properties of the corresponding parent POMs. In addition, organoimido derivatives of POMs with a remote active functional group can be exploited as building blocks to conveniently and controllably fabricate the complicated covalently-linked POM-based organic-inorganic hybrids, including the nano-dumbbells, polymeric chains and even networks of POMs [56]. This modular building block approach brings rational design and structure regulation into the synthesis of organic-inorganic hybrid molecular materials and has just been reviewed by one of our co-workers at UMKC [56]. In this chapter, concentrating on the novel DCC-dehydrating protocol to prepare organoimido derivatives of POMs recently developed in this laboratory and others, various synthetic approaches to organoimido derivatives of POMs using different organic reagents will be reviewed. As it can be expected, the reaction chemistry of organoimido derivatives of POMs stands for the fascinating future of the chemistry of organoimido derivatives of POMs since it opens not only a new road to the chemical modification of POMs, but also an exciting research arena where a variety of hybrid materials containing covalently bonded POM clusters and organic conjugated segments can be prepared in more controllable and rational manner. Hence, in the rest of this chapter, Pd-catalyzed coupling reactions of the functionalized organoimido derivatives of POMs will be introduce in details. Furthermore, various rationally assembled hybrid molecules and polymers where the organic molecules and inorganic clusters are bonded together by an imido nitrogen atom will then be highlighted.

2. Synthetic Chemistry of Organoimido Derivatives of POMs

As an important subclass of oxo-imido complexes of early transition metals, organoimido derivatives of POMs have only been synthesized and characterized very recently. They were mentioned first by J. Zubieta *et al* in an outstanding communication in 1988 [57], however, there was lack of a reliable and detailed structural characterization in this paper, probably due to the purification problem of their products. Three years later, using a different synthetic methodology, E. A. Maatta's group finished the synthesis and single crystal X-ray structural determination of the first organoimido derivative of POMs [58]. Since then, the organoimido derivatives of POMs have been extensively investigated and a number of organoimido derivatives of the Lindqvist hexamolybdate ion, $[Mo_6O_{19}]^{2-}$, have been reported [53]. However, little work has been conducted on other POMs including the other Lindqvist POMs, the Keggin POMs, and especially, the recently discovered nanoscale huge clusters of POMs such as compounds 1-3 [6-28], the synthesis of their organoimido derivatives is still a formidable challenge. Furthermore, although the bridging oxo groups of POMs can be replaced, in principle, with isoelectronic organoimido ligands, the substitution of terminal oxo groups features largely in the reported organoimido derivatives of POMs, in which the imido groups attached to only one metal center in a linear fashion and the N-atom in the imido groups can be considered sp hybridized as σ-π donor ligand to form one σ and two π bonds with the bound metal atom.

In general, organoimido derivatives of POMs can be obtained in the following ways [53]:

- Imido metathesis via Wittig-like (net) [2 + 2] exchange reactions of M=O bonds with imido-releasing agents, including isocyanates, phosphinimines, or sufinylamines.
- The α-hydrogen transfer reactions with amines.
- The displacement of labile ligands from an imido-containing metal complex by active oxometalates, followed by aggregation, or by coordination-unsaturated lacunary POMs.
- The addition or displacement reactions of active functional groups at the imido ligands from existing organoimido precursors of POMs.

Apparently, the first two strategies create the new metal-nitrogen multiple bonds and the latter ones deal with other chemical bonds from known metal imido complexes. In this section of the chapter, the synthetic methodologies in respect to the former three strategies will be surveyed. The fourth strategy will be introduced in Section 3 because it is involved in the reaction chemistry of organoimido derivatives of POMs.

2.1. Organoimido Derivatives of the Lindqvist Hexamolybadte

The hexamolybdate ion, $[Mo_6O_{19}]^{2-}$, is amongst the most well-known POM clusters for its thermal and chemical robust and ease to be prepared [59], which has the so-called Lindqvist structure [60]. As it is depicted in figure 4, the Lindqvist structure consists of a central oxyanion surrounded in an octahedral cage formed by six metal atoms. All the six metal atoms also have an octahedral environment. Besides the central oxygen atom, they each coordinate triply to one terminal oxygen atom, forming a terminal metal-oxo group (M≡O), and share an additional four doubly-bridging oxygen atoms (μ_2-O atoms) with neighboring metal atoms. Generally speaking, a Lindqvist ion has a superoctahedral structure approach to O_h point group and features its six terminal metal-oxo groups aligned along the Cartesian axes. For the hexamolybdate, these molybdyl groups (Mo≡O) are reactive enough for the terminal oxygen atoms to be directly replaced by various nitrogenous species, for example, diazenido [61], diazoalkyl [62], and imido groups [58]. In the case of imido derivatives of the hexmolybdate, so far, three types of reactions have been developed. These include reactions with phosphinimines [Eq. (1)] [58, 63], isocyanates [Eq. (2)] [64, 65], and primary amines [Eq. (3)] [66- 68], which are given in the following equations, respectively.

$$(n\text{-}Bu_4N)_2[Mo_6O_{19}] + x\ Ph_3P{=}NAr \longrightarrow (n\text{-}Bu_4N)_2[Mo_6O_{19\text{-}x}(NAr)_x] + x\ Ph_3P{=}O \quad (1)$$

$$(n\text{-}Bu_4N)_2[Mo_6O_{19}] + x\ RNCO \longrightarrow (n\text{-}Bu_4N)_2[Mo_6O_{19\text{-}x}(NAr)_x] + x\ CO_2 \quad (2)$$

$$(n\text{-}Bu_4N)_2[Mo_6O_{19}] + x\ ArNH_2 \longrightarrow (n\text{-}Bu_4N)_2[Mo_6O_{19\text{-}x}(NAr)_x] + x\ H_2O \quad (3)$$

With these approaches, the six terminal oxo (in certain case, some bridging-oxo groups) in the hexamolybdate cluster can be partially or completely substituted with organoimido ligands. Nowadays, a large number of monosubstituted [58, 63- 78], disubstituted [63, 65, 66, 71, 76, 79- 83] and polysubstituted [65, 71, 84, 85] organoimido derivatives of the hexamolybdate have been synthesized and structurally characterized.

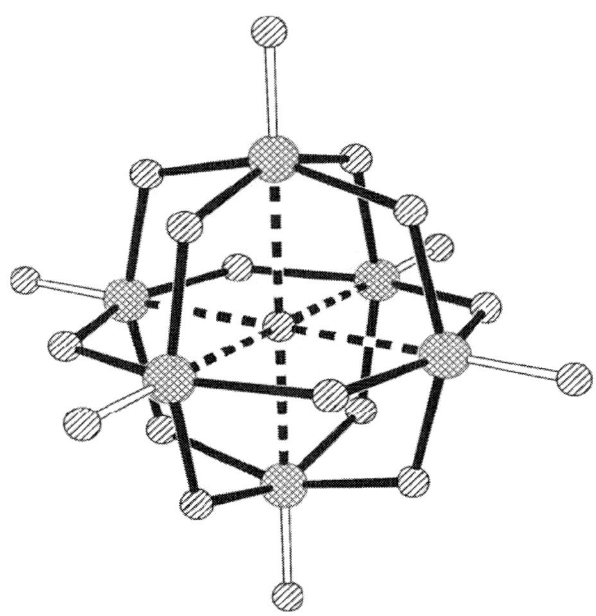

Figure 4. Structure of the Lindqvist-type hexamolybdate ion, [Mo6O19]2-.

2.1.1. Monosubstituted Derivatives

Monofunctionalized organoimido derivatives of the hexamolybdate were firstly obtained by E. A. Maatta *et al* with phosphinimines as imido-releasing reagents [Eq. (1)] in 1991 [58]. They discovered that the reaction of equimolar amounts of $(n\text{-}Bu_4N)_2[Mo_6O_{19}]$ (6) and $Ph_3P{=}Ntol$ (prepared from the reaction of p-tolyl azide and Ph_3P in ether) in strictly anhydrous pyridine at 85 °C yields $(n\text{-}Bu_4N)_2[Mo_5O_{18}(MoNC_6H_4CH_3)]$ (7) within 2 days. Two years later, A. Proust and co-workers founded that this type of reaction also works well in refluxed anhydrous acetonitrile under dried nitrogen gas for 60 hours and synthesized the monophenylimido derivative, $(n\text{-}Bu_4N)_2[Mo_6O_{18}(NPh)]$ (8) [63]. Although these reactions involve the not-easily accessible phosphinimine reagents, proceed slow, and the triphenylphosphine oxide byproduct is very difficult to separate from the desired reaction products, they nonetheless demonstrate that direct functionalization can be carried out on the hexamolybdate salt itself in a straightforward manner. Moreover, phosphinimines are powerful reagents to introduce C=N bond into organic compounds from carbonyl substrates via the famous aza-Wittig reactions in classic organic syntheses [86], the reactions thus imply that the molybdyl in the hexamolybdate behaves like a carbonyl of aldehydes, ketones or esters in some extent and may have the chemistry similar to that of a carbonyl.

The phosphinimine approach is very useful for the synthesis of arylimido monosubstituents of the hexamolybdate cluster. Recently, an interesting styrylimido-hexamolybdate complex, $(n\text{-}Bu_4N)_2[Mo_6O_{18}(NC_6H_4CH{=}CH_2)]$ (9), shown in figure 5, which is the important building block in fabricating polystyrene compositions with the POM as a pendant group, has also been prepared in *ca* 90% high yield by this approach [72]. However, attempts to make the alkylimido analogues using this approach are unsuccessful. For instance, the reaction of $(n\text{-}Bu_4N)_2[Mo_6O_{19}]$ (6) with $Bu^n{}_3P{=}NBu^t$ failed to give the tertbutylimdo derivative [64] because phosphinalkylimines are more reactive than the corresponding phosphinarylimines and result in the reduction and rearrangement of the hexamolybdate cluster.

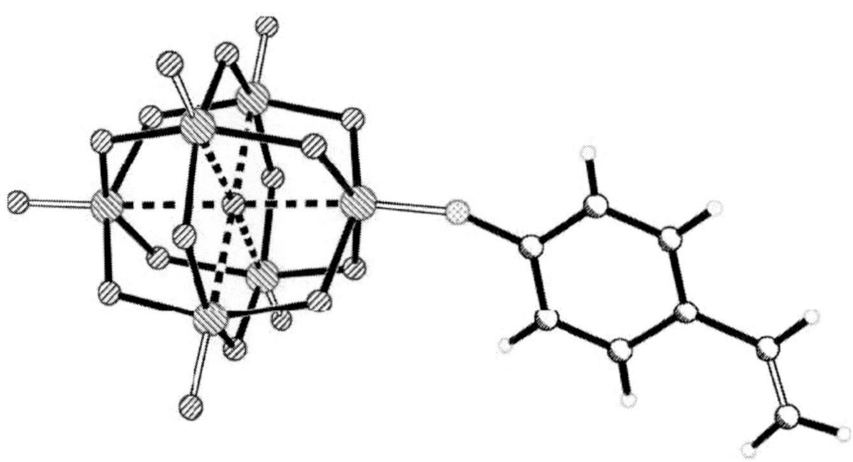

Figure 5. Ball-and-stick model of the styrylimido-hexamolybdate complex 9.

The second approach [Eq. (2)] to produce organoimido derivatives of the hexamolybdate was first set up by R. J. Errington and co-workers a little over a decade ago [64], who have shown that both of the corresponding monoimido-substituted derivatives could be afforded by reactions of alkyl and aryl isocyanates with $[Mo_6O_{19}]^{2-}$ at elevated temperatures in acetonitrile or benzonitrile, and reported the first structurally characterized tertbutylimdo derivative of the hexamolybdate, $(n\text{-}Bu_4N)_2[Mo_6O_{18}(NBu^t)]$ (10) [64]. Maatta and co-workers have subsequently developed this approach successfully using dry pyridine as the solvent [65]. They discovered that alkyl isocyanates are more reactive with the hexamolybdate clusters in dry pyridine and the reaction between them can occur at room temperature and finish in 2 to 3 days [69, 71]. Nevertheless, the reaction of aryl isocyanates with $[Mo_6O_{19}]^{2-}$ in dry pyridine still needs to be carried out at high temperature (110 °C) and must be refluxed under strictly anhydrous conditions for a prolonged time [69, 71]. It is worthwhile to point out here that this reaction is also analogous to that of isocyanates with aldehydes or ketones to form imines [87], which again suggests the comparability between the molybdyl and the carbonyl.

Using this improved route, E. A. Maatta *et al* have further synthesized a lot of monosubstituted organoimido derivatives [65, 69- 71], including the charge-transfer ferrocenylimido complex, $(n\text{-}Bu_4N)_2[(FcN)Mo_6O_{18}]$ (11), FcN = $(\eta^5\text{-}C_5H_5)Fe(\mu\text{-}\eta^5\text{:}\sigma\text{-}C_5H_4N)$, (figure 6) [70] and the organodi-imido-bridged bis(hexamolybdate) derivative, $(n\text{-}Bu_4N)_2$ $[(Mo_5O_{18})Mo{\equiv}N\text{-}Z\text{-}N{\equiv}Mo\ (O_{18}Mo_5)]$ (12), Z = 1,4-cyclo-C_6H_{10}, (figure 7) [69]. Compound 11 features the covalent attachment of an oxidizable ferrocenyl donor to a reducible POM acceptor system. Compound 12 stands for the fist example of the hexamolybdate complexes of diimido ligands, which are coordinated to two hexamolybdates and links them as a bridge.

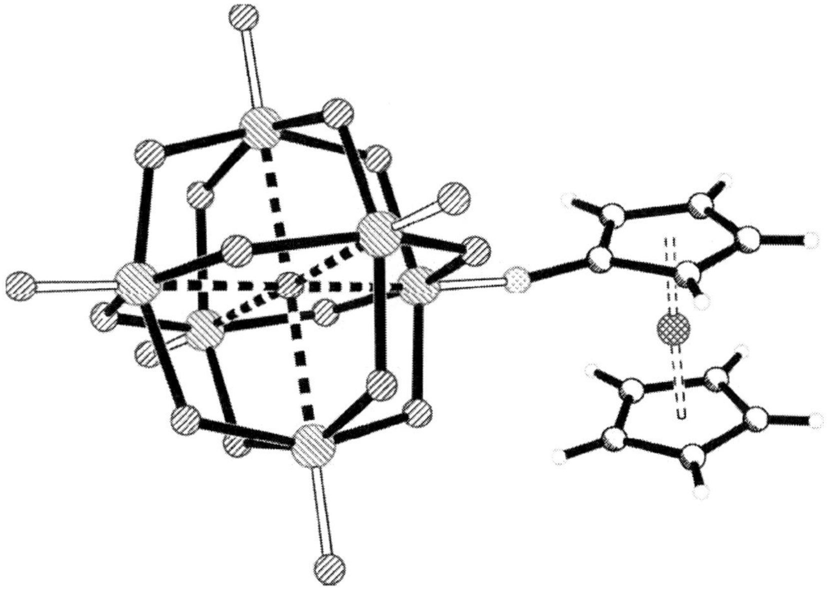

Figure 6. Ball-and-stick model of the charge-transfer ferrocenylimido-hexamolybdate complex 11.

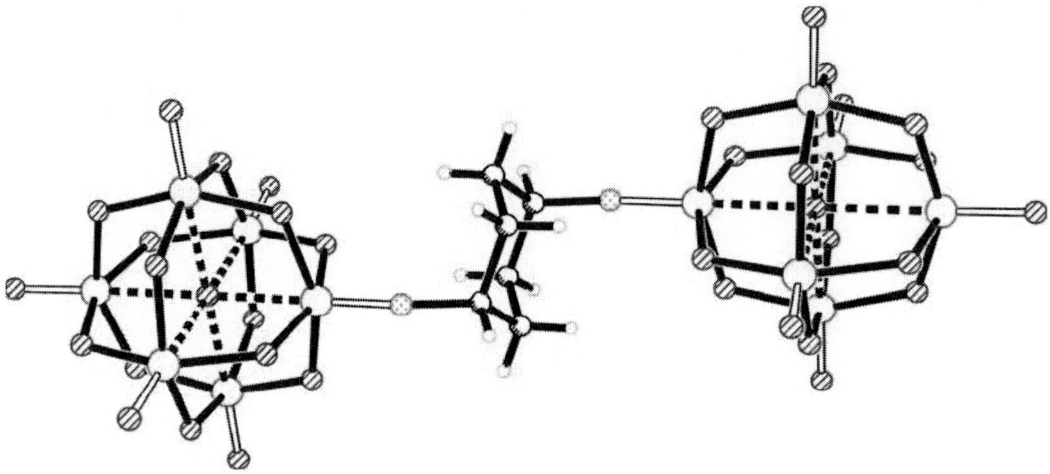

Figure 7. Ball-and-stick model of the organodi-imido-bridged bis(hexamolybdate) derivative 12.

The last approach given in Eq(3) was also put forward by R. J. Errington and co-workers firstly. They observed that in the presence of triethylamine, an aromatic primary amine, for example 1,4-diaminobenzene, reacted directly with the hexamolybdate ions in benzonitrile solvent at *ca* 150 °C [66]. Albeit this reaction occurs under rather harsh conditions and only a mixture that includes mono- and di-functionalized derivatives, and various oligomers is obtained, from which pure imido derivatives are difficult to isolate from each other, it indeed provides us a chance to prepare organoimido derivatives with cheaper and more accessible arylamines as the imido-releasing reagents. Additionally, this reaction confirms further the similarity between the molybdyl and carbonyl groups since it extremely resembles the Schiff-base reaction taken place between primary amines with aldehydes or ketones, which involves of nucleophilic addition of the amine to the carbonyl followed by elimination of water [88].

Several years later, R. A. Roesner et al improved this approach with dry pyridine as the solvent instead of benzonitrile. However, only aromatic amines with the strongly electron-donating substituents, such as the alkoxy groups, promise this reaction, under heating and dry N_2-gas protecting conditions, to create the corresponding mono-imido derivatives of the hexamolybdate in moderate yield [67].

We have now a long time interest in the chemistry of derivatives of POMs [6, 8, 9, 68, 73- 84], including new synthetic methodologies for the organoimido derivatives [68, 73- 83]. In the light of the pioneering investigations of E. A. Maatta and R. J. Errington's groups on this field, we strongly feel that the molybdyl is, under certain cases, one of the inorganic analogues, in some extent, of the carbonyl in organic chemistry [88], and some organic synthetic techniques can be extended into the chemistry of derivatives of POMs. For the primary amine approach, for example, if the resulting water is able to remove from the reactive mixture successively, just like common dehydrating reactions in organic chemistry, the reaction of Eq(3) will proceed rapidly and irreversibly, and complete thoroughly at the end.

Five years ago, together with our co-workers at UMKC, we discovered that using carbodiimides, which are among the most widely-used reagents in common organic synthesis [89-91], such as DPC (N,N'-diisopropylcarbodiimide) and DCC (N,N'-dicyclohexylcarbodiimide), as the dehydrating reagent, the reaction of $[Mo_6O_{19}]^{2-}$ with aromatic primary amines can be dramatically promoted, and have developed a novel DCC-dehydrating synthetic protocol to afford arylimido derivatives of POMs in the context of the primary amine approach [68]. As shown in Eq(4), in the presence of one equivalent of DCC, one equivalent of aromatic primary amines reacts smoothly with an excess of $(n\text{-}Bu_4N)_2[Mo_6O_{19}]$ (6) under dry N_2 gas and refluxing acetonitrile, pyridine or hot N,N'-dimethylformamide, to give the corresponding mono-substituted imido derivatives usually in excellent yield of more than 90%. Not only a variety of aromatic primary amines occur this reaction, but also it is usually completed in less than 12 hours, and pure *crystalline* products can be readily obtained with convenient bench manipulations. Moreover, this reaction can be carried out in the open air without nitrogen protection if slightly over one equivalent of DCC (1.2 equivalent) is added. Best of all, the reaction also tolerates additional remote functional groups on the aromatic ring, including iodo, ethynyl and terpyridyl groups. Figures 8-10 show the X-ray structures of three imido derivatives containing such three functional groups, $(n\text{-}Bu_4N)_2[Mo_6O_{18}(NAr)]$, Ar = 4-iodo-2,6-dimethyl-C_6H_2 (13) [68], 4-ethynyl-2,6-dimethyl-C_6H_2 (14) [68], 4-(4'-terpyridylethynyl)-2,6-dimethyl-C_6H_2 (15) [73], respectively, of which compounds 13 and 14 both have two stable conformational isomers differing by a rotation of the aromatic ring relative to the hexamolybdate skeleton. All the three complexes are valuable building blocks to construct POM-based organic-inorganic hybrid molecular materials in our controllable molecule-module-assembly strategy [56].

$$[Mo_6O_{19}]^{2-}\text{ (excess)} + ArNH_2 + DCC \longrightarrow [Mo_6O_{18}(NAr)]^{2-} + (cyclo\text{-}C_6H_{11}NH)_2CO \quad (4)$$

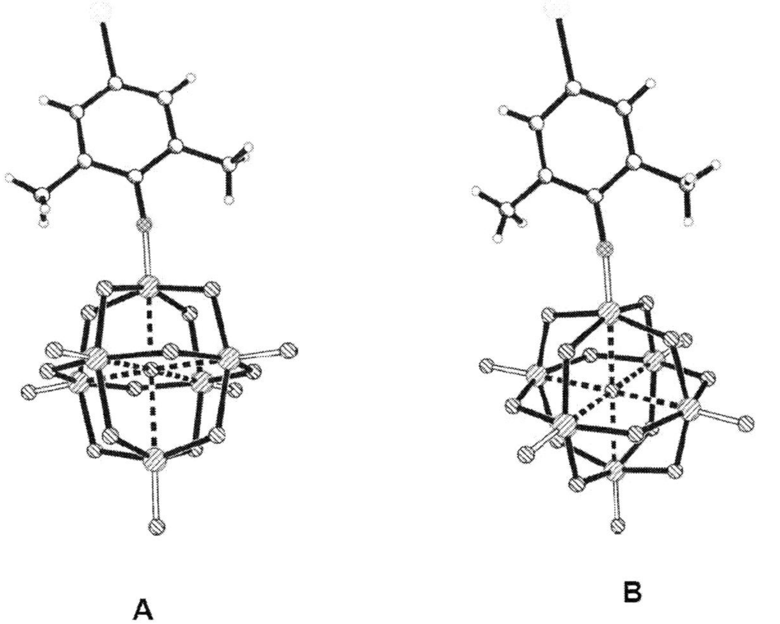

Figure 8. Structure of iodo-substituted derivative 13. A and B stand for its two conformational isomers differing by a rotation of the aromatic ring relative to the hexamolybdate skeleton.

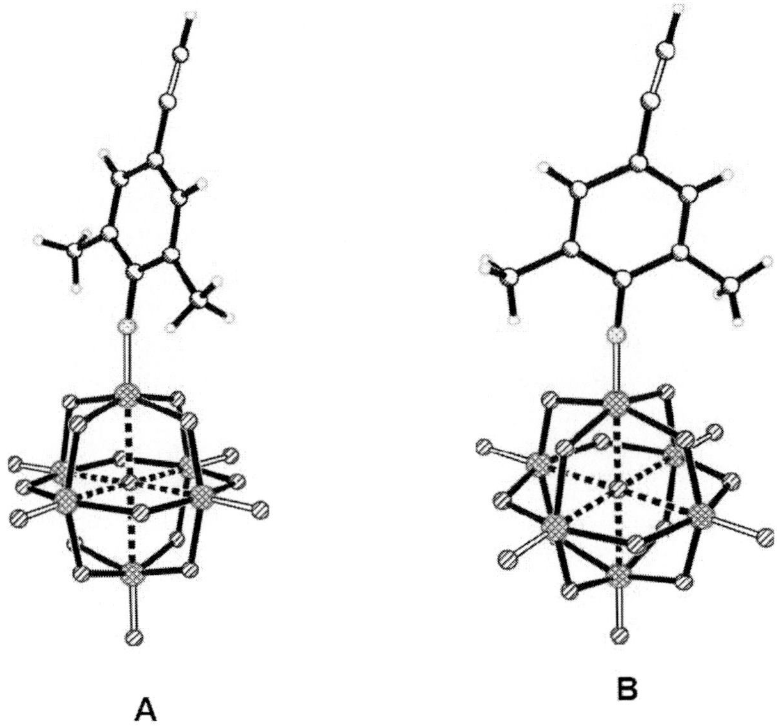

Figure 9. Structure of ethynyl-functionalized derivative 14. A and B stand for its two conformational isomers differing by a rotation of the aromatic ring relative to the hexamolybdate skeleton.

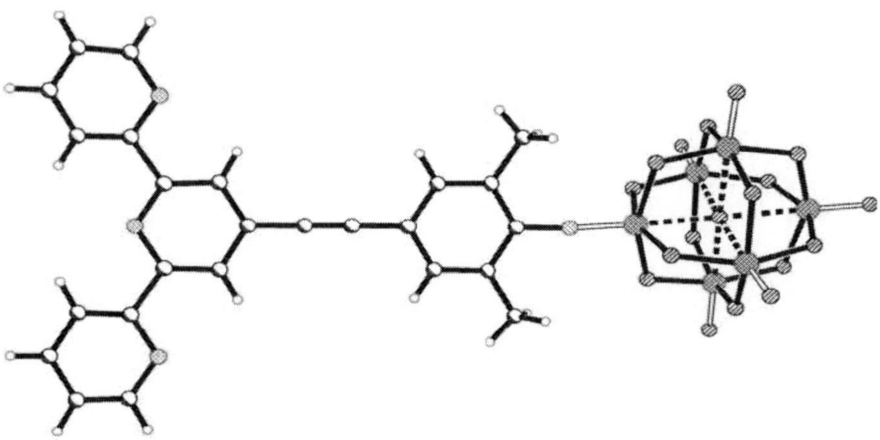

Figure 10. Ball-and-stick drawing of the organoimido hexamolybdate compound 15 containing a terpyridyl group.

In our investigations, other drying or dehydrating agents have also been tested instead of carbodiimides, including magnesium sulfate, calcium chloride, 4Å zeolite, phosphorus pentoxide, *N'N*-carbonyldiimidazole or even calcium hydride. However, they do not show similar activation effect like that of DCC [68]. This suggests that DCC does not simply serve as a dehydration agent to remove the water generated in the reaction. The likely role of DCC is presumedly to activate the terminal Mo-O bond and hence to increase the electrophilicity of the molybdenum atom, similar to its activating effect on the carboxyl group in amide or peptide synthesis [90]. In addition, an extremely excess of DCC is harmful to this reaction because DCC can undergo the side reactions directly with the molybdate ions and the initially formed products; for example, addition of eight equivalent of DCC, reaction of 2,6-dimethylaniline with $(n\text{-}Bu_4N)_2[Mo_6O_{19}]$ (6) under 12 hours of refluxing in dry acetonitrile affords, in place of the expected imido products, only an octamolybdtate compound, $(cyclo\text{-}C_6H_{11}NH_3)_4$ β-$[Mo_8O_{26}]$ (16), which is usually the major byproduct found in this reaction [68].

Recently, we have succeeded in extending this synthetic route from aniline homologs and their derivatives to other aromatic primary amines containing fused rings, including naphthylamine, anthramine, phenanthramine and aminopyrene. For instance, 1-naphthylamine undergoes this reaction effectively, in *ca* 70% yield, to afford the corresponding imido derivative (figure 11), $(n\text{-}Bu_4N)_2[Mo_6O_{18}(NC_{10}H_7\text{-}1)]$ (17) [74].

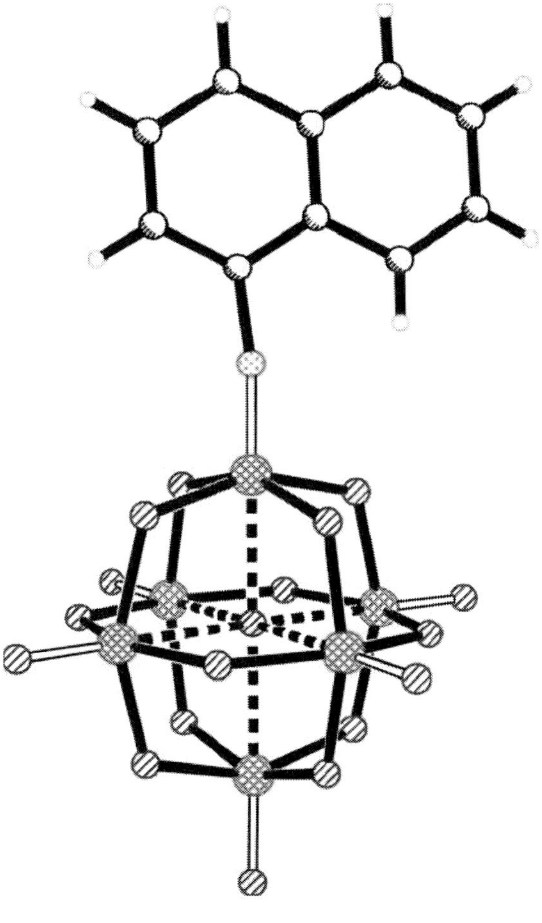

Figure 11. Ball-and-stick showing of the 1-naphthylimido hexamolybdate 17.

It should be pointed out here that, aromatic amines only with electron-withdrawing substituents such as nitro, fluoro and chloro groups do not undertake this reaction, neither do the aminopyridines, due to the weak nucleophilicity or basicity and easy oxidization of these kinds of amines [75]. We have also failed to prepare the alkylimido derivatives using this route with the corresponding aliphatic primary amines, such as benzylamine, since they violently reduce the hexamolybdate ion, usually resulting the single electron reduction produce of the parent cluster, $[Mo_6O_{19}]^{3-}$ [68, 75]. However, much effort to broaden the applicability of the DCC-dehydrating protocol to a broad spectrum of amines is still being taken in our laboratory.

Moreover, in our recent attempts to functionalize the octamolybdtates with organoimido groups, we discovered that in the presence of DCC, a proton can dramatically speed up the reaction of α-$[Mo_8O_{26}]^{4-}$ with aromatic primary amines under much milder conditions and, in the meanwhile, monofunctionalized organoimido derivatives of $[Mo_6O_{19}]^{2-}$ are selectively synthesized in high purity and moderate yield with easy workup [75]. This acid-assisting route has allowed the facile synthesis of a large number of mono-imido hexamolybdates, particularly, including those even bearing electron-withdrawing substituents such as bromo, chloro and fluoro groups [75- 78]. As shown in Eq(5), on addition of DCC (2 mmol), a variety of halogenated aromatic amines and their hydrochloride salts, in equivalent amounts (1.34 mmol), react with $(n\text{-}Bu_4N)_2$·α-$[Mo_8O_{26}]$ (1.0 mmol) smoothly, under nitrogen in hot acetonitrile (10 mL), obtaining the corresponding monosubstituted imido derivative in *ca* 50% yield after about 6 hours refluxing. Two structurally characterized halogenated imido derivatives, $(n\text{-}Bu_4N)_2[Mo_6O_{18}(NAr)]$, Ar = *p*-chlorophenyl (18) [75], *p*-bromophenyl (19) [77], respectively, are exemplified in figures 12 and 13. Nitroanilines also undergo this reaction, however, we have not successfully separated the corresponding derivatives from the reaction mixtures yet due to their decomposition in solution.

$$[Mo_8O_{26}]^{4-} + ArNH_2 + ArNH_3Cl \xrightarrow[\text{Refluxing, 6h}]{\text{2DCC, } CH_3CN} [Mo_6O_{18}(NAr)]^{2-} \quad (5)$$

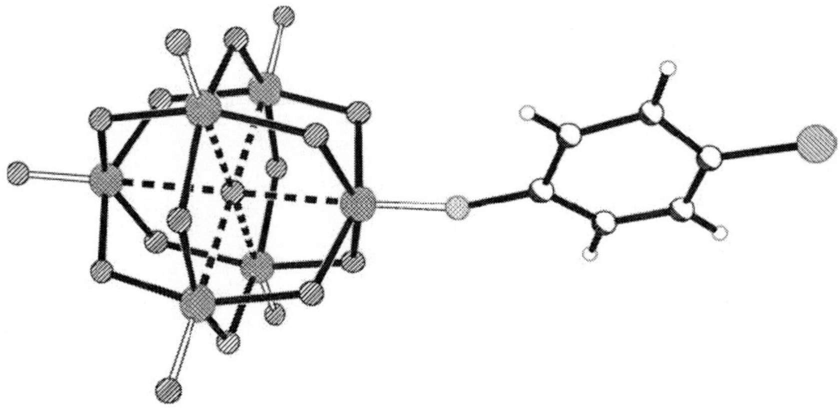

Figure 12. Ball-and-stick showing of the *p*-chlorophenylimido hexamolybdate 18.

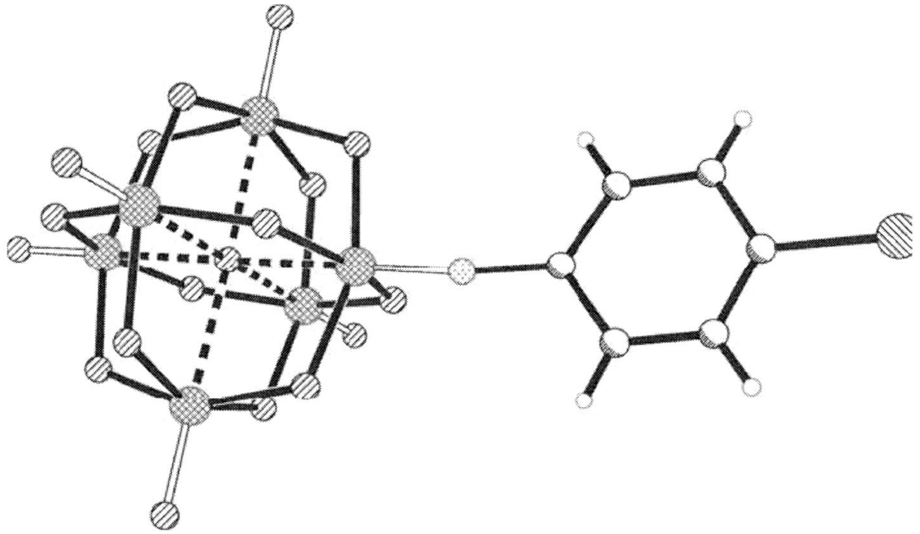

Figure 13. Ball-and-stick showing of the *p*-Bromophenylimido hexamolybdate 19.

It is worth pointing out that, the proton is crucial to the above route for the formation of monofunctionalized derivatives. It was observed that, in the absence of hydrochloride salts– which offer the protons, activated aromatic amines containing electron-donating substituents such as alkyl, alkoxy and alkylamino groups, react with α-$[Mo_8O_{26}]^{4-}$ to yield only the bifunctionalized arylimido derivatives of hexamolybdate. On the other hand, under the same conditions, no dehydrating reaction occurs between α-$[Mo_8O_{26}]^{4-}$ and inert aromatic amines with electron-withdrawing substituents such as chloro, fluoro and nitro groups, and the amines are oxidized instead. Furthermore, increasing the amount of hydrochloride salts can speed up the reaction, and correspondingly, it can be carried out even at room temperature. While more work needs to be done to shed light on the detailed reaction mechanism, the imaginable role of the proton is to complex with DCC and hence to strengthen the electrophilic ability of DCC to attack the oxo group of Mo-O group. Additionally, it also promotes the conversion of the α-octamolybdate into the hexamolybdate through a degradation and re-assembly process since in an acidic organic solvent, the hexamolybdate ions and their derivatives are much more stable than the α-octamolybdate ions.

There are a lot of spectroscopic techniques having been used to characterize the organoimido derivatives. In the low wavenumber region of the IR spectra ($\widetilde{\nu}$ <1000 cm^{-1}), the monofunctionalized organoimido derivatives display similar characteristic patterns of the parent hexamolybdate anion: there are two very strong bands of the Mo-O_t stretching and Mo-O_b-Mo asymmetric stretching vibrations at *ca* 955 (±5) and 795 (±5) cm^{-1}, respectively. Different from the parent cluster, there is also a strong shoulder peak appeared near 975 (±5) cm^{-1}, around the Mo-O stretching vibrations. This band may derive from the Mo-N stretching vibration [92] and is a diagnostic character for mono-organoimido substitution.

Compared to the ^1H NMR spectra of corresponding free amines, the protons attached to the α-carbon atoms or benzene rings in the monofunctionalized alkylimido and arylimido derivatives, respectively, all exhibit significantly downfield chemical shifts, indicating that the shielding nature of $[Mo_5O_{18}(Mo{\equiv}N\text{-})]^{2-}$ is much weaker than that of the amino group NH_2- and in agreement with the strong electron acceptability of POM clusters.

While the UV/Vis spectra of monofunctionalized alkylimido derivatives resemble that of the hexamolybdate parent intimately in the low energy region, those of monofunctionalized arylimido derivatives show considerable changes very distinctly due to their different molecular and electronic structures. The lowest energy electronic transition at 325nm in $[Mo_6O_{19}]^{2-}$, which was assigned to a L→M charge-transfer transition from the terminal oxygen non-bonding π-type HOMO to the molybdenum π-type LUMO [92], is bathochromically shifted by more than 20 nm in the monofunctionalized arylimido derivatives and becomes significantly more intense. The observed bathochromic shifts imply that the Mo-N π-bond is formed and delocalized over the organic conjugated π systems in such arylimido derivatives. In other words, there are strong electronic interactions between the hexamolybdate cluster and the conjugated organic ligands. In addition, the extent of bathochromic shifts in the reported mono-arylimido derivatives is also consistent with the length of conjugated segments and the trend in the π-donor ability of arylimido ligands.

Electrochemical data and ^{95}Mo, ^{17}O, and ^{14}N NMR data have also been recorded for a range of mono-imido derivatives of the hexamolybdate [63, 71]. The following trends have been observed [53]: (i) All species display a one-electron reduction process, the potential of which undergoes a cathodic shift in the order O < NAr < NR and, for arylimido derivatives, in the order of increasing electron-donating properties of the substituent. (ii) The $[^{95}Mo{\equiv}NR]$ resonance is shifted upfield in the order O < NAr < NR, and in the series of arylimido complexes, in the order of increasing electron-attracting properties of the substituent. (iii) The ^{17}O NMR spectra each are globally shielded with respect to that of $[Mo_6O_{19}]^{2-}$. These data are consistent with the trend in π-donor ability of the ligands, O < NAr < NR. For arylimido derivatives, molybdenum and nitrogen shieldings increase with increase in the $(\sigma + \pi)$ acceptor ability of the substituent.

The X-ray crystal structures of all the mono-organoimido derivatives have some common features: They resemble that of their hexamolybdate parent, but one terminal oxo group has been replaced by a organoimido substituent. The short Mo-N bond lengths range from 1.68-1.75 Å and nearly linear linkages of Mo-N-C are consistent with a substantial degree of triple bond character, although the angles sometimes deviate from linearity more than expected due to the force of the crystal packing [75]. The central oxygen atoms become close to the imido-bearing molybdenum atoms since the *trans* influence of an imido group is weaker compared to a terminal oxo group [76]. While the bond length involving the bridging-oxygen atoms shows considerable changes, there are no appreciable variations in the terminal bond lengths, raging from 1.66 to 1.72 Å. The pattern of short/long trans bond alternation in the three mutually orthogonal Mo_4O_4 belts observed in the original parent cluster still remains in the Mo_4O_4 belts perpendicular to the Mo-N bonds. In addition, the supramolecular "pairing self-assembly" of the cluster anions through π-π stacking between parallel phenyl rings on two neighboring clusters has also been observed in some mono-arylimido derivatives [75-77], which results the formation of dimers in their crystalline states (figure 14).

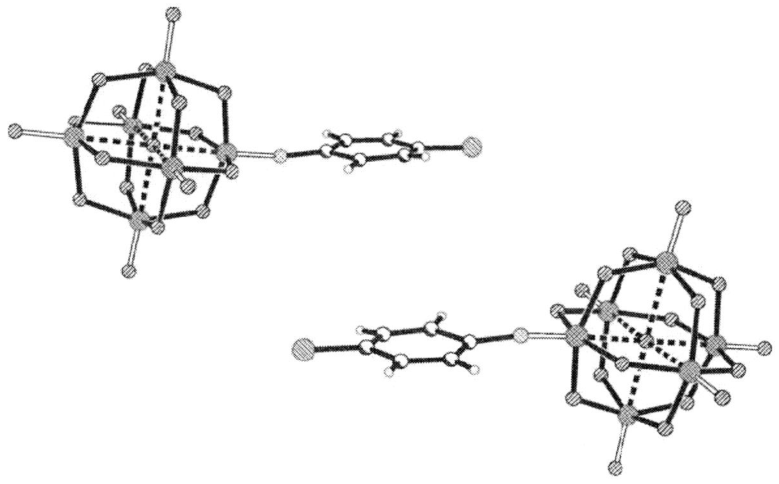

Figure 14. Dimer formation of the cluster anion in the solid of compound 18.

2.1.2. Disubstituted Derivatives

A characteristic feature of the derivatization of the hexamolybdate with organoimido ligands is the capacity for polyfunctionalization. Of which the difunctionalization is the easiest and most common. A. Proust et al noticed that, the *cis*-diimido derivative formed even when only one equivalent of $Ph_3P{=}NPh$ was exploited to react with $[Mo_6O_{19}]^{2-}$ [53]. However, using two equivalent of $Ph_3P{=}NPh$, it still gave a mixture of un-reacted hexamolybdate anions and, mono- and *cis*-diimido derivatives. The isolation of the successive derivatives is also difficult because of they cocrystallize usually in the crystalline state [63].

The first example of well characterized di-organoimido derivative was reported by E. A. Maatta and co-workers [65, 71]. They found that the reaction of $(n\text{-}Bu_4N)_2[Mo_6O_{19}]$ (6) with two equivalent of 2,6-diisopropylphenyl isocyanate in refluxing pyridine for 8 days occurs to form the *cis*-disubstituted imido hexamolybdate, $(n\text{-}Bu_4N)_2[cis\text{-}Mo_6O_{17}(NAr)_2]$, Ar = 2,6-diisopropylphenyl (20), in 67% yield. It is obvious that this is a poor route in efficiency. Moreover, they also mentioned that the crude products contain an admixture of higher- and lower-substituted hexamolybdates, which makes the separation of large amounts of the desired difunctionalized product tedious and impractical, if not impossible.

In the past years, we also checked the possibility of our DCC-dehydrating protocol in the preparation of di-organoimido derivatives of the hexamolybdate. In most case, attempts to use the same route as for the mono-substituted derivatives but with stoichiometric ratio of two equivalent aromatic primary amines failed to obtain the disubstituted derivatives in high yields. Instead, a mixture of various polysubstituted hexamolybdates are generated since there are six equally reactive sites on a hexamolybdate ion, and thus repeated recrystallization is needed for purification of the anticipant product. However, an attempt to functionalize the α-isomer of octamolybdate ion, $\alpha\text{-}[Mo_8O_{26}]^{4-}$, with this reaction routine resulted in the selective synthesis of difunctionalized arylimido derivatives of hexamolybdate with good yield and high purity [79].

This novel route is shown in Eq(6), which allows the efficient preparation of a number of di-imido hexamolybdates more conveniently with easy and simple bench operations [79-81]. For example, the refluxing reaction of $(n\text{-}Bu_4N)_2{\cdot}\alpha\text{-}[Mo_8O_{26}]$ with two equivalent of DCC and two equivalent of 4-iodo-2,6-diisopropylaniline or 4-ethynyl-2,6-dimethylaniline in dry acetonitrile for 6 to 12 hours affords the corresponding di-organoimido hexamolybdates, (n-$Bu_4N)_2$[*cis*-$Mo_6O_{17}(NAr)_2$], Ar = 4-iodo-2,6-diisopropyl-C_6H_2 (21) [79], 4-ethynyl-2,6-diisopropyl-C_6H_2 (22) [81], respectively, in good yields of 60-80%. The X-ray structures of compounds 21 and 22 are illustrated in figures 15 and 16, respectively. They both have the *cis* disubstituted structures with two activate remote functional groups at the end, which make them very important building bocks in fabricating novel organic-inorganic hybrids.

$$[Mo_8O_{26}]^{4-} + 2ArNH_2 \xrightarrow[\text{Refluxing, 6-12h}]{\text{2DCC, }CH_3CN} [cis\text{-}Mo_6O_{17}(NAr)_2]^{2-} \qquad (6)$$

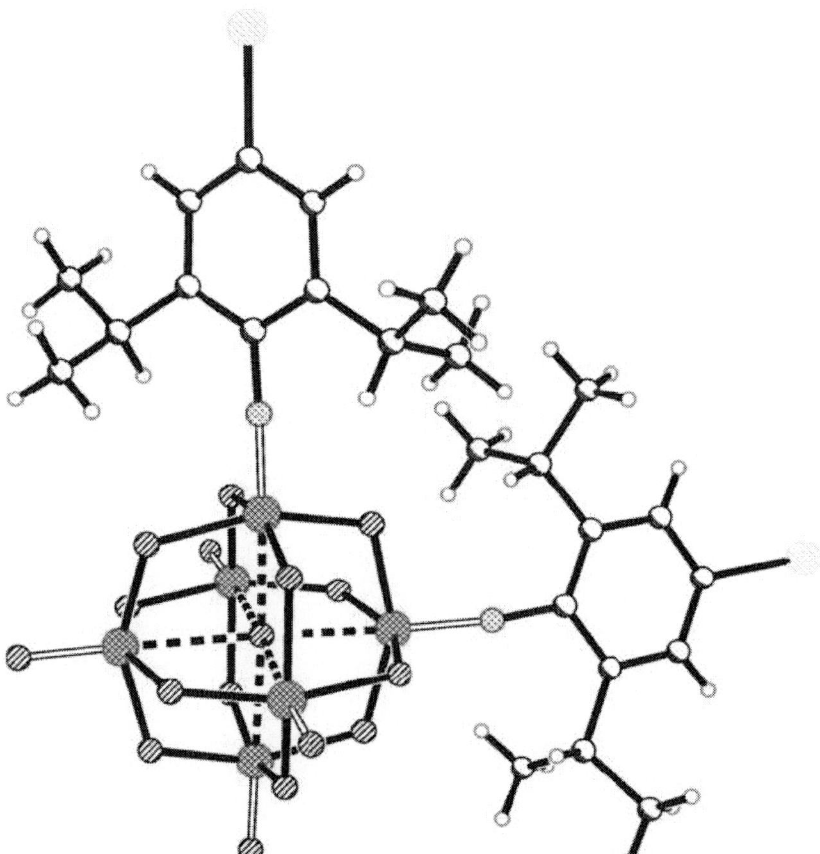

Figure 15. Ball-and-stick structure of the cis-disubstituted hexamolybdate imido derivative 21.

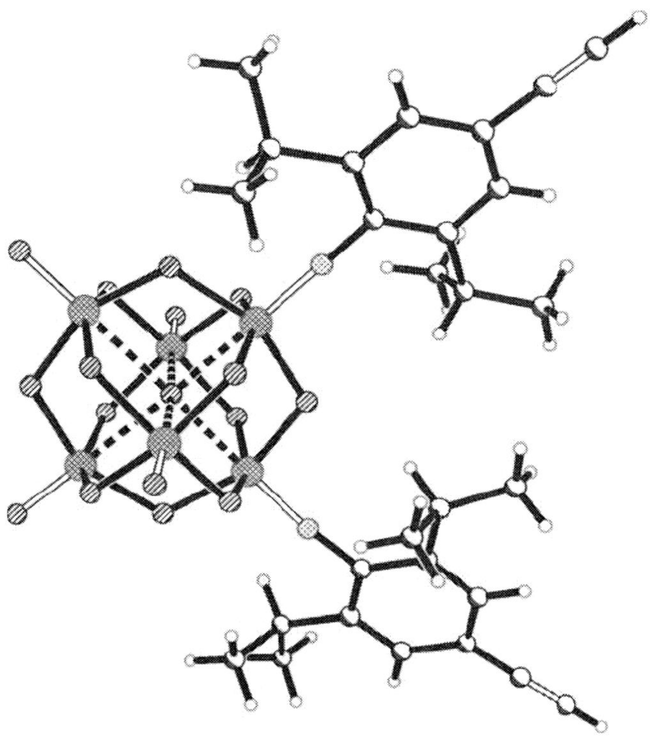

Figure 16. Ball-and-stick structure of the cis-didisubstituted hexamolybdate imido derivative 22.

In principle, there are, *cis* and *trans*, two position isomers in the terminal substituted di-imido hexamolybdates. If the steric and electronic effects of the imido groups are ignored, both of them can be formed in the same reaction, theoretically at a *cis* and *trans* ratio of 4:1, since there are four *cis* sites and only one *trans* terminal oxo reaction sites in a mono-imido derivative. While the steric repulsion is considered, which, in fact, is very common in organic substitution reactions, the *trans*-isomer will be more than 25% and becomes the principal product. But contrary to expectations, except for the *cis*-isomers, the *trans*-disubstituted hexamolybdates have scarcely been detected in the above reaction routes. It is seems that the disubstituted hexamolybdate with aryliimdo prefers to form a *cis*-isomeric derivative. Such a result is amazing, which suggests that the steric effect is less important here and on the other hand, the present of an extant imido substituent exerts an activating effect at proximal molybdyl groups. Indeed, a recent density functional theory study reveals that the *cis*-disubstituted derives are more stable in energy than the corresponding *trans*-isomers [93].

While the above routes give only the *cis*-disubstituted derives, R. J. Errington and co-workers discovered that, using their aromatic primary amines route [Eq(3)], a *trans*-disubstituted hexamolybdate, $(n\text{-}Bu_4N)_2[trans\text{-}Mo_6O_{17}(NC_6H_4NH_2)_2]$ (23), the unique example reported in the literature so far [66], was formed predominantly in the reaction between $(n\text{-}Bu_4N)_2[Mo_6O_{19}]$ (6) and two equivalent of 1,4-phenylenediamine. Moreover, ^1H NMR also confirmed the formation of the corresponding *cis*-isomer in the resulting products. Their work implies that the *trans*- disubstituted hexamolybdate is a kinetic stable product under certain supported conditions and can be isolated indeed. It also intensively suggests that there is a obvious different mechanism of the directly functionalization of the hexmolybdate with aromatic primary amines from those with phosphinimines or isocyanates and the aromatic primary amines approach is potential in the *trans*-disubstituted hexamolybdate derivatives.

Keeping the above idea in mind, we have had a renewed check very recently on our DCC-dehydrating protocol for the direct functionalization of the hexamolybdate with stoichiometric ratios of reactants [82- 83]. Indeed, for some activated aromatic primary amines containing electron-donating substitents, such as 2,6-dimethylaniline and *ortho*-anisidine, in the presence of one equivalent of DCC, the reaction of them and 0.5 equivalent of $(n\text{-}Bu_4N)_2[Mo_6O_{19}]$ (6) in heated acetonitrile can afford both the *cis*- and *trans*-disubstituted imido complexes in reasonable yields by a careful control of the refluxing time [Eq(7)].

$$[Mo_6O_{19}]^{2-} + 2ArNH_2 \xrightarrow[\text{Refluxing, 6-24h}]{\text{2DCC, }CH_3CN} [Mo_6O_{17}(NAr)_2]^{2-} \tag{7}$$

In general, a prolonged refluxing time promotes the formation of the *cis*-isomer, for example, the compound, $(n\text{-}Bu_4N)_2[cis\text{-}Mo_6O_{17}(NAr)_2]$, Ar = 2-methoxy-$C_6H_4$, (24) [82], was obtained from 24 hours refluxing reaction; On the other hand, the *trans*-isomer can be isolated from a relative short-time reaction system, for example, the compound, $(n\text{-}Bu_4N)_2[trans\text{-}Mo_6O_{17}(NAr)_2]$, Ar = 2,6-dimethyl-$C_6H_3$, (25) [83], was acquired from 6 hours refluxing reaction. In figures 17 and 18, the X-ray structures of compound 24 and 25 are shown, respectively. Compared to the parent hexamolybdate, compound 24 has a smaller optical band gap of 1.75 eV due to the incorporation of two strong electron-donating anisidido groups, which suggests that compound 24 is a potential molecular semiconductor and the optical band gap of POM-based semiconducting materials could be tuned exactly and controllably via chemical modification using suitable organic ligands. The chemical and physical properties of compound 25 are still underway in our laboratory.

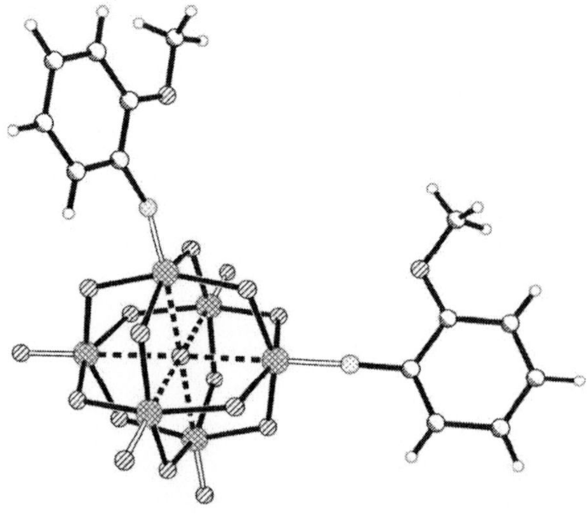

Figure 17. Ball-and-stick structure of the cis-didisubstituted hexamolybdate imido derivative 24.

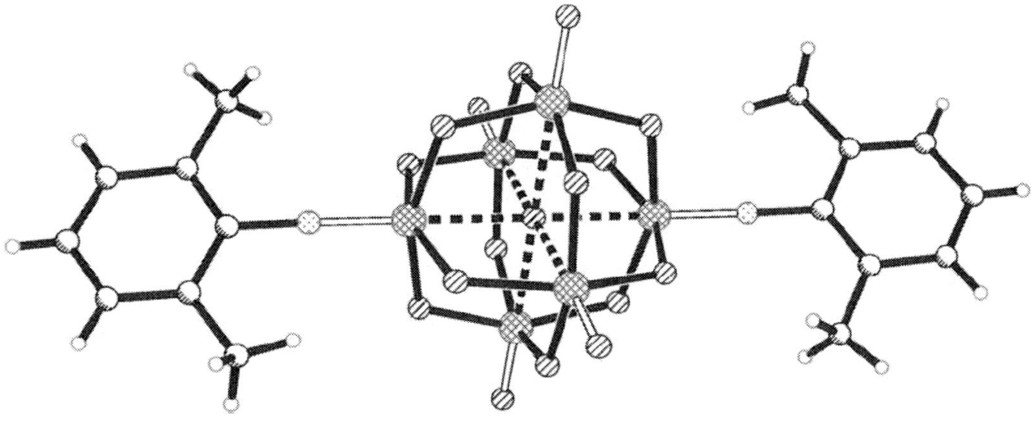

Figure 18. Ball-and-stick structure of the trans-didisubstituted imido complex 25.

Similar to their monosubstituted analogues, all the reported diimido hexamolybdates have of their own common features and some general trends in the structures and spectroscopies: (i) The bond lengths of [Mo≡NAr] triple bond are within the range of 1.71 to 1.75 Å, together with nearly linear linkages of Mo-N-C angles close to 180°. For the *cis*-isomers, one of the angles sometimes deviate from linearity more than expected due to the aggregation of the cluster anions through π-π stacking and the force of the crystal packing [80]. (ii) The central oxygen atoms inside the hexamolybdate cage are drawn closer to the imido-bearing Mo atoms, while the four bridge oxygen atoms are pushed away from the imido-bearing Mo atoms. (iii) The short-long bond alternation in the in the Mo_4O_4 belts of the parent hexamolybdate disappears in the *cis*-isomers. However, in the trans-isomers, this pattern is still kept in the belt perpendicular to the long axis of the cluster along the two Mo-N bonds. (iv) The patterns of supramolecular self-assembly due to π-π stacking or/and C-H...O hydrogen bonding are very popular. In particular, the I-O interaction has also been observed in compound 21 (figures 19). (v) In their IR spectra, the Mo-N stretching vibrations at 968 (±2) cm^{-1}, appear as a strong shoulder peak of the Mo-O stretching vibrations at 945 (±2) cm^{-1}. Compared to the corresponding mono-imido derivatives, there are obviously blue-shifts of *ca* 5 and 10 cm^{-1} for the Mo-N and Mo-O stretching vibrations, respectively. Correspondingly, the Mo-O_b-Mo asymmetric stretching vibrations at *ca* 795 (±5) cm^{-1} of the mono-imido and parent hexamolybdates are significantly blue-shifted more than 20 cm^{-1} and appear as a broad and strong band at *ca* 775 cm^{-1}. (vi) Compared to the corresponding mono-imido derivatives, the lowest energy electronic transition of L→M charge-transfer is further red-shifted by *ca* 5 nm due to the incorporation of the other arylimido ligand. In the meanwhile, ^{1}H NMR of the aromatic protons shows visible upfield shifts, and the reduction potentials are also observed to shift cathodically due to the increase of electron density in the cluster anions accompanying the incorporation of the second arylimido ligand.

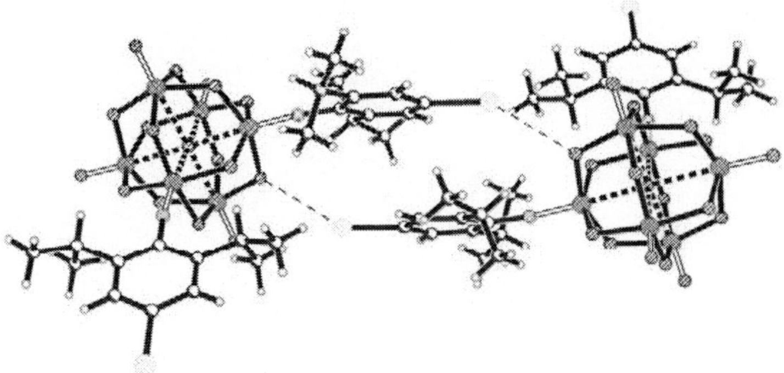

Figure 19. A view of the I-O interaction in a dimer of compound 21 in the crystalline state.

2.1.3. Polysubstituted Derivatives

The polysubstituted hexamolybdates, where three to six terminal oxo groups replaced by 2,6-diisopropylphenylimido ligands, have also been reported by E. A. Maatta and co-workers for nearly a decade [65, 71, 84]. They discovered that, by carefully controlling the molar ratios of 2,6-diisopropylaniline isocyanate over the hexamolybdate cluster , multiple-functionalized hexamolybdates, $[Mo_6O_{(19-x)}(NAr)_x]^{2-}$, Ar = 2,6-diisopropylphenyl, x = 3 (26) [71], 4 (27) [65, 71], 5 (28) [65, 71], 6 (29) [84] could be synthesized in refluxing pyridine after a prolonged reaction time of 13, 12, 11 and 21 days for 26, 27, 28, and 29, respectively. Of which the X-ray molecular structures of compounds 26 and 29 are exemplified in figures 20 and 21, respectively. While all these polysubstituted derivatives have been isolated and characterized, the nonselective nature of the reaction, which always results in an admixture of various multiple functionalized products, makes their separation process rather difficult. Apparently, the rational synthesis of polysubstituted hexamolybdates is a formidable challenge for a chemist.

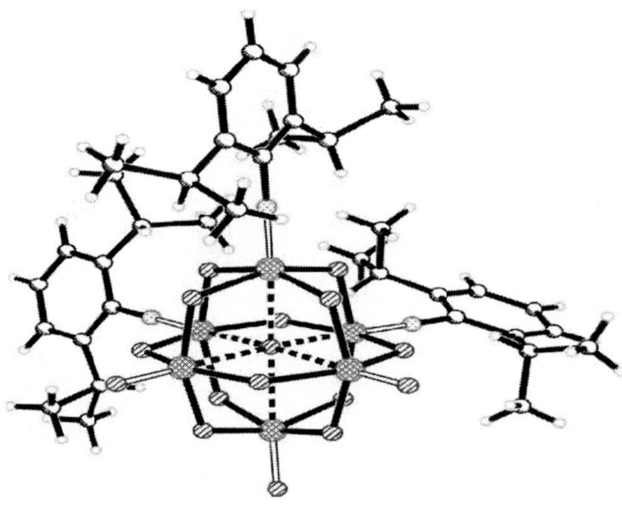

Figure 20. X-ray molecular structure of the trisubstituted hexamolybdate 26.

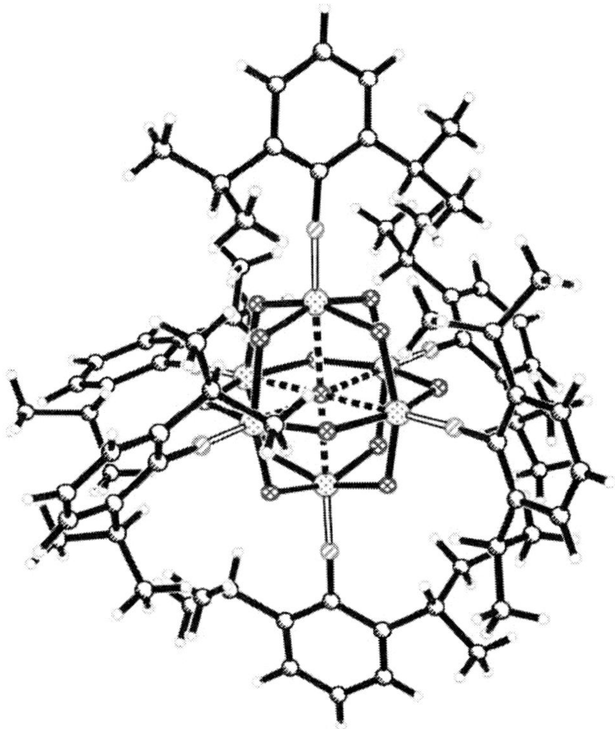

Figure 21. X-ray molecular structure of the hexasubstituted hexamolybdate 29.

We have also interested in the preparation of the polysubstituted hexamolybdates in the context of our DCC-dehydrating protocol using aromatic primary amines as the imido-releasing agents. In a recent attempt to synthesizing the *fac*-tri-2,6-dimethylanilido hexamolybdate [Eq(8)], an analogue of compound 26, contrary to our expectation, the as-resulted tri-imido derivative, $(n\text{-}Bu_4N)_2[cis\text{-}Mo_6O_{16}\ \mu_2\text{-}(NAr)(NAr)_2]$, Ar = 2,6-dimethyl-$C_6H_3$, (30) is not exclusively terminal-oxo replaced complex, and instead, one of bridging-oxo groups is substituted.

$$[Mo_6O_{19}]^{2-} + 3ArNH_2 \xrightarrow[\text{Refluxing, 24h}]{\text{3DCC, CH}_3\text{CN}} [Mo_6O_{16}(NAr)_3]^{2-} \quad (8)$$

As is illustrated in figure 22, the compound 30 has a feature that the μ_2-bridging oxygen atom sharing by the two imido-bearing Mo atoms in a *cis*-diimido hexamolybdate is substituted with a μ_2-bridging organoimido ligand. It suggests that such oxygen atoms in the *cis*-diimido hexamolybdates have been doubly activated by the neighboring imido groups and become more negative (nucleophilic) than other oxygen atoms, resulting in the easy electrophilic attack by DCC. Compound 30 stands for the first example of a bridging oxygen atom in POMs replaced by the bridging imido groups. It breaks the myth that only the terminal oxo groups can be directly replaced with imido ligands before and brings us a wonderful prospect in the chemistry of organoimido derivatives of POMs.

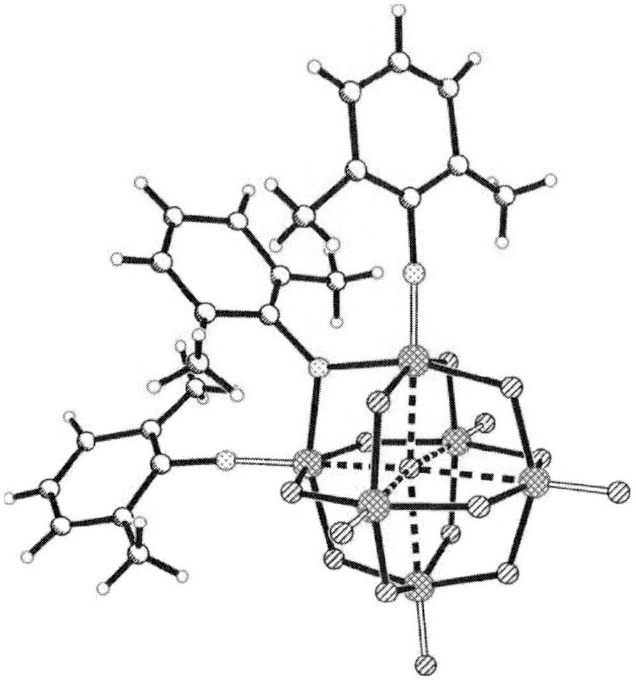

Figure 22. Ball-and-stick model of the trisubstituted hexamolybdate 30.

2.2. Organoimido Derivatives of other Lindqvist Poms

While there is fruitful in the synthetic chemistry of hexamolybdate, related studies on other POMs keep scarcely well developed. So far, there only has been obtained quite limited success in extending the above approaches to the other Lindqvist POMs.

Attempts to directly functionalize the hexatungstate anion, $[W_6O_{19}]^{2-}$, another important Lidquvist homopolyoxometalate besides the hexamolybdate, have failed to utilize all the above-mentioned approaches, since $[W_6O_{19}]^{2-}$ does not react at all with phosphinimines, isocyanates or aromatic primary amines under the present developed conditions. It seems that compared to the molybdyl groups in the hexamolybdate, the tungstyl groups in the hexatungstate is rather unreactive. The only one example of imidosubstituted hexatungstates, $(n\text{-}Bu_4N)_2[W_6O_{18}(NAr)]$, Ar = 2,6-diisopropyl-$C_6H_3$, (31) was obtained about ten years ago by E. A. Maatta et al via the reaction of $(n\text{-}Bu_4N)_2[WO_4]$ with 2,6-diisopropylphenyl isocyante in either 1,2-dichloroethane or pridine, but the yield is very low (*ca* 10% based on $n\text{-}Bu_4N)_2[WO_4]$) [94]. They thought that the plausible mechanism involves the initial formation of the imdo-tungstate complex, $[WO_3(NAr)]$, which then aggregates with additional $[WO_4]^{2-}$, in a self-assembly manner, to yield the target product. In addition, it is probably the proton in acidic contaminant, not water mentioned by Maatta *et al*, that involves in the assembly of the hexanuclear cluster, just like the above-mentioned acid-assisting route developed by us for the mono-imido hexamolybdates [75]. The deliberate addition of a suitable acid will perhaps bring us an improvement of this route.

Although the hexatungstate cluster is less labile, replacement of tungstyl groups with reactive molybdyl groups will exert an activate effect on the cluster. Surely, we discovered that the pentatungstenmolybdate ion, $[MoW_5O_{19}]^{2-}$, a Lidquvist heteropolyoxometalate, can be directly functionalized with aromatic primary amines using the DCC-dehydrating protocol [95].

$$[MoW_5O_{19}]^{2-} + ArNH_2 \xrightarrow[\text{Refluxing, 12h}]{\text{1.5DCC, CH}_3\text{CN}} [W_5O_{18}(MoNAr)]^{2-} \quad (9)$$

As shown in [Eq(9)], on addition of one and a half equivalents of DCC, the reaction of one equivalent of 2,6-dimethylaniline or 4-iodo-2,6-dimethylaniline with one equivlant of $(n\text{-}Bu_4N)_2[MoW_5O_{19}]$ does occur in hot acetonitrile under nitrogen, and in *ca* 60% yields affords the mono-imido derivatives [95], $(n\text{-}Bu_4N)_2[W_5O_{18}(MoNAr)]$, Ar = 2,6-dimethyl-$C_6H_3$, (32), 4-iodo-2,6-dimethyl-C_6H_2, (33), respectively. The X-ray crystal structure of compound 33 is given in figure 23. Indeed as expected, the terminal oxygen atom bonded to the molybdenum atom is selectively replaced by an imido substituent. This result opens a general road for the synthesis of organo imido derivatives of less reactive POM clusters, namely, by replacement one of the inert terminal metal-oxo groups with a functionalizable Mo-O group.

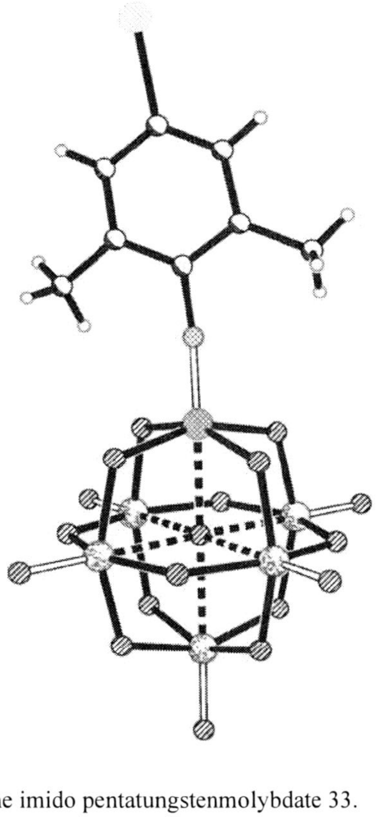

Figure 23. Ball-and-stick model of the imido pentatungstenmolybdate 33.

2.3. Organoimido Derivatives of Keggin POMs

Keggin-type POMs have a general formula of $[XMO_{40}]^{n-}$ and feature a central tetrahedral oxyanion $[XO_4]^{m-}$ encapsulated in a cage made up of twelve MO_6 octahedra by sharing corners, which represent the most popular and largest subclasses of POMs with a variety of transformations in structure and composition of the incorporated hetero-atom in the interior or metal atoms on the surface [1]. While the surface oxygen atoms of the α-phosphomolybdate ion, $[\alpha\text{-}PMo_{12}O_{34}]^{3-}$, and α-phosphotungstate ion, $[\alpha\text{-}PW_{12}O_{34}]^{3-}$, are readily alkylated with alkylation agents such as methyl sulfate and trimetyloxonium tetrafluoborate [54,55], they are unable to be replaced directly by any an imdio delivery reagent so far. For example, reaction of $[\alpha\text{-}PMo_{12}O_{34}]^{3-}$ with *p*-methyphenyl isocyanate in hot pyridine failed to afford the expected imidosubstituted α-phosphotungstate, but a degradated decanuclear imido complex instead [96].We also tried the reaction of several Keggin-structure POMs containing molybdyl groups, including the α-phosphomolybdate, with aromatic primary amines in the presence of DCC. However, in most case, aromatic amines were oxidized instead of acting as imido delivery reagents due to the high oxidation nature of the Keggin POMs of interest.

The only successful case in the synthesis of Keggin-type imido derivatives was just reported last year by A. Proust and coworkers [97]. They obtained an imido complex, $[Bu_4N]_4$ $[PW_{11}O_{39}(ReNC_6H_5)]$ (34), by the substitution reaction of the monovacant Keggin compound $[Bu_4N]_4[H_3PW_{11}O_{39}]$, a penta-dentate ligand, with the mononuclear complex $[Re(NPh)Cl_3(PPh_3)_2]$ containing a metal-imido functional group in acetonitrile [Eq(10)]. Compound 34 have clearly demonstrated by various techniques including ^1H and ^{14}N NMR, electrochemistry, and especially by ESI mass spectrometry. However, its X-ray structure has not yet been completed.

$$[H_3PW_{11}O_{39}]^{4-} + [Re(NPh)Cl_3(PPh_3)_2] \xrightarrow[\text{Refluxing, 1h}]{Et_3N,\ CH_3CN} [PW_{11}O_{39}(ReNC_6H_5)]^{4-} \quad (10)$$

In all, the functionalization of Keggin-type POMs with imido groups faces still formidable challenges and much more attentions should be paid into this interesting field in the future since the potential of Keggin POM-based molecular materials is expected to be greater than that of Lindqvist derivatives.

3. Reactive Chemistry of Organoimido Derivatives of POMs

While organoimido derivatives of POMs have been discovered for more than one decade [58] and a variety of synthetic approaches have been developed as mentioned above, the reactive chemistry of them has just been investigated very recently and only the organoimido derivatives of hexamolybdate have been involved [72, 73, 81, 98-100]. Both the inorganic cluster skeleton and the incorporated organic ligand in an organoimido POM can act as the active sites for their own enjoyed reactions. The Mo-N bond is also a possible reaction site, the correlative reactions such as hydrolysis, however, are trivial due to break of the Mo-N bond and leaving of the organic ligand from the cluster in most cases. So far, the reaction chemistry of organoimido derivatives of POMs has focused mainly on the incorporated organic ligands because it provides great potentials in controllable synthesis of organic-inorganic hybrids exploiting well-developed common organic synthesis techniques. In particular, for the imidosubstituted hexamolybdates, several important inorganic and organic reactions, which do not cleave the Mo-N bond, have been set up. These includes coordination [73], addition polymerization [72] and carbon-carbon coupling reactions [98], of which the carbon-carbon coupling has been proved to be a powerful tool for our molecule-module-assembly strategy in fabricating POM-based organic-inorganic hybrid materils [56].

3.1. Coordination

Compared to the corresponding parent cluster, the surface oxygen atoms on the cluster skeleton of an organically derivatived POM should be more reactive to coordinate with a metal ion due to the activation effect of incorporated organic ligands [53], nevertheless, such reactions have not been realized in the imidosubstituted hexamolybdates. On the other hand, if the incorporated imido groups bearing ligatable substituents such as hydroxyl, carboxyl and pryidyl, such derivatives can in principle behave as POM-functionalized organic ligands. According to this idea, we synthesized the compound 15 with a terpyridyl group (see figure 10) [73]. The coordination reaction of this hexamolybdate-functionalized ligand 15 with some bivalent transitional metal ions such as Zn^{2+} and Ru^{2+} has been confirmed by preliminary *in situ* UV studies. For example, on adding an excess of $Ru(tPy)Cl_3$ to a solution of functionalized ligand 15 in DMSO, the intense absorption at 392 nm of ligand 15 diminishes significantly, in the meanwhile, a new absorption band at 504 nm appears gradually. The new absorption is a characteristic M→L charge-transfer (MLCT) transition of the Ru-bisterpyridine complexes, which implies definitely that the new hexamolybdate-functionalized bisterpyridine-Ru complex 35 has formed by the complexation reaction shown in [Scheme 1]. Isolation of the complex 35 and its X-ray single crystal structure determination are currently in progress.

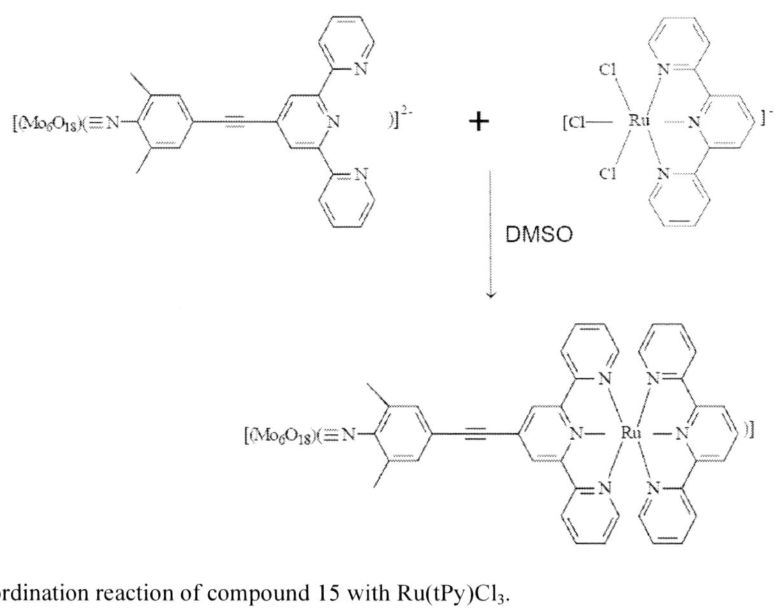

Scheme 1. Coordination reaction of compound 15 with $Ru(tPy)Cl_3$.

3.2. Addition Polymerization

The addition polymerization of organoimido derivatives of POMs bearing an unsaturated functional group such as the C=C double bond was just reported at the beginning of this new centenary [72]. In this report, E. A Maatta and co-workers discovered that, the *p*-styrenylimido hexamolybdate **9** and 4-methylstyrene could undergo conventional free radical-induced copolymerization initiated by AIBN [2,2'-azobis(2-methylpropionitrile)] in 1,2-dichloroethane and afforded a copolymer 36 with organoimido-POMs as pendants [Scheme 2]. The formation of copolymer 36 was confirmed by its IR spectrum, where the band of C=C stretching disappeared, but the characteristic doublet pattern of the terminal Mo-O and Mo-N vibrations in an monoimido hexamolybdate were still kept, compared to the precursor **9**. By comparison of the relative intensities of ^{1}HNMR due to the aryl groups, the α-CH_2 groups of the [n-NBu_4]$^+$ cations, and the styrenic CH_3 group, copolymer 36 was suggested to have a composition of nearly three 4-methyl styrene units per unit of **9**. This work demonstrates that the styrylimido ligand has similar chemical reactivity to that of the corresponding free stryrene and can be employed as a reactive site capable of delivering its metal complex into a conventional polymeric environment. Such a methodology should have broadly applicable potentials in the synthesis of novel inorganic/organic hybrid polymeric materials which incorporate covalently attached organoimido-POM complexes.

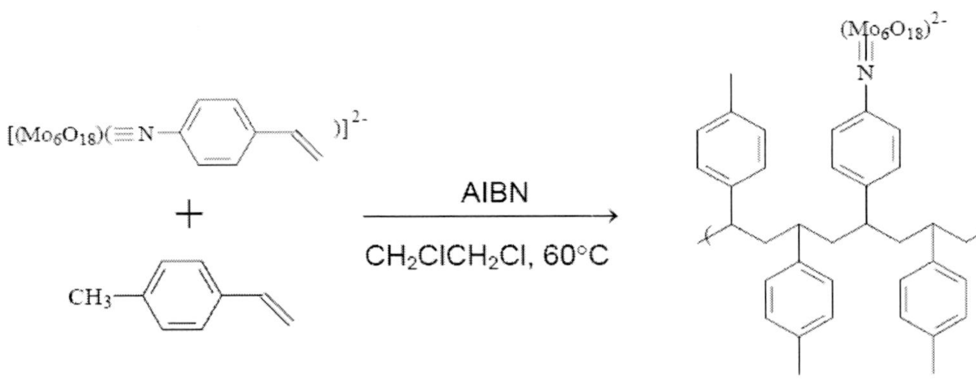

Scheme 2. Free radical-induced copolymerization Between compound **9** and 4-methylstyrene.

3.3. Carbon-Carbon Coupling and Its Application in POM-Based Organic-Inorganic Hybrids

It is well known that aryl halides and terminal acetylenes can undergo Pd catalyzed carbon-carbon coupling reactions [101-103]. The arylimido derivatives of POMs with a remote iodo or ethynyl groups, such as compounds 13 and 14, are in fact, POM-functionalized aryl halides and terminal acetylenes. If the iodo or ethynyl groups on functionalized POM clusters can undergo common Pd-catalyzed coupling reactions, and the POM clusters survive and the imido linkage does not break under the coupling reaction conditions, such organically-functionalized clusters would be invaluable building blocks for the rational synthesis of novel molecular POM–organic hybrids.

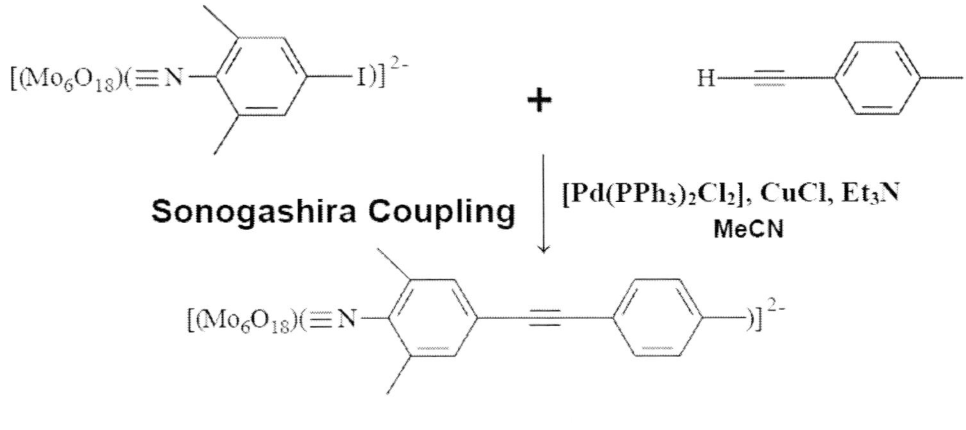

Scheme 3. Sonogashira coupling of 13 with 4-methyl-1-ethynylbenzene in acetonitrile at room temperature.

The Sonogashira coupling reaction of 13 with 4-methyl-1-ethynylbenzene [Scheme 3] or 3,5-di-tertbutyl-1-ethynylbenzene has been carried out in acetonitrile at room temperature under the protection of N_2 gas [98]. Compared to classic coupling reactions in organic synthesis [101-103], this reaction is extremely fast and finishes within only 20 minutes as monitored by thin-layer chromatography. As confirmed by the single X-ray diffraction study (figure 24), the hexamolybdate cluster survived indeed in the coupling product 37 under the reaction conditions, and obviously it activates the coupling reaction. The enhanced reactivity of the iodo function in 13 is presumably due to the electron-withdrawing nature of the hexamolybdate cluster. Prolonged reaction time and excess base were found to be harmful to the desired coupling products 37 and 38, as they slowly decomposed to the parent hexamolybdate cluster under the reaction conditions. Nevertheless, pure products can be usually obtained in more than 80% excellent yields.

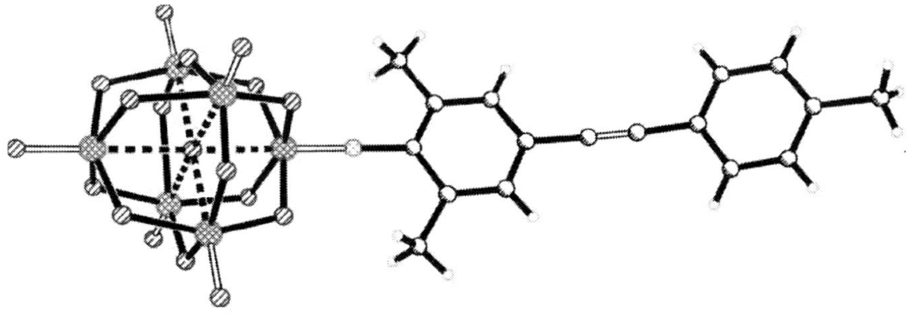

Figure 24. Ball-and-stick structure of the coupling product 37.

Hybrid Molecular Nanodumbbells

With our success in which iodo- or ethynylphenylimido-functionalized POM clusters can undergo Pd-catalyzed carbon-carbon coupling reactions, a great number of POM-based hybrids with various linking geometries, including covalently linked POM networks, can be rationally designed and easily prepared. For example, hybrid molecular nanodumbbells 39 and 40 with two hexamolybdate clusters linked by a rigid conjugated organic rod with different lengths in nanosize, shown in Scheme 4, have been synthesized by the Sonogashira coupling reaction between 13 and 14 in good yields [99].

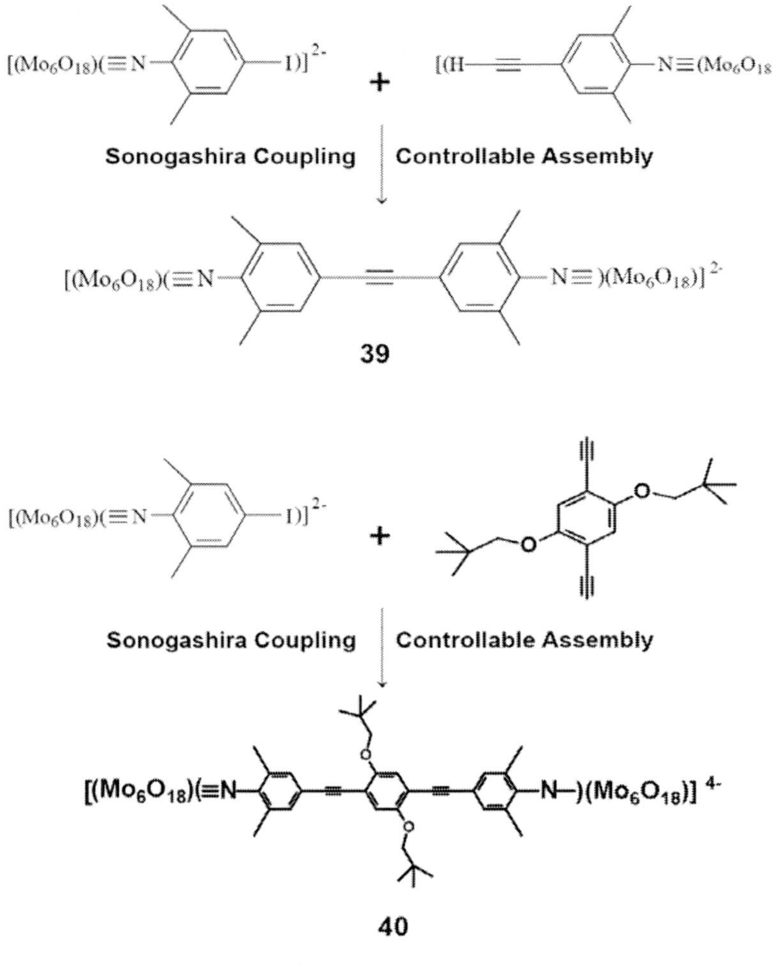

Scheme 4. Controllable assembly of hybrid nanodumbbells.

The X-ray crystal structure of nanodumbbell 39 has been determined (figure 25). Cyclic-voltammetry studies reveal a reversible one-electron reduction wave at -891 mV (vs Ag/Ag^+) for both nanodumbbells 39 and 40, which is very close to the reduction potential of other momosunstituted imido derivatives such as 13 (-909 mV), 14 (-893 mV), 37 (-893 mV), and 38 (-895 mV) [56]. UV/Vis absorption spectroscopy studies show that the lowest energy electronic transition is significantly bathochromically shifted from compounds 13 (362 nm) and 14 (368 nm), to hybrid dumbbells 39 (426 nm) and 40 (438 nm), thus indicating that the Mo≡N π bond is delocalized within the organic conjugated π system. Compared to the original oligo(phenylene ethynylene) ligands, which are strongly fluorescent molecules [104], both of two hybrid dumbbells show no fluorescence under excitation from 200 nm to 500 nm, which implies clearly that the conjugatedly-attached hexamolybdate cluster acts as an efficient fluorescence quencher in the dumbbells. These results suggest that strong electronic interactions through the conjugated organic bridge exist between the two clusters in a nanodumbbell.

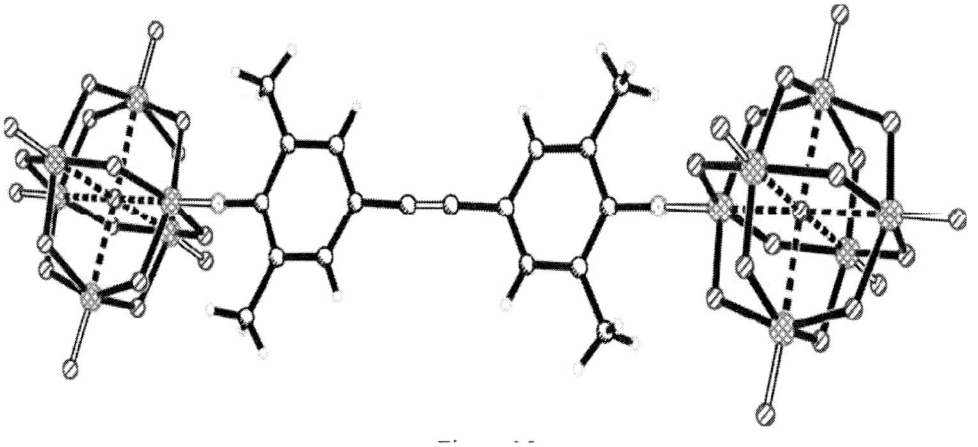

Figure 25. Ball-and-stick structure of the hybrid nanodumbbell 39.

Molecular Charge Transfer Hybrids

Charge-transfer hybrids based on organic donors and inorganic acceptors have being focused by chemists for several decades [100, 107, 108], due to their potential applications in solar energy conversion and storage such as photovoltaic cells, conducting materials including novel molecular conductors and superconductors, and nonlinear optical devices. POMs are one of the most important electron acceptors [50, 105], which have been widely exploited in preparing charge-transfer salts with organic cation donors, or charge-transfer supramolecular aggregates with neutral organic π donors through cocrystallization [106]. Efforts have recently been made to develop intramolecular charge transfer hybrids based on POMs[100]. Compound 11 is one of such examples, in which a ferrocenyl donor is bound covalently to a hexamolybdate acceptor and a charge-transfer transition from the Fe^{2+} center to the cluster is observed at 536nm [70]. However, this hybrid was obtained in only 19% yield through a direct functionalization reaction on the hexamolybdate cluster. Additionally, the short separation distance between the ferrocenyl donor and the cluster acceptor also renders the charge recombination facile as a coherent superexchange mechanism dominates [109]. These problems, however, are easy to be overcome with iodophenylimido-functionalized POM cluster building blocks through the Pd-catalyzed carbon-carbon coupling reactions.

For example, two novel charge-transfer hybrids 41 and 42, shown in Scheme 5, in which one or two ferrocenyl groups are bonded indirectly to a hexamolybdate cluster through an extended π-conjugated bridge, have been prepared in over 65% good yields from the Sonogashira coupling reactions of 13 and 21 with ethynylferrocene [100], respectively. The X-ray molecular structure of 41 was shown in figure 26.The electronic spectra of these hybrids show a broad absorption tail extending beyond 550 nm, indicating the existence of charge-transfer transition from the ferrocenyl donor to the cluster acceptor. The observation of the clear charge-transfer transition indicates the contribution of charge-transfer resonance to the ground state in both 41 and 42 even though the donor-acceptor separation distance of 11.29 Å is rather long, signaling a through-bond charge-transfer nature made possible by the organic π-conjugated bridge. Cyclic voltammetry studies reveal a single-electron oxidation wave at 0.328 V and a single electron reduction wave at -0.831 V for 41 with one ferrocenyl unit. For the hybrid with two ferrocenyl units compound 42, a lower single electron reduction potential at -1.055 V and a two-electron oxidation wave at 0.331 V are observed. The reduction process in 41 and 42 is associated with the cluster skeleton. The observed cathodical shift of reduction potential from 41 to 42 is in agreement with the increasing number of the electron-donating imido ligand in 42. The oxidation wave in these two hybrids is attributed to the ferrocenyl unit. The almost same oxidation potential of them indicates negligible electronic interactions between the two ferrocenyls in 42.

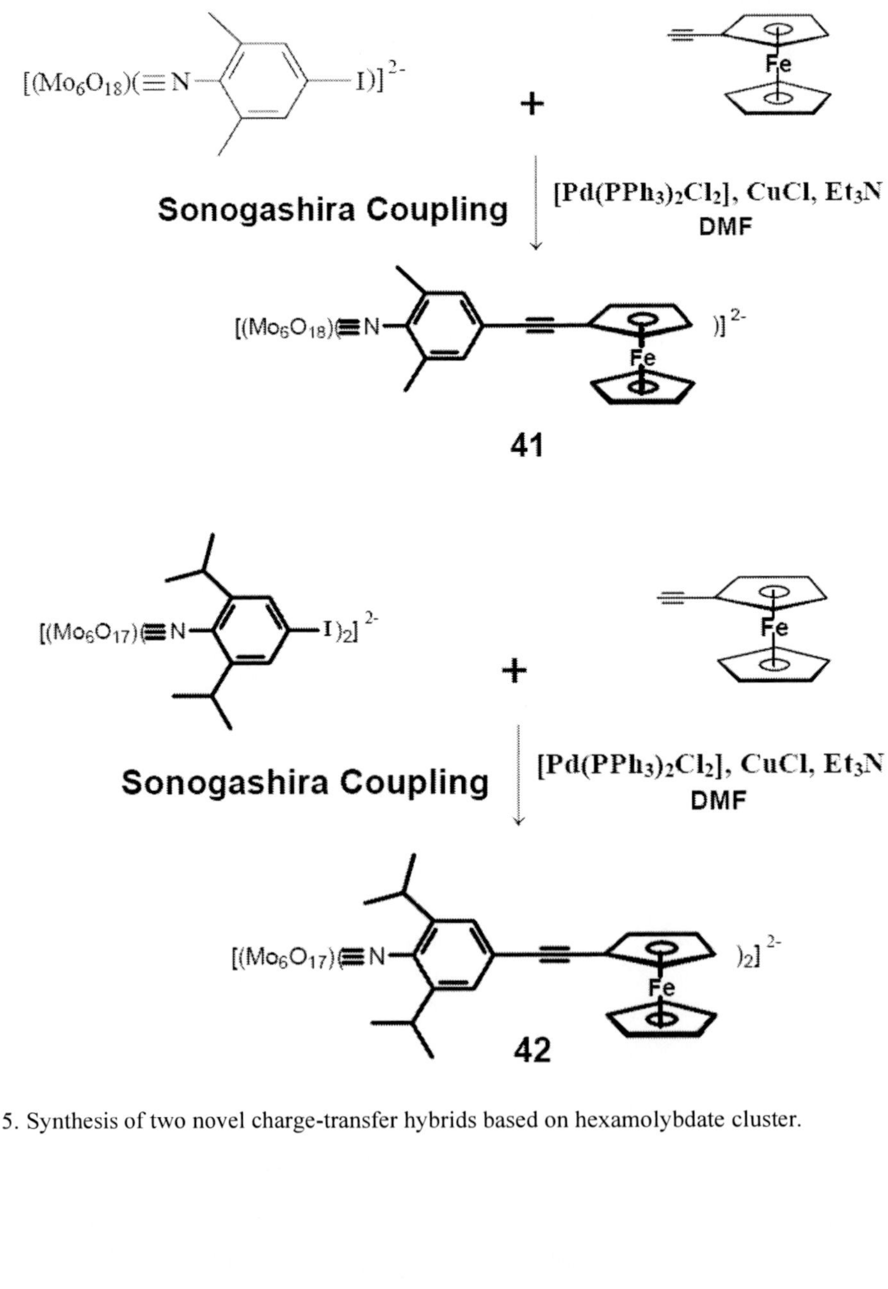

Scheme 5. Synthesis of two novel charge-transfer hybrids based on hexamolybdate cluster.

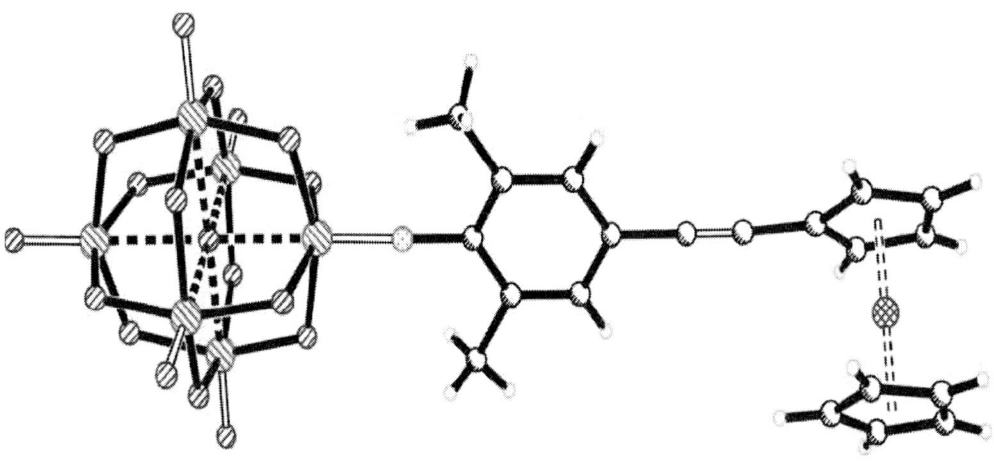

Figure 26. Ball-and-stick structure of the charge-transfer hybrid 41.

Hybrid Ploymers

Hybrid polymers containing POMs have been intensively studied for seeking new conducting materials. Conjugated polymeric hybrids containing covalently linked POM clusters, either in the main chain or in the side chains, at present, can be synthesized conveniently with the Pd-catalyzed coupling reactions. Starting from difunctionalized cluster 21 or 22, main-chain POM-containing polymers have been prepared as shown in Scheme 6. These polymers have an average molecular weight of about 160,000. Their electronic spectra are nearly identical to that of hybrid dumbbell 39, indicating that the organic π electrons delocalize only within the two imidobearing Mo atoms and the electronic interaction between the two orthogonally jointed organic π systems is limited. Cyclicvoltammetry studies reveal a reversible single-electron reduction wave at -1.2 V (vs Ag/Ag+) which is consistent with bis(arylimido)-substituted hexamolybdates. These polymers also show intense absorption in the visible range but with little fluorescence emissions, suggesting efficient fluorescence quenching of the embedded POM cluster on the organic phenylene acetylene units. Peng et al has studied their applications in photovoltaic cells and a power conversion efficiency of 0.15% was obtained, which is significantly higher than photovoltaic cells fabricated with other conjugated polymers in the same device configuration.

Recently, the hexamolybdate cluster has been incorporated into poly(phenylene ethynylene) as conjugated side-chain pendants through the Pd-catalyzed coupling reactions in Peng's groups too. These polymeric hybrids have much more potential for applications in photovoltaic cells, owing to efficient fluorescence quenching in such materials to make free charge carriers reside in different structural units [110].

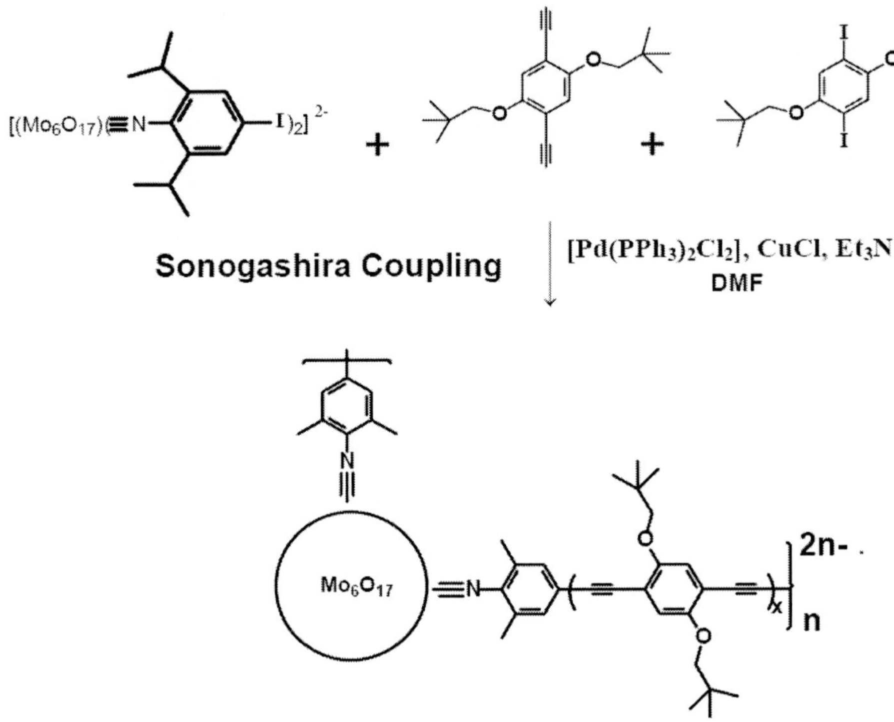

Scheme 6. Synthesis of polymeric hybrids based on hexamolybdate cluster.

4. Conclusion and Outlook

The chemistry of organoimido derivatives of polyoxometalates, especially their reactive chemistry, is still in its infancy. Although classically organic synthetic techniques have given a fascinating promise in this field, to date, only a limited number of methods have been developed for the preparation of covalently bonded organoimido derivatives of POMs and their reactions have just stepped mainly on the Pd-catalyzed coupling reactions. Nonetheless, this new chemistry indeed opens a wide road to make novel molecular hybrids and polymeric hybrids based on POMs in a much more controlled manner. With the development of this chemistry, a large number of covalently bonded POM–organic hybrids, particularly those involving organic π-conjugated systems, including molecular, dendritic, and polymeric systems are sure to be prepared in the near future. In addition, efforts on exploring the electrical, optical and photophysical properties of this new class of hybrids are expected.

Acknowledgements

Y. G. Wei sincerely thanks Professor Zhonghua Peng for professional support to start the project when he was at UMKC. He also gratefully thanks Professors Hongyou Guo and Sijia Xue for helps in continuing this research in china. The contributions of undergraduate and graduate students: Ning Ge, Na Wang, Qiang Li, Yun Xia, Zicheng Xiao, Yi Zhu and

Shaoyong Ke, are also gratefully acknowledged. This work is sponsored by NFSC-20201001, SRF for ROCS of SEM, MOST-TG2000077503 and the university funding from Hubei University of Technology.

REFERENCES

[1] Pope, M. T. *Heteropoly and Isopoly Oxometalates*. Berlin: Springer; 1983.

[2] Berzelius, J. J. Beitrag zur näheren Kenntniss des Molybdäns. Poggend. *Ann. Phys. Chem.* 1826, 6, 369–392.

[3] Pope, M. T.; Müller, A. Editors. Polyoxometalates: *From Platonic Solids to Anti-Retroviral Activity*. Dordrecht: Kluwer; 1994.

[4] Hill, C. L. Guest Editor. A Special Thematic Issue on Polyoxometalates. *Chem. Rev.* 1998, 98, 1-390.

[5] Pope, M. T.; Müller, A. Editors. Polyoxometalate Chemistry: *From Topology via Self-Assembly to Application.* Dordrecht: Kluwer; 2001.

[6] Zhang, S. W.; Wei, Y. G.; Yu, Q.; Shao, M. C.; Tang, Y. Q. Toward Quantum Line-Self-Assembly Process of Nanomolecular Building Blocks Leading to a Novel One-Dimensional Nanomolecular Polymer: Single-Crystal X-ray Structure of $\{[H_3O]^+_{12}$ $\{(H_2O)MoO_{2.5}[Mo_{36}\ O_{108}(NO)_4(H_2O)_{16}]O_{2.5}Mo(H_2O)\}^{12-}\}$. *J. Am. Chem. Soc.* 1997, 119, 6440-6441.

[7] Jiang, C. C.; Wei, Y. G.; Liu, Q.; Zhang, S. W.; Shao, M. C.; Tang, Y. Q. Self-assembly of a Novel Nanoscale Giant Cluster: $[Mo_{176}O_{496}(OH)_{32}(H_2O)_{80}]$. *Chem. Commun.* 1998, 1937-1938.

[8] Liu, G.; Wei, Y. G.; Yu, Q.; Liu, Q.; Zhang, S. W. Polyoxometalate Chain-like Polymer: the Synthesis and Crystal Structure of a 1D Compound, $[Mo_{36}O_{108}(NO)_4(MoO)_2La_2(H_2O)_{28}]_n{\cdot}56nH_2O$. *Inorg. Chem. Commun.* 1999, 2, 434-437.

[9] Wei, Y. G..; Zhang, C.; Zhang, S. W.; Shao, M. C. A Spherical Nanascale Polyoxomolybdate Encapsulating Thirty Coordinated Butyrate: Synthesis and Structure Characterization of $(NH_4)_{42}[Mo_{132}O_{372}(CH_3CH_2CH_2COO)_{30}(H_2O)_{72}]$. Unpublished result.

[10] Müller, A.; Krickemeyer, E.; Meyer, J.; Bögge, H.; Peters, F.; Plass, W.; Diemann, E.; Dillinger, S.; Nonnenbruch, F.; Randerath, M.; Menke, C. $[Mo_{154}(NO)_{14}O_{420}(OH)_{28}(H_2O)_{70}]^{(25\pm5)-}$: A Water-Soluble Big Wheel with More than 700 Atoms and a Relative Molecular Mass of About 24000. *Angew. Chem. Int. Ed.* 1995, 34, 2122-2124.

[11] Müller, A.; Krickemeyer, E.; Bögge, H.; Schmidtmann, M.; Peters, F.; Menke, C. Meyer, J. An Unusual Polyoxomolybdate: Giant Wheels Linked to Chains. *Angew. Chem. Int. Ed.* 1997, 36, 484-486.

[12] Müller, A.; Krickemeyer, E.; Bögge, H.; Schmidtmann, M.; Beugholt, C.; Kögerler, P.; Lu, C. Z. Formation of a Ring-Shaped Reduced "Metal Oxide" with the Simple Composition $[(MoO_3)_{176}(H_2O)_{80}H_{32}]$. *Angew. Chem. Int. Ed.* 1998, 37, 1220-1223.

[13] Müller, A.; Shah, S. Q. N.; Bögge, H.; Schmidtmann, M. Molecular Growth from a Mo-176 to a Mo-248 Cluster. *Nature.* 1999, 397, 48-50.

[14] Müller, A.; Serain C. Soluble Molybdenum Blues-"des Pudels Kern". *Acc. Chem. Res.* 2000, 33, 2-10.

[15] Müller, A.; Krickemeyer, E.; Bögge, H.; Schmidtmann, M.; Peters, F. Organizational Forms of Matter: An Inorganic Super Fullerene and Keplerate Based on Molybdenum Oxide. *Angew. Chem. Int. Ed.* 1998, 37, 3359-3363.

[16] Müller, A.; Sarkar, S.; Shah, S. Q. N.; Bögge, H.; Schmidtmann, M.; Sarkar, S.; Kögerler, P.; Hauptfleisch, B.; Trautwein, A. X.; Schünemann, V. Archimedean Synthesis and Magic Numbers: "Sizing" Giant Molybdenum-Oxide-Based Molecular Spheres of the Keplerate Type. *Angew. Chem. Int. Ed.* 1999, 38, 3238-3241.

[17] Müller, A.; Polarz, S.; Das, S. K.; Krickemeyer, E.; Bögge, H.; Schmidtmann, M.; Hauptfleisch, B. "Open and Shut" for Guests in Molybdenum-Oxide-Based Giant Spheres, Baskets, and Rings Containing the Pentagon as a Common Structural Element. *Angew. Chem. Int. Ed.* 1999, 38, 3241-3245.

[18] Müller, A.; Koop, M.; Bögge, H.; Schmidtmann, M.; Peters, F.; Kogerler, P. Building blocks as disposition in solution: $[\{Mo^{VI}O_3(H_2O)\}_{10}\{V^{IV}O(H_2O)\}_{20}\{(Mo^{VI}/Mo^{VI}{}_5O_{21})(H_2O)_3\}_{10}(\{Mo^{VI}O_2(H_2O)_2\}_{5/2})_2(\{NaSO_4\}_5)_2]^{20-}$, a giant spherical cluster with unusual structural features of interest for supramolecular and magneto chemistry. *Chem. Commun.* 1999, 1885-1886.

[19] Müller, A.; Shah, S. Q. N.; Bögge, H.; Schmidtmann, M.; Kogerler, P.; Hauptfleisch, B.; Leiding, S.; Wittler, K. Thirty Electrons "Trapped" in a Spherical Matrix: A Molybdenum Oxide-Based Nanostructured Keplerate Reduced by 36 Electrons. *Angew. Chem. Int. Ed.* 2000, 39, 1614-1616.

[20] Müller, A.; Das, S. K.; Bögge, H.; Schmidtmann, M.; Botar, A.; Patrut, A. Generation of Cluster Capsules (I_h) from Decomposition Products of a Smaller Cluster (Keggin-T_d) while Surviving Ones Get Encapsulated: Species with Core-shell Topology formed by a Fundamental Symmetry-driven Reaction. *Chem. Commun.* 2001, 657-658.

[21] Müller, A.; Kogerler P.; Dress A. W. M. Giant Metal-oxide-based Spheres and Their Topology: from Pentagonal Building Blocks to Keplerates and Unusual Spin Systems. *Coord. Chem. Rev.* 2001, 222, 193-218.

[22] Müller, A.; Beckmann, E.; Bögge, H.; Schmidtmann, M.; Dress A. Inorganic Chemistry Goes Protein Size: A Mo_{368} Nano-Hedgehog Initiating Nanochemistry by Symmetry Breaking. *Angew. Chem. Int. Ed.* 2002, 41, 1162-1167.

[23] Müller, A.; Roy S. En Route from the Mystery of Molybdenum Blue via Related Manipulatable Building Blocks to Aspects of Materials Science. *Coord. Chem. Rev.* 2003, 245, 153-166.

[24] Müller, A.; Kuhlmann C.; Bögge, H.; Schmidtmann M.; Baumann M.; Krickemeyer E. $Mo^{V}{}_2O_4{}^{2+}$ directs the formation and subsequent linking of potential building blocks under different boundary conditions: A related set of novel cyclic polyoxomolybdates. *Eur. J. Inorg. Chem.* 2001, (9), 2271-2277.

[25] Yamase T.; Prokop P. V. Photochemical Formation of Tire-shaped Molybdenum Blues: Topology of a Defect Anion, $[Mo_{142}O_{432}H_{28}(H_2O)_{58}]^{12-}$. *Angew. Chem. Int. Ed.* 2002, 41, 466-469.

[26] Yang, W. B.; Lu C. Z.; Lin X.; Zhuang H. H. A Novel Acetated 54-member Crown-shaped Polyoxomolybdate with Unprecedented Structural Features: $Na_{26}[\{Na(H_2O)_2\}_6(\{\mu_3\text{-}OH\}_4Mo^{V}{}_{20}Mo^{VI}{}_{34}O_{164}(\mu_2\text{-}CH_3COO)_4)]\cdot\approx 120H_2O$. *Chem. Commun.* 2000, 1623-1624.

[27] Yang, W. B.; Lu C. Z.; Lin X.; Zhuang H. H. Novel Acetate Polyoxomolybdate "Host" Accommodating a Zigzag-Chainlike “Guest” of Five Edge-Shared Sodium Cations: $Na_{21}\{[Na_5(H_2O)_{14}]\subset[Mo_{46}O_{134}(OH)_{10}(\mu\text{-}CH_3COO)_4]\}\cdot CH_3COONa\cdot\approx 95H_2O$. *Inorg. Chem.* 2002, 41, 452-454.

[28] Wassermann K.; Dickman M. H.; Pope M. T. Self-Assembly of Supramolecular Polyoxometalates: The Compact, Water-Soluble Heteropolytungstate Anion $[As_{12}^{III}Ce_{16}^{III}(H_2O)_{36}W_{148}O_{524}]^{76-}$. *Angew. Chem. Int. Ed.* 1997, 36, 1445-1448.

[29] Liu T. B. Supramolecular Structures of Polyoxomolybdate-Based Giant Molecules in Aqueous Solution. *J. Am. Chem. Soc.* 2002, *124,* 10942-10943.

[30] Liu T. B.; Diemann E.; Li H. L.; Dress A. W. M., Müller, A. Self-assembly in Aqueous Solution of Wheel-shaped Mo_{154} Oxide Clusters into Vesicles. *Nature.* 2003, 426, 59-62.

[31] Liu T. B. An Unusually Slow Self-Assembly of Inorganic Ions in Dilute Aqueous Solution. *J. Am. Chem. Soc.* 2003, 125, 312-313.

[32] Liu, G.; Cai, Y. G.; Liu, T. B. Automatic and Subsequent Dissolution and Precipitation Process in Inorganic Macroionic Solutions. *J. Am. Chem. Soc.* 2004, 126, 16690-16691.

[33] Bredas, J. L. Silbey, R. Editors. *Conjugated Polymers.* Dordrecht/ The Netherlands: Kluwer; 1991.

[34] Nalwa, H. S. Editor. Handbook of Organic Conductive Molecules and Polymers I – IV. Chichester/ England: Wiley; 1997.

[35] Pope, M.; Swenberg, C. E. Editors. *Electronic Processes in Organic Crystals and Polymers*. 2nd edition. Oxford: Oxford University Press; 1999.

[36] Heeger, A. J. Semiconducting and Metallic Polymers: The Fourth Generation of Polymeric Materials. *J. Phys. Chem. B.* 2001, *105*, 8475-8491.

[37] Kraft, A.; Grimsdale, A. C.; Holmes A. B. Electroluminescent Conjugated Polymers - Seeing Polymers in a New Light. *Angew. Chem. Int. Ed.* 1998, 37, 402-428.

[38] Kelly, S. M. *Flat Panel Displays: Advanced Organic Materials.* Cambridge: Royal Society of Chemistry; 2002.

[39] Katz, H. E.; Bao, Z. The Physical Chemistry of Organic Field-Effect Transistors. *J. Phys .Chem. B.* 2000, 104, 671-678.

[40] Katz, H. E.; Bao, Z.; Gilat, S. L. Synthetic Chemistry for Ultrapure, Processable, and High-Mobility Organic Transistor Semiconductors. *Acc. Chem. Res.* 2001, 34, 359-369.

[41] Katz, H. E. Recent Advances in Semiconductor Performance and Printing Processes for Organic Transistor-Based Electronics. *Chem. Mater.* 2004, *16*, 4748-4756.

[42] Newman, C. R.; Frisbie, C. D.; da Silva Filho, D. A.; Bredas, J.-L.; Ewbank, P. C.; Mann, K. R. Introduction to Organic Thin Film Transistors and Design of n-Channel Organic Semiconductors. *Chem. Mater.* 2004, *16*, 4436-4451.

[43] Hide, F.; Diaz-Garcia, M. A.; Schwartz, B. J.; Heeger, A. J. New Developments in the Photonic Applications of Conjugated Polymers. *Acc. Chem. Res.* 1997, 30, 430-436.

[44] McGehee, M. D.; Heeger, A. J. Semiconducting (Conjugated) Polymers as Materials for Solid-State Lasers. *Adv. Mater.* 2000, 12, 1655-1668.

[45] Dagani, R. Nobel Prize in chemistry. *Chem. Eng. News.* 2000, 78(42), 4-4.

[46] Shirakawa H. The Discovery of Polyacetylene Film: The Dawning of an Era of Conducting Polymers (Nobel Lecture). *Angew. Chem. Int. Ed.* 2001, 40, 2574-2580.

[47] MacDiarmid, A. G. “Synthetic Metals”: A Novel Role for Organic Polymers (Nobel Lecture). *Angew. Chem. Int. Ed.* 2001, 40, 2581-2590.

[48] Heeger, A. J. Semiconducting and Metallic Polymers: The Fourth Generation of Polymeric Materials (Nobel Lecture). *Angew. Chem. Int. Ed.* 2001, 40, 2591-2611.

[49] Pope, M. T.; Müller, A. Polyoxometalate Chemistry: An Old Field with New Dimensions in Several Disciplines. *Angew. Chem. Int. Ed.* 1991, 30, 34-48.

[50] Yamase T. Photo- and Electrochromism of Polyoxometalates and Related Materials. *Chem. Rev.* 1998, 98, 307-326.

[51] Sanchez, C.; Soler-Illia, G. J. A. A.; Ribot, F.; Lalot, T.; Mayer, C. R.; Cabuil, V. Designed Hybrid Organic-Inorganic Nanocomposites from Functional Nanobuilding Blocks. *Chem. Mater.* 2001, *13*, 3061-3083.

[52] Katsoulis D. E. A Survey of Applications of Polyoxometalates. *Chem. Rev.* 1998, 98, 359-388.

[53] Gouzerh, P. Proust, A. Main-Group Element, Organic, and Organometallic Derivatives of Polyoxometalates. *Chem. Rev.* 1998, 98, 77-112.

[54] Knoth, W. H. Harlow, R. L. Derivatives of heteropolyanions. 3. O-alkylation of $Mo_{12}PO_{40}^{3-}$ and $W_{12}PO_{40}^{3-}$. *J. Am. Chem. Soc.* 1981, *103*, 4265-4266.

[55] Knoth, W. H. Farlee, R. D. Esters of phosphotungstic acid. Anhydrous phosphotungstic acid. *Inorg. Chem.* 1984, *23*, 4765-4766.

[56] Peng, Z. H. Rational Synthesis of Covalently Bonded Organic-Inorganic Hybrids. *Angew. Chem. Int. Ed.* 2004, 43, 930-935.

[57] Kang, H.; Zubieta, J. Co-ordination complexes of polyoxomolybdates with a hexanuclear core: synthesis and structural characterization of $(NBu^n_4)_2[Mo_6O_{18}(NNMePh)]$, *Chemical Communications*, 1988, (17), 1192-1193.

[58] Du, Y.; Rheingold, A. L.; Maatta, E. A. A Polyoxometalate Incorporating an Organoimido Ligand: Preparation and Structure of $[Mo_5O_{18}(MoNC_6H_4CH_3)]^{2-}$ *J. Am. Chem. Soc.* 1992, *114*, 345-346.

[59] Hur, N. H.; Klemperer, W. G.; Wang, R.-C. Tetrabutylammonium Hexamolybdate(VI) *Inorg. Synth.* 1990, 27, 77-78.

[60] Allcock, H. R.; Bissell, E. C.; Shawl, E. T. Crystal and molecular structure of a new hexamolybdate-cyclophosphazene complex. *Inorg. Chem.* 1973, 12, 2963-2967.

[61] Hsieh, T.- C.; Zubieta, J. A. Synthesis and Characterization of Oxomolybdate Clusters Containing Coordinatively Bound Organo-Diazenido Units: The Crystal and Molecular Structure of the Hexanuclear Diazenido-Oxomolybdate, $(NBu^n_4)_3[Mo_6O_{18}(N_2C_6H_5)]$ *Polyhedron.* 1986, 5, 1655-1657.

[62] Kwen, H.; Young, V. G., Jr.; Maatta, E. A. A Diazoalkane Derivative of a Polyoxometalate: Preparation and Structure of $[Mo_6O_{18}(NNC(C_6H_4OCH_3)CH_3)]^{2-}$ *Angew. Chem. Int. Ed.* 1999, 38, 1145-1146.

[63] Proust, A.; Thouvenot, R.; Chaussade, M.; Robert, F.; Gouzerh, P. Phenylimido Derivatives of $[Mo_6O_{19}]^{2-}$: Syntheses, X-ray Structures, Vibrational, Electrochemical, ^{95}Mo and ^{14}N NMR Studies. *Inorg. Chim. Acta* 1994, 224, 81-95.

[64] Errington, R. J.; Lax, C.; Richards, D. G.; Clegg, W.; Fraser, K. A. New Aspects of Non-Aqueous Polyoxometalate Chemistry: 3. Suface Oxide Reactivity. In: Pope, M. T.; Müller, A. Editors. *Polyoxometalates: From Platonic Solids to Anti-Retroviral Activity.* Dordrecht: Kluwer; 1994; p 112-113.

[65] Strong, J. B.; Ostrander, R.; Rheingold, A. L.; Maatta, E. A. Ensheathing a Polyoxometalate: Convenient Systematic Introduction of Organoimido Ligands at Terminal Oxo Sites in $[Mo_6O_{19}]^{2-}$. *J. Am. Chem. Soc.* 1994, 116, 3601-3602.

[66] Clegg, W.; Errington, R. J.; Fraser, K. A.; Holmes, S. A.; Schäfer, A. Functionalisation of $[Mo_6O_{19}]^{2-}$ with Aromatic Amine: Synthesis and Structure of a hexamolybdate Building Block with Linear Difunctionality. *Chem. Commun.* 1995, 455-456.

[67] Roesner, R. A.; McGrath, S. C.; Brockman, J. T.; Moll, J. D.; West, D. X.; Swearingen, J. K.; Castineiras, A. Mono- and di-functional aromatic amines with *p*-alkoxy substituents as novel arylimido ligands for the hexamolybdate ion. *Inorg. Chim. Acta.* 2003, 342, 37-47.

[68] Wei, Y. G.; Xu, B. B.; Barnes, C. L.; Peng, Z. H. An Efficient and Convenient Reaction Protocol to Organoimido Derivatives of Polyoxometalates. *J. Am. Chem. Soc.* 2001, *123*, 4083-4084.

[69] Stark, J. L.; Rheingold, A. L.; Maatta, E. A. Polyoxometalate clusters as building blocks: preparation and structure of bis(hexamolybdate) complexes covalently bridged by organodiimido ligands. *Chem. Commun.* 1995, 1165-1166.

[70] Stark, J. L.; Young, V. G., Jr.; Maatta, E. A Functionalized Polyoxometalate Bearing a Ferrocenylimido Ligand: Preparation and Structure of $[(FcN)Mo_6O_{18}]^{2-}$. A. *Angew. Chem. Int. Ed.* 1995, 34, 2547-2548.

[71] Strong, J. B.; Yap, G. P. A.; Ostrander, R.; Liable-Sands, L. M.; Rheingold, A. L.; Thouvenot, R.; Gouzerh, P.; Maatta, E. A. A New Class of Functionalized Polyoxometalates: Synthetic, Structural, Spectroscopic, and Electrochemical Studies of Organoimido Derivatives of $[Mo_6O_{19}]^{2-}$. *J. Am. Chem. Soc.* 2000, *122*, 639-649.

[72] Moore, A. R.; Kwen, H.; Beatty, A. M.; Maatta, E. A. Organoimido-polyoxometalates as polymer pendants. *Chem. Commun.* 2000, 1793-1794.

[73] Xu, B. B.; Peng, Z. H.; Wei, Y. G.; Powell, D. R. Polyoxometalates covalently bonded with terpyridine ligands. *Chem. Commun.* 2003, 2562-2563.

[74] Zhu, Y.; Wu, P. F.; Ge, N.; Wei, Y. G.; Wang, Y. The First Naphthylimido Derivatives of the hexamolybdate: Synthesis and Characterization of $(Bu_4N)_2[Mo_6O_{18}(NC_{10}H_7\text{-}1)]$. Submitted to publication.

[75] Wu, P. F.; Li, Q.; Ge, N.; Wei, Y. G.; Wang, Y.; Wang, P.; Guo, H. Y. An Easy Routine to Monofunctionalized Organoimido Derivatives of the Lindqvist Hexamolybdate. *Eur. J. Inorg. Chem.* 2004, 2819-2822.

[76] Li, Q.; Wu, P. F.; Wei, Y. G.; Xia, Y.; Wang, Y.; Guo, H. Y. Organic-Inorganic Hybrids: Preparation and Structural Characterization of $(Bu_4N)_2[Mo_6O_{17}(NAr)_2]$ and $(Bu_4N)_2[Mo_6O_{18}(NAr)]$ ($Ar = o\text{-}CH_3C_6H_4$). *Z. Anorg. Allg. Chem.* 2005, 631, in Press.

[77] Li, Q.; Wu, P. F.; Wei, Y. G.; Wang, Y.; Wang, P.; Guo, H. Y. An organic-inorganic hybrid of polyoxoanion cluster incorporating para-bromophenylimido ligand: Preparation, spectroscopic studies and crystal structure of $(Bu_4N)_2[Mo_6O_{18}(NC_6H_4Br\text{-}p)]$. Submitted to publication.

[78] Xue, S. J.; Ke, S. Y.; Wei, Y. G. Synthesis, spectroscopic characterization and crystal structure of an arylimido derivative of with electron-withdrawing fluoro substituents. Unpublished result.

[79] Xu, L.; Lu, M.; Xu, B. B.; Wei, Y. G.; Peng, Z. H.; Powell, D. R. Towards Main-Chain-Polyoxometalate-Containing Hybrid Polymers: A Highly Efficient Approach to Bifunctionalized Organoimido Derivatives of Hexamolybdates. *Angew. Chem. Int. Ed.* 2002, 41, 4129-4132.

[80] Li, Q.; Wu, P. F.; Wei, Y. G.; Wang, Y.; Wang, P.; Guo, H. Y. Synthesis, structure and supramolecular assembly in the crystalline state of a bifunctionalized arylimido derivative of hexamolybdate. *Inorg.Chem. Commun.* 2004, 7, 524-527.

[81] Lu, M.; Xie, B.; Kang, J.; Chen, F.-C.; Yang, Y.; Peng, Z. H. Synthesis of Main-Chain Polyoxometalate-Containing Hybrid Polymers and Their Applications in Photovoltaic Cells. *Chem. Mater.* 2005, *17*, 402-408.

[82] Xia, Y.; Wu, P. F.; Wei, Y. G.; Wang, Y.; Guo, H. Y. A Hybrid Molecular Semiconductor: Synthesis and Structural Characterization of the Bifunctionalized Organoimido Derivative of Hexamolybdate, $(n\text{-}Bu_4N)_2[Mo_6O_{17}(\equiv NAr)_2]$ (Ar = o-$CH_3OC_6H_4$). Submitted to publication.

[83] Xia, Y.; Wu, P. F.; Wei, Y. G.; Guo, H. Y. Synthesis, Spectroscopic and Structure of an Unexpected Trans- Disubstituted Arylimido Derivative of the Hexamolybdate, $(n\text{-}Bu_4N)_2[Mo_6O_{17}(\equiv NAr)_2]$ (Ar = 2,6-$(CH_3)_2C_6H_4$). Submitted to publication.

[84] Strong, J. B.; Haggerty, B. S.; Rheingold, A. L.; Maatta, E. A. A Superoctahedral Complex Derived from a Polyoxometalate: the Hexakis(arylimido)hexamolybdate Anion $[Mo_6(NAr)_6O_{13}H]^-$. *Chem. Commun.* 1997, 1137-1138.

[85] Xia, Y.; Wu, P. F.; Wei, Y. G.; Guo, H. Y. The Unprecedent Bridging Oxo Replacement in a Trifunctionalized Organoimido Derivative of the Hexmolybate, $[Mo_6O_{16}(\equiv NAr)_2(\mu_2\text{-}NAr)]^{2-}$ (Ar = 2,6-$(CH_3)_2C_6H_4$). Submitted to publication.

[86] Johnson, A. W. *Ylides and Imines of Phosphorus*. Wiley: New York; 1993.

[87] Saunders, J. H.; Slocombe, R. J. The Chemistry of the Organic Isocyanates. *Chem. Rev.* 1948, *43*, 203-218.

[88] Patrick G. Section J– Aldehydes and Ketones: J6. Nucleophilic Addition– Nitrogen Nucleophiles. *Instant Notes in Organic Chemistry*. Reprint Edition; BIOS Scientific Publishers Limited: Beijing; 2002; p 183-184.

[89] Williams, A.; Ibrahim, I. T. Carbodiimide Chemistry: Recent Advances. *Chem. Rev.* 1981, *81*, 589-636.

[90] Kurzer, F.; Douraghi-Zadeh, K. Advances in the Chemistry of Carbodiimides. *Chem. Rev.* 1967, *67*, 107-152.

[91] Khorana, H. G. the Chemistry of Carbodiimides. *Chem. Rev.* 1953, *53*, 145-166.

[92] Nugent, W.; Mayer, J. E. *Metal-Ligand Multiple Bonds*. Wiley: New York; 1988; p122-125.

[93] Yan, L.-K.; Su, Z.-M.; Guan, W.; Zhang, M.; Chen, G.-H.; Xu, L.; Wang, E.-B. Why Does Disubstituted Hexamolybdate with Arylimido Prefer to Form an Orthogonal Derivative? Analysis of Stability, Bonding Character, and Electronic Properties on Molybdate Derivatives by Density Functional Theory (DFT) Study. *J. Phys. Chem. B.* 2004, *108*, 17337-17343.

[94] Mohs, T. R.; Yap, G. P. A.; Rheingold, A. L.; Maatta, E. A. An Organoimido Derivative of the Hexatungstate Cluster: Preparation and Structure of $[W_6O_{18}(NAr)]^{2-}$ (Ar = 2,6-(i-Pr)$_2C_6H_3$). *Inorg. Chem.* 1995, *34*, 9-10.

[95] Wei, Y. G.; Lu, M.; Cheung, C. F.- C.; Barnes, C. L.; Peng, Z. H. Functionalization of $[MoW_5O_{19}]^{2-}$ with Aromatic Amines: Synthesis of the First Arylimido Derivatives of Mixed-Metal Polyoxometalates. *Inorg. Chem.* 2001, *40*, 5489-5490.

[96] Proust, A.; Taunier, S.; Artero, V.; Robert, F.; Thouvenot, R.; Gouzerh, P. The Unexpected Reactivity of *p*-Tolylisocyanate towards the Keggin Anion $\alpha\text{-}[PMo_{12}O_{40}]^{3-}$. *Chem. Commun.* 1996, 2195 – 2196.

[97] Dablemont, C.; Proust, A.; Thouvenot, R.; Afonso, C.; Fournier, F.; Tabet, J.-C. Functionalization of Polyoxometalates: From Lindqvist to Keggin Derivatives. 1. Synthesis, Solution Studies, and Spectroscopic and ESI Mass Spectrometry Characterization of the Rhenium Phenylimido Tungstophosphate $[PW_{11}O_{39}\{ReNC_6H_5\}]^{4-}$. *Inorg. Chem.* 2004, *43*, 3514-3520.

[98] Xu, B. B.; Wei, Y. G.; Barnes, C. L.; Peng, Z. H. Hybrid Molecular Materials Based on Covalently Linked Inorganic Polyoxometalates and Organic Conjugated Systems. *Angew. Chem. Int. Ed.* 2001, 40, 2290-2292.

[99] Lu, M.; Wei, Y. G.; Xu, B. B.; Cheung, C. F.- C.; Peng, Z. H.; Powell, D. R. Hybrid Molecular Dumbbells: Bridging Polyoxometalate Clusters with an Organic π-Conjugated Rod. *Angew. Chem. Int. Ed.* 2002, 41, 1566-1568.

[100] Kang, J.; Nelson, J. A.; Lu, M.; Xie, B.; Peng, Z. H.; Powell, D. R. Charge-Transfer Hybrids Containing Covalently Bonded Polyoxometalates and Ferrocenyl Units. *Inorg. Chem.* 2004, *43*, 6408-6413.

[101] Sonogashira, K.; Tohda, Y.; Hagihara, N. Convenient synthesis of acetylenes. Catalytic substitutions of acetylenic hydrogen with bromo alkenes, iodo arenes, and bromopyridines. *Tetrahedron Lett.* 1975, 50, 4467-4470.

[102] deSousa, P. T. The Palladium-catalysed Cross Coupling Reaction of Acetylenic Compounds with Aryl Halides and Related Compounds. *Quimica Nova.* 1996, 19, 377-382.

[103] Wang, Y. F.; Deng, W.; Liu, L.; Guo, Q. X. Recent Progress in Sonogashira Reaction. *Chin. J. Org. Chem.* 2005, 25, 8-24.

[104] Ziener, U.; Godt, A. Synthesis and Characterization of Monodisperse Oligo(phenyleneethynylene)s. *J. Org. Chem.* 1997, 62, 6137-6143.

[105] Sadakane, M.; Steckhan, E. Electrochemical Properties of Polyoxometalates as Electrocatalysts. *Chem. Rev.* 1998, 98, 219-238.

[106] Coronado, E.; Gómez-García, C. J. Polyoxometalate-Based Molecular Materials. *Chem. Rev.* 1998, 98, 273-296.

[107] Gust, D.; Moore, T. A.; Moore, A. L. Molecular Mimicry of Photosynthetic Energy and Electron Transfer. *Acc. Chem. Res.* 1993, 26, 198-205.

[108] Wasielewski, M. R. Photoinduced Electron Transfer in Supramolecular Systems for Artificial Photosynthesis. *Chem. Rev.* 1992, 92, 435-461.

[109] Weiss, E. A.; Ahrens, M. J.; Sinks, L. E.; Gusev, A. V.; Ratner, M. A.; Wasielewski, M. R. Making a Molecular Wire: Charge and Spin Transport through *para*-Phenylene Oligomers. *J. Am. Chem. Soc.* 2004, *126*, 5577-5584.

[110] Xu, B. B.; Lu, M.; Kang, J.; Wang, D. G.; Brown, J.; Peng, Z. H. Synthesis and Optical Properties of Conjugated Polymers Containing Polyoxometalate Clusters as Side-Chain Pendants. *Chem. Mater.* 2005, *17*, 2841-2851.

In: Research in Hybrid Materials
Editors: Simon J. Brunner and Julian W. Egger
ISBN 978-1-60456-718-2

Chapter 5

A NEW APPROACH TO HYBRID SYSTEMS OF RENEWABLE ENERGY UTILIZATION

Yu.V. Vorobiev*, J. Gonzalez-Hernandez, P. Gorley, P. Horley and L. Bulat******

*CINVESTAV, Unidad Queretaro, Queretaro 76230, Mexico
**CIMAV, Chihuahua 31109, Mexico
Department of Electronic and Energy Engineering,
Chernivtsi National University, Chernivtsi 58013, Ukraine
***St. Petersburg State University of Refrigeration and Food Eng.,
St. Petersburg 191002, Russia

ABSTRACT

A general analysis is given of hybrid systems consisting of different combinations of 4 devices frequently employed for renewable energy utilization: Photovoltaic Solar Panel (PV), Solar Thermal Plane Collector (ST), Wind Generator (WG) and Heat-to-Electric/Mechanic Energy Convertor (HE); some of the combinations include radiation energy flux concentrators of different degrees. The main result of the consideration made is that the hybrid systems are more efficient than the sum of the constituents and more stable in relation to spontaneous variations of the renewable energy source potential (like wind velocity, insolation, etc.). However, to realize the possibilities mentioned, all the elements of a given hybrid system have to be especially designed and made for this specific system. For example, the PV panel for the hybrid PV/Thermal system ought to have a substrate with high thermal conductivity, to allow for heat extraction from the panel by the adjacent Solar Thermal Plane Collector, and practically no commercial panels with these characteristics are available. Besides, the PV panel as a part of the hybrid system will demand a special choice of semiconductor material and surface treatment which could be different from those for conventional panels. The limiting efficencies for some hybrid systems are estimated; these efficiencies exceed the efficiencies of separate use of the devices discussed. The most promising hybrid system

---

* On sabbatical leave from CINVESTAV-Queretaro

is the PV panel made as spectrum splitter in combination with HE converter, of which total efficency could be around 50 %.

INTRODUCTION

Among many devices employed for the renewable energy generation, a special place is occupied by photovoltaic solar cells, for many reasons (like low maintenance, long lifetime, etc.). The key problem in solar cells science and engineering is the restricted efficiency of Solar Energy conversion resulting from the impossibility of efficient utilization of a wide solar spectrum with one semiconductor material. The possibilities of using more than one material (multijunction cells, spectral splitting with many cells [1-6]) or hypothetical materials with very specific parameters [7,8] do not give a simple solution to the problem because of technological problems and very high cost.

On the other hand, the hybrid systems of a different kind recently became very popular (for example, [9-14]); some of them, like PV/Thermal, are developed up to commercial stage. It is evident that these hybrid systems *a priori* could provide a higher efficiency of utilization of renewable energy source resulting from the smaller total area than that of the sum of separate parts of a system, having at the same time a relatively low cost.

In addition the hybrid systems have advantages compared to their elements, since each one of the separate devices forming the system has its own working conditions which quite often are contradictory (for example, the windy weather is profitable for wind generators, but it reduces efficiency of solar thermal collectors), therefore the hybrid system can be made almost independent upon the variation of these conditions, and thus more stable and reliable.

The common way to build a hybrid system is to use a combination of existing devices, which is definitely the simplest and the cheapest approach. However, this way may not be the most efficient one. The present paper intends to show that to optimize the construction and performance of any hybrid system, one has to make a special analysis of the coupling conditions of the devices used, and, in the majority of cases, to design each device accordingly to the specific system's demands. The result will be the highest efficiency and stability of the system.

THEORETICAL ANALYSIS

Separate Performance of the System's Elements

Here we give the basic information necessary to discuss the different combinations of the elements (for the 4 elements, there are 10 possible combinations; we shall discuss only those which we consider are the most important).

Photovoltaic Solar Panel (PV) with the sensitive area A_{PV}, efficiency η_{PV} is exposed to solar radiation of power W during Δt_1 hours per day (the last parameter is the averaged considering the hourly changes in the solar position). Thus, the total energy generated by a PV per day is

$$E_{PV} = \eta_{PV}\, W\, A_{PV}\, \Delta t_1 \qquad (1)$$

For a typical Si solar photovoltaic module with an efficiency of 12 % and area of 1 m^2, under solar illumination corresponding to AM1.5 conditions (844 W/m^2), the power generated will be approximately 100 W, and the energy corresponding to the average 5.5 sun hours per day (data taken for the state of Queretaro, Mexico, no Sun tracking is assumed), in agreement with (1) is 550 Wh.

Solar Thermal Plane Collector (ST) is converting the sun radiation into heat stored in water or air circulating within the collector. Its efficiency η_T is usually 50 – 70 % ([9-11]), and is very dependent upon the working conditions like the temperature difference between input and output, cooling conditions, etc., so thermal energy generated by ST with area A_T during the sun hours will be

$$E_T = \eta_T\, W\, A_T\, \Delta t_1 \qquad (2)$$

and for each square meter of area, the daily thermal energy produced is around 3 kWh.

Wind Generator (WG) utilizes a secondary product of solar radiation – the wind, and its performance is determined, first of all, by the wind velocity v. For a WG "controlling" the area A_A of air flux, the power W_W converted to electricity is the part η_W (WG efficiency) of the corresponding air flux kinetic energy per unit time which is equal to 0.5 $\rho V v^2$, ρ being the air density and V – volume per second of the air flux controlled. For the latter we write $V = A_A\, v$, and thus obtain

$$W_W = 0.5\, \eta_E\, \rho\, A_A\, v^3, \qquad (3)$$

so the total energy E_W produced during the working time interval Δt_2 will be

$$E_W = 0.5\, \eta_E\, \rho\, A_A\, v^3\, \Delta t_2 \qquad (4)$$

Here the corresponding time interval can be 24 hours per day, although the wind velocity could vary considerably during the day.

Taking the flux controlled area equal again to 1 m^2, air density $\rho = 1.29\ kg/m^3$, wind velocity 8 m/s which is considered as the minimum for a good WG performance, and the effectivity 0.25 (some average value), we get from (3) the power 82.6 W; for 24 hours per day performance, according to (4) it gives approximately 2 kWh of energy.

The estimations made show that each of the devices considered produces considerable amount of energy per day (from 0.55 to 3 kWh), which could be utilized for domestic or other applications. The analysis below will show the advantages in combining these elements in hybrid systems, and the specific demands related with the array.

Heat-to-Electric/Mechanic Energy Convertor (HE). In this part, some heat engines could be considered (like Stirling Engine frequently used in the field discussed, with extensive literature). Here we present information about thermoelectric generators (TEG) which is less familiar to the general audience. The thermoelectric method is a direct method of energy conversion similar to the photovoltaic one, so it is natural to combine these two methods in a hybrid system. It is also important that thermoelectric devices (again, like the photovoltaic ones) are highly reliable and could work 10 – 30 years practically without any technical service (see [15-19]) which unfortunately could not be said in relation to heat engines.

The TEG efficiency. The efficiency of a thermoelectric generator TEG is determined by so called thermoelectric figure of merit

$$Z = \frac{\alpha^2}{\rho\,\kappa}, \tag{5}$$

where ρ is the electric resistivity, κ - the thermal conductivity and α - the thermo-emf coefficient of a thermoelectric (semiconductor) used.

A modern thermoelectric module (a unit thermoelectric converter) is a battery – a number of alternate n- and p-type semiconductor branches. They are connected electrically in series with metallic connection strips, sandwiched between two electrically insulating and thermally conducting ceramic plates to form a module. So, any TEG contains two different materials (corresponding indexes "1" and "2"), and the optimized figure of merit for a module can be presented in the form

$$Z_0 = \frac{(\alpha_1 - \alpha_2)^2}{\left(\sqrt{\kappa_1\rho_1} + \sqrt{\kappa_2\rho_2}\right)^2} \tag{6}$$

The efficiency of TEG can be characterized in two ways:

[1] Efficiency corresponding to the classic condition of the maximum power output $R = r$ (the external electric resistance R equal to the TEG internal resistance r)

$$\eta_1 = \eta_0 \left(\frac{4}{Z_0 T_h} + 2 - \frac{T_h - T_c}{2T_h} \right)^{-1} \tag{7}$$

[2] Since the thermoelectric devices (the same as photovoltaic ones) are not characterized by a constant internal resistance, there exist another condition for the maximum TEG efficiency corresponding to the relation $\left(R/r\right)_{opt} = M = \sqrt{1 + Z_0 T_M}$, where T_M is the average temperature $T_M = 0.5\,(T_h + T_c)$. In this case

$$\eta_{max} = \eta_0 \frac{M - 1}{M + \frac{T_c}{T_h}}. \tag{8}$$

The coefficient $\eta_0 = \dfrac{T_h - T_c}{T_h}$ = $\Delta T/T_h$ in (7) and (8) is the ideal thermodynamic efficiency of a heat engine (the Carnot efficiency). Usually η_{max} is a little higher than η_1; as a rule, the difference between the values of the TEG efficiency given by the two expressions above does not exceed 4% [15]. Our purpose is to estimate the highest efficiency of the hybrid system, therefore we use for calculations the expression (8).

Thermoelectric materials. To proceed, we have to specify value of the figure of merit Z. It is necessary to point out that the figure of merit of semiconductors depends on temperature, and different kinds of semiconductor materials should be selected for different operating temperatures. The industrial thermoelectric materials can be divided in dependence of operating temperature into the following groups [15-18]:

- for temperatures up to 500K - the solid solutions based on bismuth telluride
- (Bi_2Te_3, Bi_2Te_3-Bi_2Se_3) are used;
- for temperatures up to 800K - the PbTe;
- for space applications (T>900K) - the solid solutions based on Ge-Si.

The main problem of thermoelectric material research is how to increase the figure of merit Z. Despite the efforts of many research groups in different countries, during 1950 - 2000 years the increase achieved in the dimensionless thermoelectric coefficient ZT was only from 0.75 to 1.0 (the data refer to room temperature). But an essential progress in the field was made during the last few years. The remarkable successes have been achieved on different directions:

1. Usage of the new physical ideas in nano-scale microstructures. A high quality 2D quantum superlattice with nano-scale films based on p-Bi_2Te_3/Sb_2Te_3 having $ZT = 2.4$ at room temperature was obtained [19-21]. A nanomaterial with quantum dots (1D structure) based on PbSeTe has the value of $ZT = 2.0$ at room temperature [22]. The increase of the figure of merit was also obtained in a special structure with cold points (like contacts between a plate and edge of a cone). The measured figure of merit of the last structure based on p-$Bi_{0.5}Sb_{1.5}Te_3$ and n- $Bi_2Te_{2.9}Se_{0.1}$ corresponds to $ZT = 1.7$ at room temperature [23,24].
2. Alongside with semiconductor TEG there exist thermionic devices that also can be used for direct generation of electricity. However, the traditional thermionic converters work only at very high temperatures, above 800 K, necessary for electrons to overcome the potential barrier. But recently a theory has been offered [25] of a thermionic converter with a nanometers barrier thickness; here electrons can overcome the potential barrier by quantum tunneling. Apparently such thermotunnel converter has been realized [26]. The parameter equivalent to thermoelectric figure of merit of this converter is $ZT = 4$.
3. Another option is the application of some new bulk materials as thermoelectrics – so called skutterrudites and clathrates. It was shown that the thermal conductivity can be reduced in such materials [27], therefore the figure of merit will increase. The examples of skutterrudites are $CoSb_3$ (n-type) and Zn_4Sb_3 (p-type) and of the

clathrates - $CoFe_4Sb_{12}$. These materials have good figure of merit in a wide temperature range and therefore can be used for TEGs.

Thus, for calculation of the efficiency of hybrid system in the temperature range 300 – 600 K we can use the following thermoelectric parameters: $ZT = 1$, value that have modern industrial thermoelectrics; $ZT = 2$, this value has thermoelectrics received in laboratories (nano-scale microstructures), and $ZT = 4$, value for thermionic converter with quantum tunneling. Using the expression (8) to estimate the efficiency and taking the temperature difference between hot and cold TEG sides equal to 300 K, we obtained for the three cases mentioned the corresponding efficiency of 10, 15 and 20 %.

Hybrid Systems (Different Combinations of Elemental Devices)

Thermal-Wind (ST+WG). This complex may be useful for thermic applications (for example, to stabilize the temperature regime of a greenhouse or small living house in case of great daily variations of temperature, which is typical for Mexico), the energy generated during the day by ST is collected by hot water which is stored in thermally insulated water tank, to be used at night for heating of a building. The electricity generated by the WG will be used also for heating (through an electric heater with an efficiency practically of 100 %, or in a heat pump system where the efficiency could be almost doubled). The efficiency of the ST decreases almost linearly with the wind velocity; this decrease ought to be compensated by the effect of wind generator WG: its power is never linear with the wind velocity, so the compensation will be within some limits; actually, for a strong wing, there could be overcompensation. We can introduce an average wind velocity v_a and find the conditions of energy stabilization in relation to relatively small deviations ($\pm \Delta v$) of this velocity, taking $|\Delta v| < v_a$.

Thus, for the ST efficiency we write

$$\eta_T = \eta_T^o (1 - \alpha v) = \eta_T^o [1 - \alpha(v_a \pm \alpha\Delta v)],$$

so that

$$E_T = C [1 - \alpha(v_a \pm \alpha\Delta v)], \text{ where } C = W A_T \Delta t_1 \tag{2A}$$

and for the WG

$$E_W = C^* (v_a \pm \Delta v)^3 \approx C^* (v_a^3 \pm 3 v_a^2 \Delta v), \tag{4A}$$

$$C^* = 0.5 \eta_E \rho A_A \Delta t_2.$$

Then for the total energy produced by the system per day we have

$$E_{tot} = [C(1 - \alpha v_a) + C^* v_a^3] \pm (3 C^* v_a^2 - C\alpha) \Delta v \tag{9}$$

Now we have the condition of stability of the system energy as

$$(3\ C^*\ v_a^2 - C\alpha) = (1.5\eta_E\ \rho\ A_A\ \Delta t_2\ v_a^2 - W\ A_T\ \alpha\ \Delta t_1) = 0 \tag{10}$$

We shall see below how this condition could be applied to a real system. However, one general remark could be made now. The wind generators most frequently used (HA, or Horisontal Axis devices) demand very high wind velocity for normal performance (8 m/s or more). At these conditions, the thermal collectors have very low efficiency, and their use is not practical; the supportable wind velocity for their application is 3 – 5 m/s. It means that for the hybrid system of the kind discussed, the HA WG is not a good choice; at relatively low wind velocities, the Vertical Axis (VA) devices are more efficient.

Photovoltaic-Thermal (PV + ST) System. This system (so-called "combi-panel") has an evident advantage compared to the performance of the separated devices. In that system the ST absorbs the excessive heat of the PV thus cooling it and therefore enhances the efficiency of the PV panel. It is also clear that the total amount of energy produced per unit area increases by this combination. The detailed analysis of thermal balance in this system was published in our previous paper [28]; the results of the present analysis support the main conclusion. All the previous papers on the subject have taken for granted that the systems elements have to be of the same area. However, it is evident that the presence of the PV panel above the solar collector reduces the heat flux to the collector and thus its efficiency; on the other hand, the collector ability to extract heat from the PV panel is reduced while the water (air) inside it is heated. Therefore, the optimal case would be to make PV panel of smaller area than that of the heat collector, and to place it above the initial collector's part (i.e. that corresponding to the entrance of the cold water).

Another point is that the conventional PV usually could not provide a good thermal contact with the ST heat collector, so, this hybrid system demands a special construction of the PV with a high thermal conductivity of the substrate used; one example is given in the paper mentioned [28].

Wind-Photovoltaic (WG + PV) System. These systems are already in use; they are unique among the hybrids discussed in a sense that there are no specific demands to the devices combined. The system generates and stores electric energy day and night; since the WG has more working hours per day than the PV, low-power WG could match the higher power PV. To guarantee the normal performance of the system at different conditions of weather, the excessive battery bank ought to be used. This system is more effective in places with very high wind velocity: in addition to the WG driving, the wind stabilizes the PV temperature and therefore the efficiency of its performance.

Wind-Thermal-Photovoltaic (WG+ST+PV) System. Here we consider the PV/Thermal system discussed above in combination with the wind generator WG, the system provides electricity generated by PV panel (at daytime) and by the wind generator (day and night), as well as the hot water from the ST collector (ought to be stored in thermally insulated vessel to be used in absence of illumination by sun). The total energy (electric + thermal) produced by the system per day will be

$$E_{tot} = (\eta_{PV}\ A_{PV} + \eta_T\ A_T)W\Delta t_1 + 0.5\ \eta_E\ \rho\ A_A\ v^3\ \Delta t_2. \tag{11}$$

The specific demands to the PV based on the necessity of a good heat exchange between the PV and the ST were discussed above.

Photovoltaic-Heat-to-Electric (PV+HE) system. The scheme of a hybrid system discussed is presented in figure 1, parts A and B refer to the two different options (see below). It is evident from the figure that the two versions are very similar and contain the same basic elements (the concentrator "CONC", photovoltaic cell "PVC", High Temperature Stage "HTS" which is the HE converter, and the 2-axis Solar Tracking System "STS2"), although the type and construction of the cell (and the concentrator) could be different in the two cases. Below we discuss the working principles and possible parameters of the system in each version.

A. System with separation of "thermal solar radiation" and low temperature operation of solar cell. (figure 1 A). This system needs a solar cell which at the same time acts as a spectral filter (splitter) neither absorbing nor dissipating the solar radiation part with quantum energy $h\nu < E_g$ ("thermal solar radiation") where E_g is the cell absorbing material band gap. It is shown below that this part could be quite large, especially for semiconductor materials with relatively wide band gap. Thus, in case of a semiconductor with $E_g = 1.75$ eV, approximately 50 % of solar radiation energy corresponds to the condition $h\nu > E_g$ and is suitable for photovoltaic conversion, and another 50 %, with $h\nu < E_g$, could be used as thermal energy. In our system, this "thermal solar radiation" should be concentrated on the hot side of the HTS (thus providing its high temperature), and is converted by it into electric energy when HTS is Thermoelectric Generator (TEG), or to mechanic energy by heat engine as the HTS (like Stirling Engine [29] which has efficiency very close to that of the Carnot cycle). It is known that mechanic energy could be converted into electricity with efficiency above 90 % which allows to have rather high total energy conversion coefficient of the hybrid system.

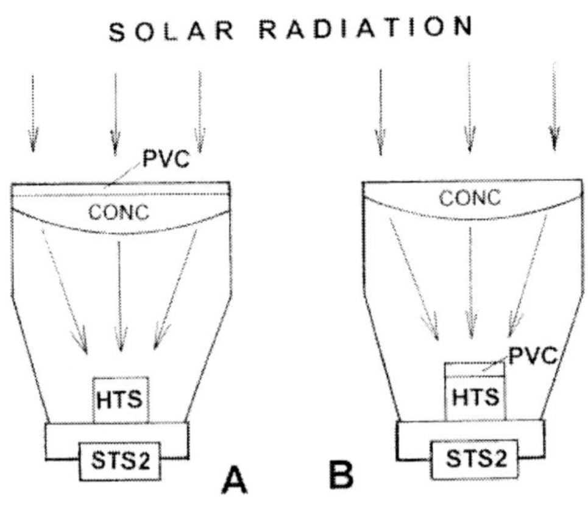

Figure 1. Schemes of hybrid system with High Temperature Stage.

To calculate the percentage of "thermal solar radiation" as a function of semiconductor band gap, we used the simple graphical analysis of solar radiation spectrum introduced by Henry [2] and included now in textbooks [30], with a numerical integration to determine the function n_{ph} (E_g), the solar flux absorbed by a semiconductor (as in Henry paper, we assume that the semiconductor is opaque for photon energies greater than E_g and transparent for energies less than E_g). This flux is given by

$$n_{ph}(E_g) = \int_{E_g}^{\infty} \frac{dn_{ph}}{d\hbar\omega} d\hbar\omega \qquad (12)$$

The function obtained is shown in figure 2: the lower curve (squares) for AM1.5 spectrum, and upper one (circles) – for AM0. Following the graphical analysis procedure [2,30]), on the basis of these curves we found the percentage ξ of "thermal solar radiation": in figure 2 it is shown by triangles (upper curve for AM0, lower for AM1.5; the corresponding values of the figure 2 ought to be multiplied by 10 to get ξ in %). One can see that ξ varies between 10 – 12 % for E_g = 0.8 eV and 78 – 80 % for E_g = 2.3 eV. The ξ vaues obtained for AM1.5 spectrum are also included in figure 3 (circles).

This thermal part of solar radiation energy has to drive the HTS (heat engine or TEG); assuming that HTS efficiency is proportional to that of Carnot cycle, with a coefficient K < 1 (the difference between K and 1 shows how close the HTS is to the ideal engine; in general, K could be temperature dependent, in particular, in the TEG case). Thus, this part of solar energy conversion is characterized by the efficiency

$$\eta_{therm} = \xi \cdot K\ \Delta T/T_h \qquad (13)$$

where T_h is the temperature of the hot side of the HTS, and ΔT the temperature difference between cold and hot sides.

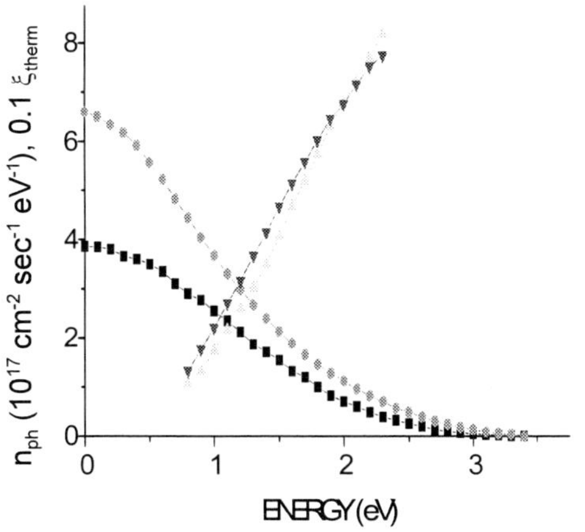

Figure 2. Graphical analysis of the efficiency of an ideal solar cell, for AM1.5 and AM0 solar spectra, with the calculated percentage of "thermal" radiation.

The solar cells for hybrid system of this kind have yet to be designed and made, but we do not see any principal obstacles to that. To estimate the cell efficiency η_{CA} as a function of band gap, we took 0.75 of the corresponding cell's ideal efficiency at non-concentrated radiation η_{id} [2,30]; since for all well developed cells (based on Ge, GaAs, $CuInSe_2$) the best efficiency obtained is well above this value, we consider this approximation reasonable. The corresponding values (η_{CA} = 0.75 η_{id}) for the AM1.5 irradiation are shown in figure 3 by squares (lower curve). We may add that if a variety of these cells-filters will be available, the cells with decreasing band gaps could be connected optically in series to increase the total efficiency, as an alternative to the multijuncion cells. An additional note is that the cell for this system should not have any surface texturizing which is typical for conventinal solar cells.

Using for the HTS efficiency the expression (13), we get the total system efficiency

$$\eta_{totA} = \eta_{CA} + \eta_{therm} = \eta_{CA} + \xi \cdot K\ \Delta T/T_h \qquad (14)$$

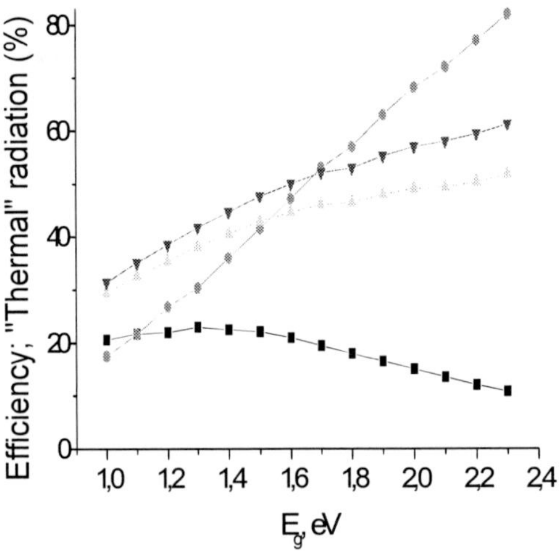

Figure 3. Percentage of "thermal" radiation in the AM1.5 spectrum, PV cell estimated efficiency and the total system efficiency as function of E_g..

The curves calculated according to (14) for K = 0.8, ΔT = 500 and 1000 K, are shown in figure 3 by up- and down-sided triangles. In this case, the total efficiency could exceed 50 - 55 %, without an employment of expensive high technologies. We must note that the temperature differences used for our calculations do not exceed those practically employed in Stirling engines (for example, Bernd Kammerich 1 kW Engine, Germany).

B. System without the solar spectrum division, a PV cell operates at high temperature. This is more straitforward system (figure 1B) than the one discussed previously; the conditions of the PV cell performance are not quite favorable here, but the system, in principle, could be constructed on the basis of the elements which already exist (although the lifetime of the PV cell in question needs special study for this working regime). The cell is subjected to concentrated sunlight, which usually enhances its efficiency; the thermal flux through the cell is transfered into the HTS by direct thermal contact, thus the cell working temperature is equal to the T_h parameter of the HTS.

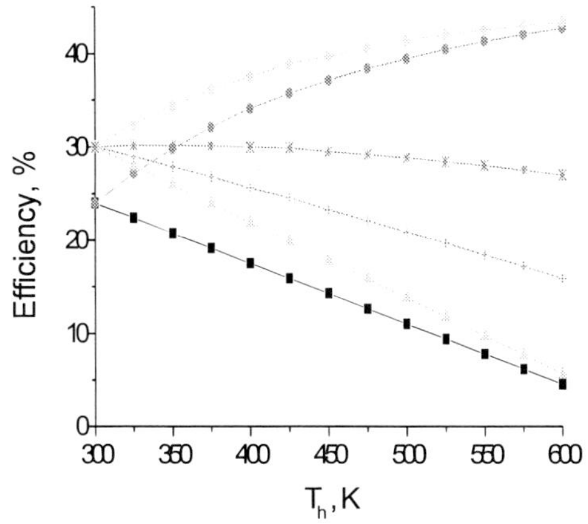

Figure 4. Calculated temperature dependencies of the PV cells efficiency (squares, triangles) and of the total hybrid system efficiency, in case of the heat engine as HTS (circles, diamonds) and the TEG (three intermediate curves, ZT = 1, 2, 4).

For calculation of the expected system efficency, two types of cells with relatively high (but not record) efficiency are considered: a GaAs single junction cell [31] with room temperature efficiency (η_o) of 24 %, and multijunction GaAs-based cell with corresponding efficiency of 30 % [3]. Both values refer to radiation concentration of approximately 50 Suns which is sufficient to achieve cell temperature higher than 450 K.

The temperature dependence of cell efficiency $\eta(T)$ is assumed linear with the coefficient $\beta = (d\eta/dT)/\eta$ equal to $-$ 2.7X10^{-3} K^{-1} [32]; it gives for the temperature 450 K the efficiency values $\eta* = 14.3$ % for single junction (SJ) cell, and 17.8 % for multijunction (MJ) cell. Thus, practically 80 % (i.e. $1 - \eta*$) of solar radiation energy will be transformed to heat within a cell, and may be used for a heat-to-electric/mechanic energy conversion by the second stage of a hybrid system – a HTS.

The total conversion efficiency of hybrid system could be written as

$$\eta_{totB} = \eta* + (1 - \eta*)\,\eta_{HTS} \tag{15}$$

Here we have $\eta* = \eta_o\,(1 - \beta\,\Delta T)$, $\Delta T = T_h - T_{room} \approx T_h - T_c$, ($T_c$ is the temperature of the cold size of the HTs which is approximately equal to the ambient temperature T_{room}), and for the HTS we take, as in part "A", that it is proportional to the Carnot engine efficiency: η_{HTS} = K $\Delta T/T_h$, which gives for the total conversion efficiency

$$\eta_{totB} = \eta_o\,(1 - \beta\,\Delta T) + [1 - \eta_o\,(1 - \beta\,\Delta T)]\,K\,\Delta T/T_h. \tag{16}$$

Figure 4 shows the calculated results for the hybrid system with the two cells mentioned above; the temperature dependence of the cells efficiency is shown by squares (SJ) and triangles (MJ). The total efficiency with the heat engine acting as HTS is represented by the growing curves, (circles for single junction cell, diamonds for multijunction cell, K = 0.8).

One could see that the second stage gives significant increase in the total efficency for relatively modest values of ΔT, and the total efficency could be as high as 35 – 40 %.

For the system "B" having the thermoelectric generator TEG as the HTS, we present in figure 4 the results of calculations of the total efficiency for three *ZT* values given above, taking as a basis the equation (15) and using η_{max} (8) instead of η_{HTS}. The PV cell is supposed to be of MJ type, with the room temperature efficiency of 30 %; the corresponding results for *ZT* = 1, 2 and 4 are shown in figure 4 by three descending curves starting from efficiency 30 % at 300 K (the larger *ZT*, the higher curve). It is seen that the TEG has considerable effect on the efficiency in all cases, although the total efficiency of the system with TEG is lower than with heat engine. Taking into account the extraordinary recent progress in the TEG field, we can expect further increase of the TEG efficency, and that of the hybrid system.

EXPERIMENTAL

The basic elements employed in experimental study were the following. The solar heat collector ST used was of the model Powermat, with water as a heat collecting agent, having a surface area of 4 m^2; approximately 90 % of the surface area is controlled by the internal tube system providing an efficient heat interchange. The black PVC absorber covering the collector is resistant to UV solar radiation (the guaranteed lifetime is 20 years) and provides a small weight of the panel (about 3 kg in empty state, and 7 kg when filled with water). The maximum heating efficiency at the absence of the wind estimated using the Hottel-Whillier model [11-13] was around 60 – 70 %, which corresponds to the best collectors known. This result is illustrated in figure 5, which gives the dependence of the ST efficiency upon the temperature difference between the water entering the collector and the ambient temperature T_a.

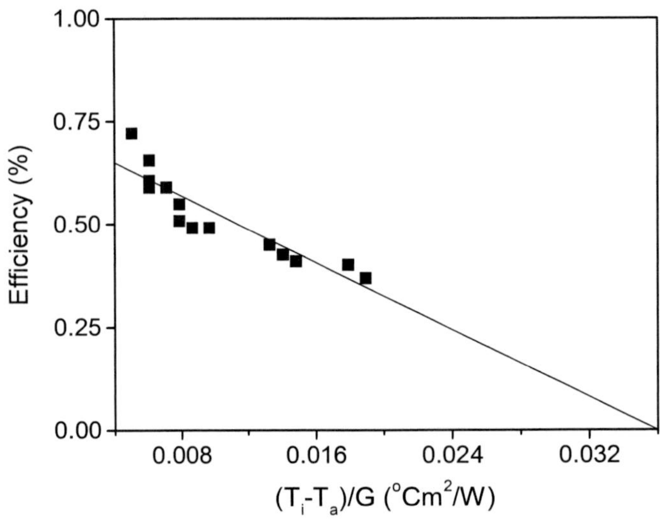

Figure 5. The temperature dependence of the thermal efficiency of the ST studied.

The dependence is practically linear, in good agreement with the Hottel-Whillier model, and the maximal value is close to 70 % indicating a good performance of the collector. The expression used to find the ST efficiency from experimental data was:

$$\eta_T = M_t\, C_c\, (T_o - T_i) / GS \qquad (17)$$

where M_t is the water flux (mass per second), G – solar radiation intensity, C_c - the water specific heat capacity, T_o is the output water temperature, and T_i - its input temperature.

The PV panels investigated and used in PV/Thermal systems were of crystalline Si type (c-Si; in particular, we used the panel made by Russian plant OKB "Krasnoe Znamya" OKBKZ M100/12 with an area of about 1 m^2 and the power generated under AM1.5 solar radiation of 100 W, and the other one assembled in our laboratory from 85×85 mm^2 cells made by the same company). Besides, in the experiments designed to achieve a good thermal contact with the collector, we used the panels made of amorphous Si (α-Si, ECD Company, Troy, Michigan, USA) and $CuInSe_2$ (CIS) commercial panel made by Siemens, USA. The electric efficiency of all the PV panels used was between 10 and 18 %. The prototype panel which we constracted to achieve the best thermal contact was made of Al substrate (thickness 2 mm) covered with 1 μm thick insulating film of PMMA/silica composite dielectric coating, with OKBKZ-made six c-Si solar cells (85X85 mm^2) area, and a glass cover.

One option to increase the efficiency of conventional solar panels is the use of bifacial panels [33,34]; to take full advantage of the rear photosensitive face, the panel ought to be placed at some distance from the white diffuse reflecting surfaces [34], or small degree radiation concentrators might be used. We employed the latter option (see photo in figure 6), the concentrator made of 5 stainless steel plates collected solar radiation from an area twice as big as the panel. It resulted in considerable heating of the panel; to overcome this effect, we plan to develop the water-running plane heat collector transparent in visible (not to affect much the rear face photovoltaic performance) and put it in thermal contact with the panel rear face. This system will produce electric energy and hot water, as any PV/Thermal system; we expect it to be useful for future domestic applications.

Figure 6. Bifacial solar cell with radiation concentrator for rear face.

The wind generator WG was developed and built in our laboratory (figure 7). The vertical axis (VA) model of Savonius type, 3 m high, with an active area $A_A = 1.5\ m^2$, having an automobile alternator with a home made gear-box as generating unit, under conditions of moderate wind, typical for the region of application (state of Querétaro, Mexico, average wind velocity about 5 m/s) provides approximately 40 W of electric energy operating at 120 rpm (see figure 8).

Figure 7. The constructed Vertical Axis Wind Generator .

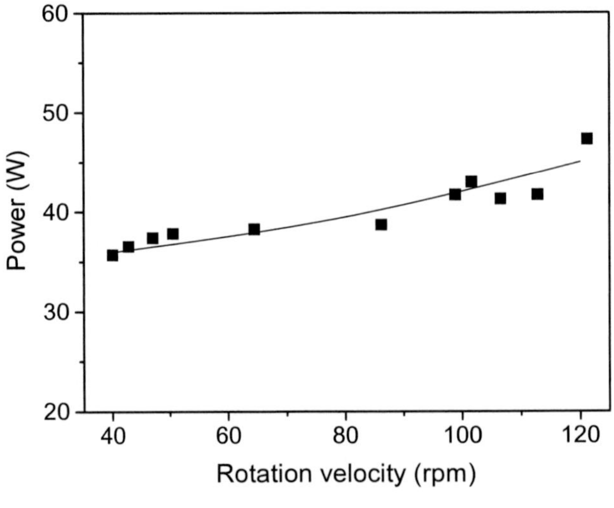

Figure 8. Characteristics of the experimental wind generator.

The PV+ST+WG system made according to the discussion above, is schematically shown in figure 9 (construction scheme – left part, electric scheme – right one). The electric part of the system includes the battery bank (4 sealed lead-acid batteries "Prism" 12 V, capacity of 105 AH each), Solar Charge Controller of the model Steca (Germany) and the DC-AC Inverter Proam (China). The arrangement made to investigate the regime of the solar thermal collector ST included two home-made electrical digital thermometers (1, 2 in figure 9) based on the temperature sensors LM335, the water flux meter CICASA Delaunet MD-15 (3 in figure 9), all three devices mentioned were connected to the computerized data acquisition system. The electrical connections scheme is made according to the generally accepted rules and notations, and does not demand comments.

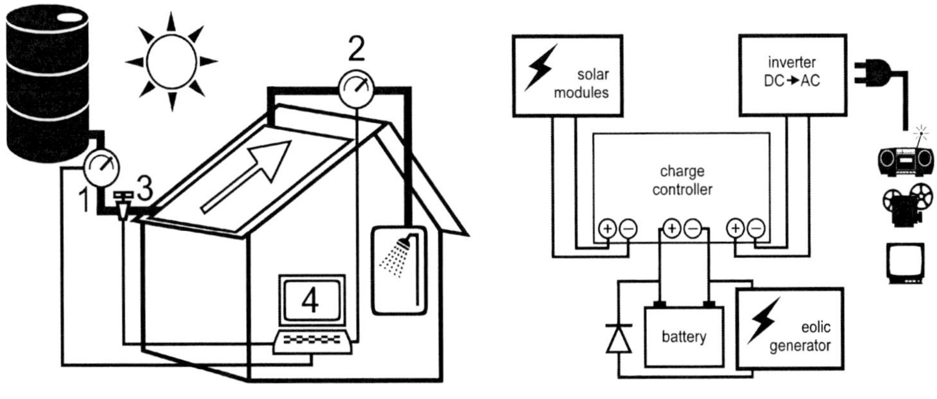

Figure 9. Construction and electric shemes of the PV/ST/WG hybrid system (see text).

The devices used for study of hybrid systems performance were capable of production per day at normal working conditions discussed above E_{PV} = 0.55 kWh (PV panel M100/12) and E_W = 0.96 kWh (home made WG) of electric energy, plus E_T = 12 kWh of thermal energy (ST Powermatt; the value given refers to the absence of wind). From (3) we find that our WG generating 40 kW at wind velocity v_a = 5 m/s has an efficiency η_E = 0.3 which could be considered as a good parameter.

At the wind velocity mentioned, the ST will produce, according to (2A), 10.1 kWh of thermal energy per day (to find this value, we determined experimentally the coefficient α in (2A), taking measurements of the ST efficiency depending on v. Thus we obtained α = 0.032 s/m).

In *Thermal-Wind Hybrid System* (ST+WG) designed for building heating purposes, according to discussion above, at average wind velocity of 5 m/s (expression (9)) it will provide approximately 11 – 12 kWh of heat energy, depending on the way of transforming of electric-to-heat energy. To discuss the heat production stability of the system in relation to small variations of wind velocity, we have to calculate the variations in the energy generation by the system's elements according to (10). For the parameters given, we obtain the first part of (10) describing the dependence of WG productivity upon the changes of v ($1.5\eta_E\ \rho\ A_A \Delta t_2\ v_a^2$) equal to 0.576 kWhs/m, and the second part (similar dependence of the ST productivity, $-\ W\ A_T\ \alpha\ \Delta t_1$) equal to – 0.594 kWhs/m. Thus we see that their sum is reasonably close to zero, i.e. our system is well balanced. According to the values found, the daily variations in energy production caused by changes of wind velocity $\Delta v = \pm 3$ m/s will be around 0.05 kWh, which is less than 0.5 % of the total energy generated per day.

In *PV/Thermal Hybrid System* (PV+ST) discussed above we obtain, in first approximation, the above mentioned amount E_{PV} of electric energy and E_T of thermal energy, with the difference that now the whole energy is provided by the area A_T, not the sum of areas $A_T + A_{PV}$. To get an additional advantage caused by the PV cooling in thermal contact with the ST, a special PV construction is needed; we have shown [28] that the actual increase of the PVM efficiency could be as high as 10 %.

Our *Wind-Photovoltaic Hybrid System* (WG+PV) produces daily approximately 1.5 kWh of electric energy at average conditions of state of Queretaro, Mexico (we should stress that our model of WG is good for moderate wind velocities; for more windy places, the HA models will be better). For actual utilization of this energy with conentional electrical devices, energy storage, control and conversion appliances are necessary, as shown in figure 8.

The *Wind-Thermal-Photovoltaic System* (WG+ST+PV) provides at average conditions of the state around 1.5 kWh of electric energy and 10 kWh of thermal energy. Depending on the necesities and the actual climatic conditions, any part of electric energy produced can be used for stabilization of thermal regime; on the other hand, this amount of energy is sufficient to operate one remote classroom equipped with receptor of educational satellite programs and TV-video set during 6 hours a day, which was proved experimentally in one of the rural schools of the state of Queretaro, Mexico (the classroom is in constant use since August 2002).

In relation to hybrid systems with High Temperature Stage (PV+HE), the preliminary investigations were made of the performance of different solar cells (those of α-Si, $CuInSe_2$, c-Si with p-n junction and with Schottky barrier) at high temperatures, using concentrated and non-concentrated solar radiation. For the former type of experiments, the concentrator with 2-axial Sun tracking system was made (figure 10), the radiation concentration degree around 30, providing the temperature at focal plane up to 200 °C. The corresponding theory for high temperature cell performance was also developed. Figure 11 shows theoretical and experimental dependencies of parameters of the c-Si solar cells with p-n junction at the temperature interval 60 – 170 °C (the detailed description of theoretical model and experimental equipment is given in [35]). The results obtained are related to type "B" of the system discussed, and show the definite possibility of solar cell to be used at high temperatures. For the realization of all the advantages of the system (both "A" and "B" types), the new cells are necessary, and their design and construction are in process now.

Figure 10. Sun-tracking radiation concentrator with c-Si solar cell in focal plane.

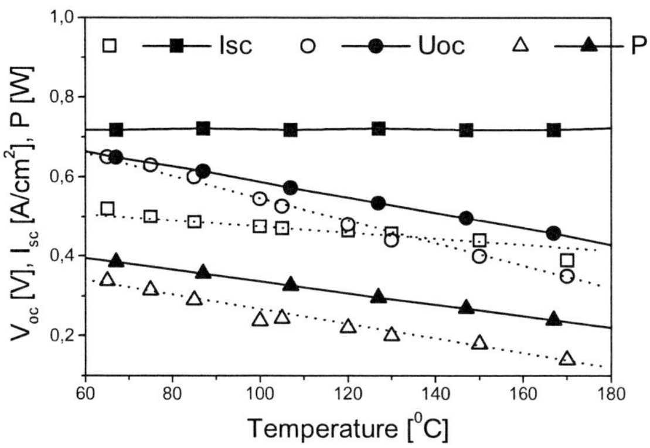

Figure 11. Temperature dependence of c-Si solar cell parameters:theory (solid symbols) and experiment (hollow symbols).

Conclusion

The recent development in solar photovoltaic cell technologies brought great progress to the field, but unfortunately, there are no clear prospects today to develop economic cells with efficiency higher than 20 %. Hybrid systems of different kinds promise much higher efficiency of utilization of solar (renewable) energy, which could be in total around 40 – 60 %, with a reasonable cost and some other advantages. However, hybrid systems, as a rule, are not efficiently working when made by a mechanical combination of the existing elemental devices. Some of them, if not all, have to be designed and constructed especially for a particlular hybrid system. Thus, a photovoltaic panel for applications in a PV/Thermal system has to be designed in an optimized way to provide a good thermal exchange with the heat collector, which needs specific materials and arrangements. The ideal for hybrid systems would be a photovoltaic panel transparent for part of the solar spectrum (infrared one) which is not absorbed by its semiconductor material, and having a good heat exchange with the heat collector. On the other hand, the solar heat collector could be made transparent in the visible region, to be able to use with bifacial panels and not affect greatly their performance. The experimental systems designed, built and studied confirm the conclusions made.

References

[1] Karam, N.H.; King, R.R., et al., *Sol. En.Mat. Sol. Cells.*, 2001, vol. 66, pp. 453-466.
[2] Henry, C.H., *J. Appl. Phys.*, 1980, vol. 51, pp. 4494-4500.
[3] Yamaguchi, M., *Sol. En. Mat. Sol. Cells.*, 2003, vol. 75, pp. 261-269.
[4] Tobias, I; Luque, A., *Progr. in Photovolt.,* 2002, vol. 10, pp. 323-329.
[5] Andreev, V.M.; Grilikhes, V.A., et al., *Sol. En. Mat. Sol. Cells.*, 2004, vol. 84, pp. 3-17.
[6] Imenes, A.G.; Mills, D.R., *Sol. En.Mat. Sol. Cells.*, 2004, vol.84, pp. 19-69.
[7] Luque, A.; Marti, A., *Phys. Rev. Lett.*, 1997, vol. 78, pp. 5014-5017.

[8] Luque, A.; Marti, A., *Progr. in Photovolt.*, 2001, vol. 9, pp. 73-86.
[9] Kern Jr., E.C.; Russel, M.C., Proc. 13 *ISES PVSC*, Washington D.C., USA, 1978, pp. 1153-1157.
[10] Huerta, J., et al., *Proc. ANES/ISES*, Mexico, 2000, pp. 529-533.
[11] Raghuraman, P., J. *Sol. En. Eng.*, 1981, vol. 103, pp. 291-298.
[12] Takashima T., et al., *Sol. En.,* 1994, vol.52, p.241-245.
[13] Florschuetz, L.W., *Sol. En.*, 1979, vol. 22, pp. 361-366.
[14] Zondag, H.A., et al., *Sol. En.*, 2002, vol. 72, pp. 113-128.
[15] Rowe, D.M., Ed.;*CRC Handbook of Thermoelectrics*, CRC Press, London, N.Y., 1995, 702 p.
[16] Anatychuk, L. I. Thermoelements and Thermoelectric Devices (Handbook), *Naukova Dumka,* Kiev, 1979, 768 p.
[17] Bulat, L.P., et al., *Thermoelectric Cooling*. St. Petersburg, SPbSURandFE, 2002, 47 p.
[18] L.P.Bulat, L.P., *Kholodilnaya Tekhnika*, 2004, No. 8, 7 p.
[19] Fleurial, J.-P., et al., *Proc. of 23rd International Conference on Thermoelectrics*, Book of Abstracts, Adelaide, Australia, 2004, p.70.
[20] Venkatasubramantan, R., et al., *Nature*, 2001, vol. 413, pp. 597-608.
[21] Venkatasubramantan, R., *US Paten*t No.: 6,300,150,B1. Oct. 9, 2001.
[22] Harman, T.C., et al., *Science*, 2002 September 27; 297: 2229.
[23] Ghoshal, U., Ghoshal, S., McDowell, C., Shi, L., *Appl. Phys. Letters*, 2002, vol.80. pp. 3006-3008.
[24] Ghoshal, U., *Proc. XXI International Conf. on Thermoelectrics*, 2002. IEEE, p.540.
[25] Hishinuma Y., Geballe, T.H., Moyzhes, B.Y, Kenny T.W., *Appl. Phys. Lett.*, 2001, vol.78, p.2572.
[26] Tavkhelidze, A., Skhiladze, G., Bibilashvili, A., Tsakadze, L., Jangadze, L., Taliashvili, Z., Cox, I., Berishvili, Z., *Proc. XXI International Conf. on Thermoelectrics*, 2002. IEEE. P.435.
[27] Caillat, T. et al., *Proc. XX International Conf. on Thermoelectrics*, 2001. IEEE, p.282.
[28] Zakharchenko, R., et al., *Sol. En. Mater. Sol. Cells.*, 2004, vol. 82, pp. 253-261.
[29] Organ, A.J., *The Regenerator and the Stirling Engine*, Harvill Press, 1997, pp. 35-88.
[30] Sze, S.M., *Physics of Semiconductor Devices*, 2nd ed., John Wiley and Sons, N.Y., 1981, pp. 830-835.
[31] Algora, C., et al., *IEEE Trans. Electron Devices*, 2001, vol. 48, pp. 840-844.
[32] Nann, S.; Emery, K., *Sol. En.Mat. Sol. Cells*, 1992, vol. 27, pp. 189-216.
[33] Luque, A., et al., *Sol. Cells*, 1980, vol. 2. pp. 151-166.
[34] Luque, A., et al., *Sol. Cells*, 1984-85, vol. 13, pp. 277-292.
[35] Meneses-Rodriguez, D., et al., *Sol. En. J.,* 2005, vol. 78, pp. 243-250.

In: Research in Hybrid Materials
Editors: Simon J. Brunner and Julian W. Egger

ISBN 978-1-60456-718-2

Chapter 6

Analysis of Hybrid Actuated Laminated Piezoelectric Sandwich Beams and Active Vibration Control Applications

S. Raja*

Aeroelasticity and Smart Structures Group, Structures Division,
National Aerospace Laboratories, Bangalore – 560017, India

Abstract

Distributed actuation and sensing are the key elements in the development of active structural control methodology. Piezoelectric materials are popularly considered as active elements (actuators or sensors) due to their good frequency bandwidth, low cost and fast energy conversion nature. As actuators, they develop isotropic or directional actuation strains, which are governed by mainly five piezoelectric constants (d_{31}, d_{32}, d_{33}, d_{15}, d_{24}). The longitudinal (d_{33}) and extension (d_{31}, d_{32}) actuations have been thoroughly studied; however shear actuation (d_{15}) is relatively a new concept but shows promising feature. It is a novel idea to combine the extension and shear actuations to develop a hybrid actuation mode for active vibration control applications, exploiting the benefits of both. The hybrid active laminate can be built, employing a transversely polarized (d_{31}) lamina and an axially polarized (d_{15}) lamina. Appropriate constitutive models are derived with an assumption that each lamina behaves as elastically orthotropic and electro-mechanically orthorhombic crystal class mm2. A two node sandwich beam element is developed using the isoparametric FE procedures to conduct numerical experiments. Active control analysis is performed using a modal control approach and the procedure is outlined to obtain the reduced order models without loosing the dynamic information of the vibrating systems.

Active stiffening (piezoelectric straining) and active damping (piezoelectric resistive force) are the two active effects systematically analyzed by numerical studies. Collocated and non-collocated actuator configurations are considered, employing extension and shear actuators in sandwich beam architectures to evaluate the performance of above mentioned active effects. In the vibration amplitude control, the shear actuation has been

* E-mail address: raja@css.cmmacs.ernet.in

found very effective, as it develops locally shear strain. Also, a sine wave actuation mode is observed when a shear actuator is activated in a Clamped-Clamped construction. Interesting deflection behaviours are observed under hybrid actuation mode for various boundary effects. The mode shape control concept using piezoelectric stiffening has been introduced, where a Clamped-Free laminated beam is taken as an illustration. It is a useful technique, as the mode shapes influence significantly the dynamic instability of thin walled composite structures.

1. Introduction

Piezoelectric materials have tremendous potential in vibration control applications as distributed sensors and actuators, which can be either surface bonded or embedded into composite substrate to build smart laminated or sandwich structures. However, to effectively use the distributed actuation property of piezoelectric materials, proper modelling techniques and new smart concepts must be developed. One such attempt is made here by combining shear and extension actuations to develop a hybrid actuation. An orthorhombic crystal system (PZT/PVDF) is considered and in particular mm2 class is modelled. The governing equations are derived first and then they are used to obtain the finite element formulations. Numerical experiments are carried out to systematically assess the merit of individual actuation performance and the hybrid actuation.

In the last 30 years, several works have been reported in the finite element formulation of piezoelectric coupling with isotropic and orthotropic composite substrates. Truss, beam, plate, solid and shell elements have been derived and are employed in different structural applications; however the main focus has been laid on structural vibration control. This is due to the fact that finite element method can handle the piezoelectric coupling in complex structural geometries effectively, using discretization process (Allik and Hughes, 1970; Nailon *et al.*, 1983; Tzou and Tseng, 1990, 1991; Ha *et al.*, 1992; Hwang and Park, 1993; Heyliger *et al.*, 1994, Saravanos and Heyliger, 1995, Saravanos *et al.*, 1997, Chang and Chopra, 1999).

Since an orthorhombic crystal system of mm2 class has five piezoelectric constants and they can couple the electric fields with strain fields, studies are made to investigate new smart structure concepts, employing them (Barrett, 1992; Zhang and Sun, 1996, 1999; Benjeddou *et al.*, 1997, 1999; Aldraihem and Khdeir, 2000, Raja *et al.*, 2002). Smart sandwich structure constructions are proposed, in which the piezoelectric materials are either surface bonded or placed as a core along with regular soft-core material (Koconis *et al.*, 1994). Analytical, numerical and experimental solutions are subsequently developed for modelling the shear actuated composite structures (Senthil and Batra, 2001; Benjeddou and Deue, 2001; Abramovich and Meyer-Piening, 2001).

Active control system (ACS) is a novel idea primarily developed to prevent catastrophic structural failures (like flutter) through feedback controls. Various control algorithms are employed to develop active vibration suppression schemes. With the emerging smart structure technology, active control solutions to structural problems have become a reality. The literature relevant to the control theory is enormous, also well established; therefore, the present discussion is limited only to the application of active controls in smart structures

(Lazarus and Crawley, 1992; Tzou and Hollkamp, 1994; Baz and Poh, 1996; Han *et al.*, 1997; Park *et al.*, 1999).

2. EXTENSION AND SHEAR ACTUATIONS

Despite the fact that an orthorhombic crystal system has five piezoelectric constants, only three piezoelectric constants, namely d_{31}/d_{32} and d_{33} and their associated actuations are much exploited in the smart structure applications. Piezoelectric constants d_{31} and d_{32} couple the transverse electric field E_3 to the normal strains ε_{11} and ε_{22} forms the basis for extension actuation mechanism (EAM). This type of actuator either develops extension or compression depending on the polarity of the applied electric field, which can be ultimately converted into bending mechanism (figure 1(a & b)). On the other hand, the piezoelectric constants d_{15} and d_{24} couple the axial electric fields E_1 and E_2 with transverse shear strains γ_{13} and γ_{23}, respectively to develop piezoelectric shear actuation (SAM). Interestingly, if the constitutive relations are transformed such that the transverse electric field E_3 couples both normal and shear strains, it is then possible to model both actuation mechanisms in a unified way. This implies that the direction of polarization should be made parallel to transverse electric field for extension actuation and perpendicular, for shear actuation (figure 2).

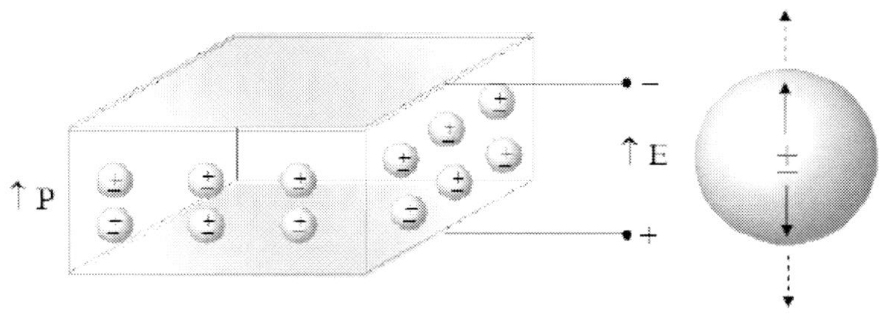

Figure 1 (a). Extension actuation mechanism (EAM)

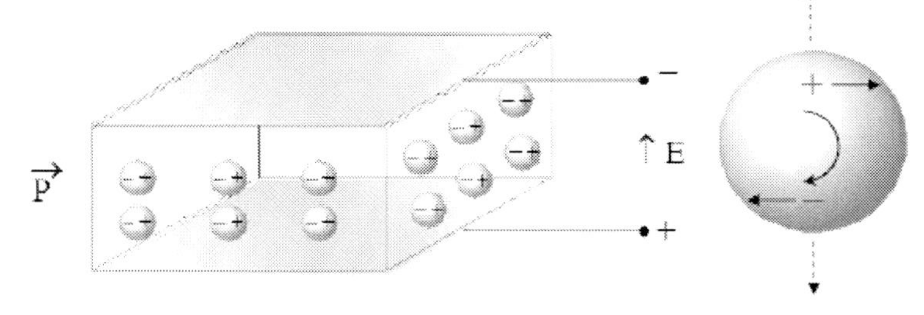

Figure 1 (b). Shear actuation mechanism (SAM)

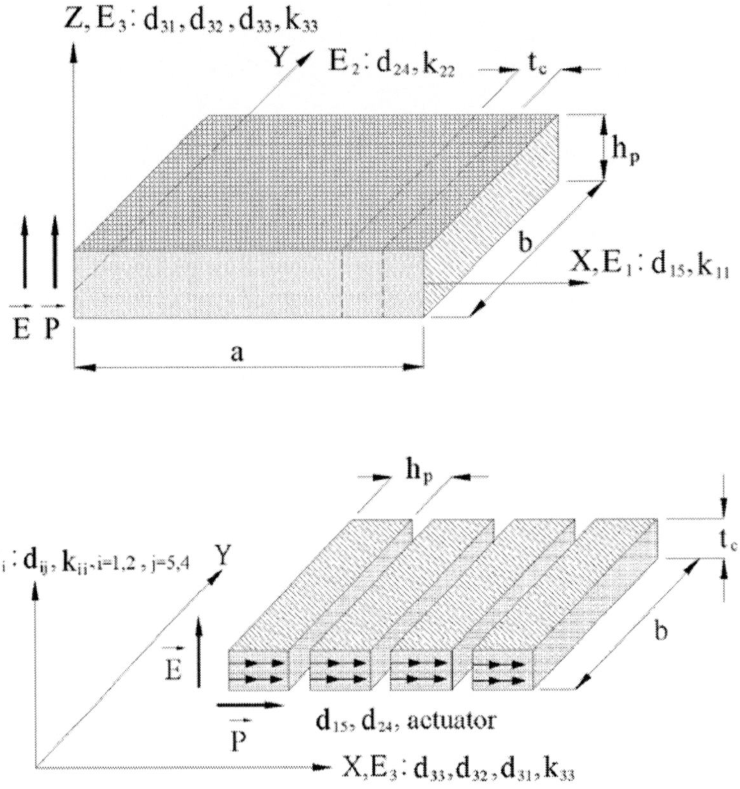

Figure 2. Extension and shear actuation

2.1 Basic Equations and Weak Formulation

The governing equations of the piezoelectric continuum are grouped under three categories.

(a) Constitutive Equations

$$\sigma_{ij} = c_{ijkl}\,\varepsilon_{kl} - d_{kij}\,E_k \quad \text{(Actuator)} \tag{1}$$

$$D_i = d_{ijk}\,\varepsilon_{kl} + \kappa_{ik}\,E_k \quad \text{(Sensor)} \tag{2}$$

(b) Differential Equations

$$\sigma_{ij,i} + f_{bj} = \rho\,\ddot{u}_j \quad \text{(Stress Equation of Motion)} \tag{3}$$

where f_{bj} is the body force.

$$D_{i,i} = 0 \quad \text{(Equation of Electrostatics)} \tag{4}$$

(c) Field Equations

$$\varepsilon_{ij} = \frac{1}{2}\left(u_{j,i} + u_{i,j}\right) \quad \text{(Kinematics)} \tag{5}$$

$$E_i = -\phi_{,i} \quad \text{(Electric Field to Potential)} \tag{6}$$

The generalized weak formulation for the coupled piezoelectric field is derived using the virtual work principle. As the piezoelectric solid is capable of working as actuator and sensor, the internal energy developed in the system is basically electro-mechanical in nature and is known as electric enthalpy. Hamilton's principle can be directly applied on this electric enthalpy to derive the weak form of governing equations for the electro-mechanical system. However, the virtual work principle is employed to make the energy formulation unique.

The virtual work done on a piezoelectric solid, if it behaves as an actuator, is derived from the stress equation of motion and is given by

$$\int_V (\sigma_{ij,i} + f_{bj} - \rho \ddot{u}_j)\,\delta u_j \, dv = 0. \tag{7}$$

Using the divergence theorem, and with the help of natural boundary conditions $\sigma_{ij}\, n_i = f_{sj}$, where n_i is the outer normal, equation (7) may be written as

$$\int_V (\rho \ddot{u}_j\, \delta u_j + \sigma_{ij}\, \delta\varepsilon_{ij})dv = \int_V (f_{bj}\, \delta u_j)dv + \int_S (f_{sj}\, \delta u_j)ds. \tag{8}$$

The virtual work done on a piezoelectric solid, if it behaves as a sensor is obtained from the equation of electrostatics and is defined as

$$\int_v (D_{i,i}\, \delta\phi)dv = 0. \tag{9}$$

Using the divergence theorem, and with the help of natural boundary conditions $D_i\, n_i = q$, equation (9) may be written as

$$\int_V (D_i\, \delta\phi_{,i})\,dv = \int_S (q\, \delta\phi)\,ds. \tag{10}$$

The generalized weak form of coupled piezoelectric problem is derived from equations (8), (10), and using the constitutive relations, for the composite sandwich structure (see figure 4) as

$$\sum_{i=t,c,b} \int_{v^{(i)}} (\{\delta u^{(i)}\}^T \rho^{(i)} \{\ddot{u}^{(i)}\} + \{\delta\varepsilon^{(i)}\}^T [c^{(i)}]\{\varepsilon^{(i)}\} + \{\delta\varepsilon^{(i)}\}^T [d^{(i)}]\{E^{(i)}\}$$

$$+\{\delta E^{(i)}\}^T [d^{(i)}]^T \{\varepsilon^{(i)}\} - \{\delta E^{(i)}\}^T [\kappa^{(i)}]\{E^{(i)}\})dv - \int_{v^{(i)}} \{\delta u^{(i)}\}^T \{f_b^{(i)}\} dv$$

$$- \int_{s^{(i)}} \{\delta u^{(i)}\}^T \{f_s^{(i)}\} ds + \int_{s^{(i)}} q^{(i)} \delta\phi^{(i)} ds = 0 \tag{11}$$

where t, c, b are the top, core and bottom sub-laminates, respectively.

2.2 Constitutive Models

The stress – strain relationship is established with the following assumptions:

1) The lamina is elastically orthotropic and piezoelectrically orthorhombic
2) The transverse electric field couples both normal and transverse shear strains
3) The adhesive layer is very thin and so shear deformation in it is neglected

Therefore, the axial electric field is transferred to the transverse direction i.e., $1 \rightarrow 3$ coordinate change (see figure 2) to maintain the same electrode pattern for both actuations. Thus, two constitutive models are employed in the finite element formulation, one for Extension Actuation (EAM) and another for Shear Actuation (SAM).

2.2.1 Stress-Strain Relations for Extension Actuated Lamina (EAM)

The active lamina properties with respect to material axes are defined as follows:

$$\begin{Bmatrix} \sigma_{11} \\ \sigma_{22} \\ \tau_{23} \\ \tau_{13} \\ \tau_{12} \\ D_1 \\ D_2 \\ D_3 \end{Bmatrix} = \begin{bmatrix} Q_{11} & Q_{12} & 0 & 0 & 0 & 0 & 0 & -Q_{PE31} \\ Q_{12} & Q_{22} & 0 & 0 & 0 & 0 & 0 & -Q_{PE32} \\ 0 & 0 & Q_{44} & 0 & 0 & 0 & -Q_{PE24} & 0 \\ 0 & 0 & 0 & Q_{55} & 0 & -Q_{PE15} & 0 & 0 \\ 0 & 0 & 0 & 0 & Q_{66} & 0 & 0 & 0 \\ 0 & 0 & 0 & Q_{PE15} & 0 & Q_{DE11} & 0 & 0 \\ 0 & 0 & Q_{PE24} & 0 & 0 & 0 & Q_{DE22} & 0 \\ Q_{PE31} & Q_{PE32} & 0 & 0 & 0 & 0 & 0 & Q_{DE33} \end{bmatrix} \begin{Bmatrix} \varepsilon_{11} \\ \varepsilon_{22} \\ \gamma_{23} \\ \gamma_{13} \\ \gamma_{12} \\ E_1 \\ E_2 \\ E_3 \end{Bmatrix} \tag{12}$$

where

$$Q_{11} = C_{11} - \frac{C_{13}C_{13}}{C_{33}};\ Q_{12} = C_{12} - \frac{C_{13}C_{23}}{C_{33}};\ Q_{22} = C_{22} - \frac{C_{23}C_{23}}{C_{33}};\ Q_{44} = C_{44};\ Q_{55} = C_{55};$$

$$Q_{66} = C_{66};\ Q_{PE31} = d_{31} - \frac{C_{13}d_{33}}{C_{33}};\ Q_{PE32} = d_{32} - \frac{C_{23}d_{33}}{C_{33}};\ Q_{PE24} = d_{24};\ Q_{PE15} = d_{15};$$

$$Q_{DE11} = \kappa_{11};\ Q_{DE22} = \kappa_{22};\ Q_{DE33} = \kappa_{33} + \frac{d_{33}d_{33}}{C_{33}}.$$

The constitutive relationship in the X-Y-Z coordinate system can be obtained as follows.

$$\begin{aligned} &\{\sigma\}_{xyz} = [\overline{Q}]\{\varepsilon\}_{xyz}, \\ &\text{where } [\overline{Q}] = [T]^T[Q][T]. \end{aligned} \tag{13}$$

and 'T' is the transformation matrix given by,

$$[T] = \begin{bmatrix} m^2 & n^2 & 0 & 0 & mn & 0 & 0 & 0 \\ n^2 & m^2 & 0 & 0 & -mn & 0 & 0 & 0 \\ 0 & 0 & m & -n & 0 & 0 & 0 & 0 \\ 0 & 0 & n & m & 0 & 0 & 0 & 0 \\ -2mn & 2mn & 0 & 0 & m^2 - n^2 & 0 & 0 & 0 \\ 0 & 0 & 0 & 0 & 0 & 1 & 0 & 0 \\ 0 & 0 & 0 & 0 & 0 & 0 & 1 & 0 \\ 0 & 0 & 0 & 0 & 0 & 0 & 0 & 1 \end{bmatrix}, m=\cos\theta;\ n=\sin\theta.$$

The transformed coefficients are then obtained and are given below.

$$\begin{aligned} \overline{Q}_{11} &= m^4Q_{11} + n^4Q_{22} + 2m^2n^2(Q_{12} + 2Q_{66}) \\ \overline{Q}_{12} &= (m^4 + n^4)Q_{12} + m^2n^2(Q_{11} + Q_{22} - 4Q_{66}) \\ \overline{Q}_{15} &= mn^3(Q_{12} - Q_{22} + 2Q_{66}) + m^3n(Q_{11} - Q_{12} - 2Q_{66}) \\ \overline{Q}_{18} &= (m^2Q_{PE31} + n^2Q_{PE32}) \\ \overline{Q}_{22} &= m^4Q_{22} + n^4Q_{11} + 2m^2n^2(Q_{12} + 2Q_{66}) \\ \overline{Q}_{25} &= mn^3(Q_{11} - Q_{12} - 2Q_{66}) + m^3n(Q_{12} - Q_{22} + 2Q_{66}) \\ \overline{Q}_{28} &= (m^2Q_{PE32} + n^2Q_{PE31}) \\ \overline{Q}_{33} &= m^2Q_{44} + n^2Q_{55} \end{aligned} \tag{14}$$

$$\overline{Q}_{34} = -mn(Q_{44} - Q_{55})$$
$$\overline{Q}_{36} = nQ_{PE15}$$
$$\overline{Q}_{37} = mQ_{PE24}$$
$$\overline{Q}_{44} = m^2 Q_{55} + n^2 Q_{44}$$
$$\overline{Q}_{46} = mQ_{PE15}$$
$$\overline{Q}_{47} = -nQ_{PE24}$$
$$\overline{Q}_{55} = m^2 n^2 (Q_{11} - 2Q_{12} + Q_{22}) + (m^2 - n^2) 2Q_{66}$$
$$\overline{Q}_{58} = mn(Q_{PE31} - Q_{PE32})$$
$$\overline{Q}_{66} = Q_{DE11}$$
$$\overline{Q}_{77} = Q_{DE22}$$
$$\overline{Q}_{88} = Q_{DE33}$$

Further the constitutive relations are reduced for a beam using the standard static condensation procedure.

$$\begin{Bmatrix} \sigma_{xx} \\ \tau_{xz} \\ D_x \\ D_z \end{Bmatrix} = \begin{bmatrix} \overline{Q}^*_{11} & 0 & 0 & -\overline{Q}^*_{PE31} \\ 0 & \overline{Q}^*_{55} & -\overline{Q}^*_{PE15} & 0 \\ 0 & \overline{Q}^*_{PE15} & \overline{Q}^*_{DE11} & 0 \\ \overline{Q}^*_{PE31} & 0 & 0 & \overline{Q}^*_{DE33} \end{bmatrix} \begin{Bmatrix} \varepsilon_{xx} \\ \gamma_{xz} \\ E_x \\ E_z \end{Bmatrix},$$

$$\overline{Q}^*_{11} = \frac{(m^2 + n^2)^4 Q_{66} (Q_{11} Q_{22} - Q_{12}^2)}{m^4 Q_{22} Q_{66} + n^4 Q_{11} Q_{66} - m^2 n^2 (2Q_{12} Q_{66} + Q_{12}^2 - Q_{11} Q_{22})}$$

$$\overline{Q}^*_{PE31} = \frac{(m^2 + n^2)^2 Q_{66} [m^2 (Q_{22} Q_{PE31} - Q_{12} Q_{PE32}) - n^2 (Q_{12} Q_{PE31} - Q_{11} Q_{PE32})]}{m^4 Q_{22} Q_{66} + n^4 Q_{11} Q_{66} - m^2 n^2 (2Q_{12} Q_{66} + Q_{12}^2 - Q_{11} Q_{22})}$$

$$\overline{Q}^*_{55} = \frac{(m^2 + n^2)^2 Q_{55} (Q_{DE22} Q_{44} + Q_{PE24}^2)}{m^2 (Q_{DE22} Q_{44} + Q_{PE24}^2) + n^2 Q_{DE22} Q_{55}} \qquad (15)$$

$$\overline{Q}^*_{PE15} = \frac{m(m^2+n^2)Q_{PE15}(Q_{DE22}Q_{44}+Q^2_{PE24})}{m^2(Q_{DE22}Q_{44}+Q^2_{PE24})+n^2Q_{DE22}Q_{55}}$$

$$\overline{Q}^*_{DE11} = Q_{DE11} + \frac{n^2Q^2_{PE15}Q_{DE22}}{m^2(Q_{DE22}Q_{44}+Q^2_{PE24})+n^2Q_{DE22}Q_{55}}$$

$$\overline{Q}^*_{DE33} = \frac{\left\{\begin{array}{l} m^4(Q^2_{PE32}Q_{66}+Q_{DE33}Q_{22}Q_{66})+n^4(Q^2_{PE31}Q_{66}+Q_{DE33}Q_{11}Q_{66}) \\ +m^2n^2[Q^2_{PE31}Q_{22}+Q^2_{PE32}Q_{11}-2Q_{PE31}Q_{PE32}(Q_{12}+Q_{66}) \\ -Q_{DE33}(Q^2_{12}-Q_{11}Q_{22}+2Q_{12}Q_{66})] \end{array}\right\}}{m^4Q_{22}Q_{66}+n^4Q_{11}Q_{66}-m^2n^2(2Q_{12}Q_{66}+Q^2_{12}-Q_{11}Q_{22})}$$

2.2.2 Stress – Strain Relations for Shear Actuated Lamina (SAM)

Since the axial electric field is transformed into transverse direction, the elastic constants are also modified accordingly. The lamina properties with respect to material axes are defined for shear actuation as follows:

$$\begin{Bmatrix} \sigma_{11} \\ \sigma_{22} \\ \tau_{23} \\ \tau_{13} \\ \tau_{12} \\ D_1 \\ D_2 \\ D_3 \end{Bmatrix} = \begin{bmatrix} Q_{33} & Q_{23} & 0 & 0 & 0 & -Q_{PE33} & 0 & 0 \\ Q_{23} & Q_{22} & 0 & 0 & 0 & -Q_{PE32} & 0 & 0 \\ 0 & 0 & Q_{66} & 0 & 0 & 0 & 0 & 0 \\ 0 & 0 & 0 & Q_{55} & 0 & 0 & 0 & -Q_{PE15} \\ 0 & 0 & 0 & 0 & Q_{44} & 0 & -Q_{PE24} & 0 \\ Q_{PE33} & Q_{PE32} & 0 & 0 & 0 & Q_{DE33} & 0 & 0 \\ 0 & 0 & 0 & 0 & Q_{PE24} & 0 & Q_{DE22} & 0 \\ 0 & 0 & 0 & Q_{PE15} & 0 & 0 & 0 & Q_{DE11} \end{bmatrix} \begin{Bmatrix} \varepsilon_{11} \\ \varepsilon_{22} \\ \gamma_{23} \\ \gamma_{13} \\ \gamma_{12} \\ E_1 \\ E_2 \\ E_3 \end{Bmatrix} \quad (16)$$

where

$$Q_{33} = C_{33} - \frac{C_{13}C_{13}}{C_{11}};\ Q_{23} = C_{23} - \frac{C_{13}C_{12}}{C_{11}};\ Q_{22} = C_{22} - \frac{C_{12}C_{12}}{C_{11}};\ Q_{44} = C_{44};\ Q_{55} = C_{55};$$

$$Q_{66} = C_{66};\ Q_{PE33} = d_{33} - \frac{C_{13}d_{31}}{C_{11}};\ Q_{PE32} = d_{32} - \frac{C_{12}d_{31}}{C_{11}};\ Q_{PE24} = d_{24};\ Q_{PE15} = d_{15};$$

$$Q_{DE11} = \kappa_{11};\ Q_{DE22} = \kappa_{22};\ Q_{DE33} = \kappa_{33} + \frac{d_{31}d_{31}}{C_{11}}.$$

The constitutive relationship in the X-Y-Z coordinate system can be obtained as follows:

$$\{\sigma\}_{xyz} = [\overline{Q}]\{\varepsilon\}_{xyz}, \quad \text{where } [\overline{Q}] = [T]^T[Q][T]. \tag{17}$$

and 'T' is the transformation matrix given by,

$$[T] = \begin{bmatrix} m^2 & n^2 & 0 & 0 & mn & 0 & 0 & 0 \\ n^2 & m^2 & 0 & 0 & -mn & 0 & 0 & 0 \\ 0 & 0 & m & -n & 0 & 0 & 0 & 0 \\ 0 & 0 & n & m & 0 & 0 & 0 & 0 \\ -2mn & 2mn & 0 & 0 & m^2-n^2 & 0 & 0 & 0 \\ 0 & 0 & 0 & 0 & 0 & m & -n & 0 \\ 0 & 0 & 0 & 0 & 0 & n & m & 0 \\ 0 & 0 & 0 & 0 & 0 & 0 & 0 & 1 \end{bmatrix}$$

Therefore, SAM lamina may be directionally placed to activate either θ_x or θ_y. The transformed coefficients are obtained and are given below.

$$\begin{aligned}
\overline{Q}_{11} &= m^4Q_{33} + 2m^2n^2(Q_{23} + 2Q_{44}) + n^4Q_{22} \\
\overline{Q}_{12} &= n^4Q_{23} + m^2n^2(Q_{33} + Q_{22} - 4Q_{44}) + m^4Q_{23} \\
\overline{Q}_{15} &= mn^3(Q_{23} - Q_{22} + 2Q_{44}) + m^3n(Q_{33} - Q_{23} - 2Q_{44}) \\
\overline{Q}_{16} &= mn^2(Q_{PE32} - 2Q_{PE24}) + m^3Q_{PE33} \\
\overline{Q}_{17} &= -(n^3Q_{PE32} + m^2n(Q_{PE33} + 2Q_{PE24})) \\
\overline{Q}_{22} &= m^4Q_{22} + 2m^2n^2(Q_{23} + 2Q_{44}) + n^4Q_{33} \\
\overline{Q}_{25} &= mn^3(Q_{33} - Q_{23} - 2Q_{44}) + m^3n(Q_{23} - Q_{22} + 2Q_{44}) \\
\overline{Q}_{26} &= mn^2(Q_{PE33} + 2Q_{PE24}) + m^3Q_{PE32} \\
\overline{Q}_{27} &= -(n^3Q_{PE33} + m^2n(Q_{PE32} - 2Q_{PE24})) \\
\overline{Q}_{33} &= m^2Q_{66} + n^2Q_{55} \\
\overline{Q}_{34} &= -mn(Q_{66} - Q_{55}) \\
\overline{Q}_{38} &= nQ_{PE15}
\end{aligned} \tag{18}$$

$$\overline{Q}_{44} = m^2Q_{55} + n^2Q_{66}$$
$$\overline{Q}_{48} = mQ_{PE15}$$
$$\overline{Q}_{55} = m^2n^2(Q_{33} - 2Q_{23} + Q_{22}) + (m^2 - n^2)^2Q_{44}$$
$$\overline{Q}_{56} = m^2n(Q_{PE33} - Q_{PE32}) + (m^2 - n^2)nQ_{PE24}$$
$$\overline{Q}_{57} = -mn^2(Q_{PE33} - Q_{PE32} + Q_{PE24}) + m^3Q_{PE24}$$
$$\overline{Q}_{66} = m^2Q_{DE33} + n^2Q_{DE22}$$
$$\overline{Q}_{67} = -mn(Q_{DE33} - Q_{DE22})$$
$$\overline{Q}_{77} = m^2Q_{DE22} + n^2Q_{DE33}$$
$$\overline{Q}_{88} = Q_{DE11}$$

Further the constitutive relations are reduced for a beam using the standard static condensation procedure.

$$\begin{Bmatrix} \sigma_{xx} \\ \tau_{xz} \\ D_x \\ D_z \end{Bmatrix} = \begin{bmatrix} \overline{Q}^*_{11} & 0 & -\overline{Q}^*_{PE33} & 0 \\ 0 & \overline{Q}^*_{55} & 0 & -\overline{Q}^*_{PE15} \\ \overline{Q}^*_{PE33} & 0 & \overline{Q}^*_{DE33} & 0 \\ 0 & \overline{Q}^*_{PE15} & 0 & \overline{Q}^*_{DE11} \end{bmatrix} \begin{Bmatrix} \varepsilon_{xx} \\ \gamma_{xz} \\ E_x \\ E_z \end{Bmatrix}, \tag{19}$$

$$\overline{Q}^*_{11} = \frac{\left\{ \begin{aligned} &(m^2+n^2)^4[m^2(Q_{33}Q_{22}Q_{44}Q_{DE22} - Q^2_{23}Q^2_{PE24} - Q^2_{23}Q_{44}Q_{DE22} \\ &+Q_{33}Q_{22}Q^2_{PE24}) + n^2Q_{44}(Q^2_{PE33}Q_{22} + Q_{33}Q^2_{PE32} - 2Q_{PE32}Q_{PE33}Q_{23} \\ &-Q^2_{23}Q_{DE33} + Q_{33}Q_{22}Q_{DE33})] \end{aligned} \right\}}{\tilde{Q}_r},$$

$$\overline{Q}^*_{PE33} = \frac{\left\{ \begin{aligned} &m(m^2+n^2)^3[m^2(Q_{44}Q_{DE22} + Q^2_{PE24})(Q_{22}Q_{PE33} - Q_{PE32}Q_{23}) \\ &+n^2(Q_{33}Q_{PE32}(Q^2_{PE24} - Q_{PE32}Q_{PE24} + Q_{44}Q_{DE22}) + Q_{PE24} \\ &(Q^2_{23}Q_{DE33} - 2Q_{PE32} + Q_{PE24}) - Q_{22}Q_{PE24}(Q^2_{PE33} + Q_{33}Q_{DE33}) \\ &+Q_{44}Q_{DE22} - Q_{PE33}Q_{23})] \end{aligned} \right\}}{\tilde{Q}_r},$$

$$\overline{Q}^*_{55} = \frac{(m^2+n^2)^2Q_{66}Q_{55}}{m^2Q_{66} + n^2Q_{55}},$$

$$\overline{Q}^*_{PE15} = \frac{m(m^2+n^2)Q_{PE15}Q_{66}}{m^2Q_{66} + n^2Q_{55}},$$

$$\bar{Q}^*_{DE33} = \frac{\left\{\begin{array}{l}(m^2+n^2)^2\ [m^4(Q^2_{PE24}+Q_{44}Q_{DE22})(Q^2_{PE32}+Q_{22}Q_{DE33})+ \\ n^4(Q_{DE33}Q_{33}+Q^2_{PE33})(Q^2_{PE24}+Q_{44}Q_{DE22})+ \\ n^2m^2(Q_{22}Q_{DE22}(Q_{DE33}Q_{33}+Q^2_{PE33})\text{-}Q_{DE22}Q^2_{23}Q_{DE33} \\ -2Q_{23}(Q_{DE33}Q_{DE22}Q_{44}+Q_{DE22}Q_{PE32}Q_{PE33}+Q_{DE33}Q^2_{PE24})\text{-} \\ Q_{PE32}(\text{-}Q_{DE22}Q_{33}Q_{PE32}+2Q_{44}Q_{DE22}Q_{PE33}+2Q_{PE33}Q^2_{PE24}))]\end{array}\right\}}{\tilde{Q}_r},$$

$$\bar{Q}^*_{DE11} = \frac{m^2Q_{DE11}Q_{66}+n^2(Q_{DE11}Q_{55}+Q^2_{PE15})}{m^2Q_{66}+n^2Q_{55}},$$

$$\tilde{Q}_r = \left\{\begin{array}{l} m^6Q_{22}(Q_{DE22}Q_{44}+Q^2_{PE24})+n^6Q_{44}(Q^2_{PE33}+Q_{33}Q_{DE33})+ \\ m^2n^4[(Q_{22}(Q^2_{PE33}+Q_{33}Q_{DE33})\text{-}Q^2_{23}Q_{DE33}-2Q_{23}(Q_{PE32}Q_{PE33}+ \\ Q_{44}Q_{DE33}-Q_{PE33}Q_{PE24})+\ Q_{33}(Q_{DE22}Q_{44}\text{-}2Q_{PE24}Q_{PE32}+Q^2_{PE32}+Q^2_{PE24}) \\ \text{-}2Q_{44}Q_{PE32}Q_{PE33}]+m^4n^2[Q_{22}(Q_{33}Q_{DE22}\text{-}2Q_{PE33}Q_{PE24}+Q_{44}Q_{DE33}) \\ \text{-}Q^2_{23}Q_{DE22}+\ Q_{23}(2Q_{PE24}Q_{PE32}\text{-}2Q^2_{PE24}\text{-}2Q_{DE22}Q_{44})+Q^2_{PE32}Q_{44}] \end{array}\right\}.$$

2.2.3 *Stress Resultants of Hybrid Actuated Sandwich Beam*

In the present finite element analysis, a sandwich model is proposed based on *sub-laminate* concept to incorporate the extension and shear actuations. The stress resultants are calculated separately for the face laminates, and the core by integrating the stresses in each layer through the sub-laminate thickness. The shear strain of the sandwich beam cross section is obtained from the shear strain of the core laminate using the small displacement approximation.

Face Laminate:

$$\begin{Bmatrix} N_x \\ M^{gb}_y \\ M^{lb}_y \\ Q_z \end{Bmatrix}_i = \begin{bmatrix} A_E & \pm d_iA_E & B_E & 0 & A_{PE} \\ \pm d_iA_E & d_i^2A_E & \pm d_iB_E & 0 & 0 \\ B_E & \pm d_iB_E & D_E & 0 & B_{PE} \\ 0 & 0 & 0 & 0 & 0 \end{bmatrix}_i \begin{Bmatrix} \varepsilon^m_{xi} \\ \varepsilon^{gb}_{xi} \\ \varepsilon^{lb}_{xi} \\ \gamma^s_{xz} \\ E_{zi} \end{Bmatrix} \quad \text{where } i = t, b \tag{20}$$

Core laminate:

$$\begin{Bmatrix} N_x \\ M_y^{gb} \\ M_y^{lb} \\ Q_{zc} \end{Bmatrix}_c = \begin{bmatrix} A_E & B_E & 0 & 0 & 0 \\ B_E & D_E & 0 & 0 & 0 \\ 0 & 0 & 0 & 0 & 0 \\ 0 & 0 & 0 & A_E & A_{PE} \end{bmatrix}_c \begin{Bmatrix} \varepsilon_{xc}^{m} \\ \varepsilon_{xc}^{gb} \\ \varepsilon_{xc}^{lb} \\ \gamma_{xzc}^{s} \\ E_{zc} \end{Bmatrix}$$

$$\text{where } A_{PE}, A_{DE}, A_E = \sum_{k=1}^{n} \left[\bar{Q}^*\right]_k b_k (h_k - h_{k-1});$$

$$B_{PE}, B_E = \frac{1}{2}\sum_{k=1}^{n} \left[\bar{Q}^*\right]_k b_k (h_k^2 - h_{k-1}^2); \tag{21}$$

$$D_{DE}, D_E = \frac{1}{3}\sum_{k=1}^{n} \left[\bar{Q}^*\right]_k b_k (h_k^3 - h_{k-1}^3),$$

E – elastic, PE – piezoelectric, DE – dielectric.

After estimating the core shear stress resultant Q_{zc}, the shear stress resultant of the entire sandwich section is determined as:

$$Q_{zs} = \frac{d_s}{t_c} Q_{zc}. \tag{22}$$

2.2 Electric Field to Potential Relations

In the present formulation there are three sub-laminates considered, namely, top, core and bottom faces. One multifunctional layer is modeled in each laminate and it can be located anywhere in the sub-laminate thickness direction. The total electric potential in the kth active layer of the sub-laminate is

$$\phi_i(x,z) = \phi_{0i}(x) + \frac{(Z_i - h_{k-1}^{(i)})}{(h_k^{(i)} - h_{k-1}^{(i)})}\phi_{1i}(x), \quad i = t, b, c \tag{23}$$

Here $\phi_{0i} = \dfrac{\phi_i^{\,+} + \phi_i^{\,-}}{2}$ is the mean potential, $\phi_{1i} = \phi_i^{\,+} - \phi_i^{\,-}$ is the difference of potential (figure 3) and h_k is the distance of kth layer from the midplane of the sub-laminate.

Subsequently, the two electric fields taken in the constitutive relations are,

$$E_{xi} = -\left(\frac{\partial \phi_{0i}}{\partial x} + \frac{(Z_i - h_{k-1}^{(i)})}{(h_k^{(i)} - h_{k-1}^{(i)})}\frac{\partial \phi_{1i}}{\partial x}\right), \quad i = t, b, c. \tag{24}$$

$$E_{zi} = -\frac{1}{(h_k^{(i)} - h_{k-1}^{(i)})}\phi_{li}, \quad i = t, b, c. \tag{25}$$

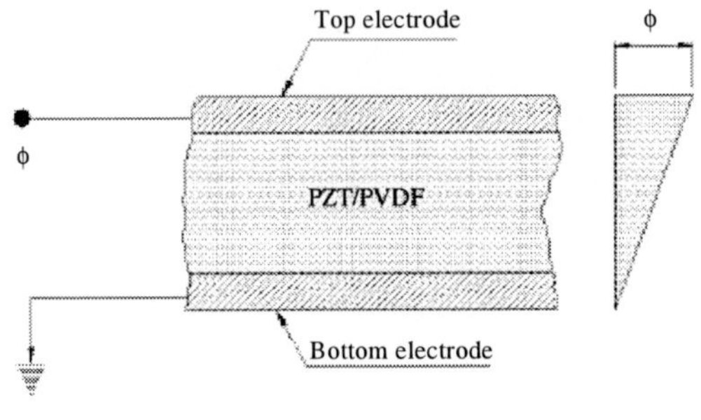

Figure 3. Electric potential variation across the thickness of active lamina

2.4 Finite Element Formulation for Hybrid Actuated Beam

The finite element formulation is done based on the following assumptions:

- Face laminates are thin and behave as Euler-Bernoulli beams.
- Core is thick, laminated, and characterised by a first order shear deformation theory.
- The shear rotation of the core is transferred as a global curvature to the face laminates.
- Transverse displacement is assumed to be constant across the thickness.
- Electric potential variation across the active lamina is also assumed to be linear.

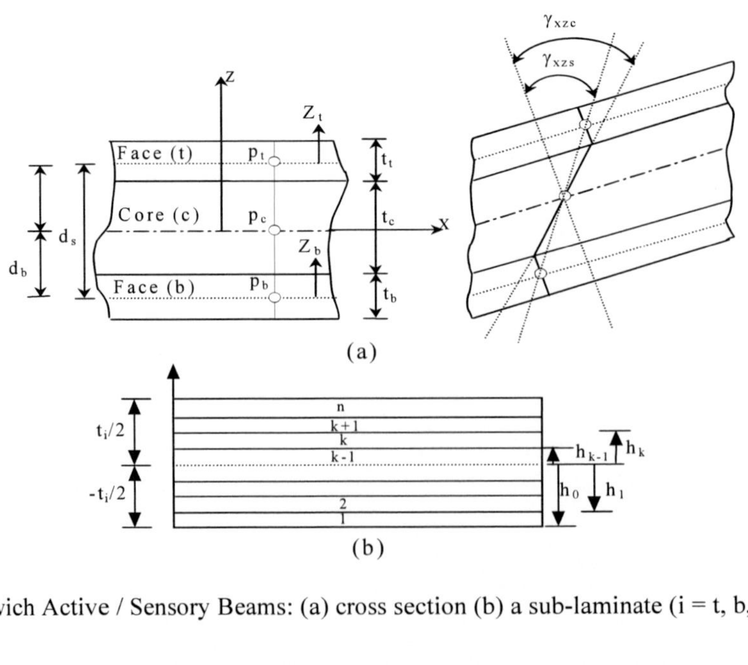

Figure 4. Sandwich Active / Sensory Beams: (a) cross section (b) a sub-laminate (i = t, b, c).

2.4.1 Displacement Field Equations and Kinematic Relations

The displacements are defined at the midplane of the core as u_0, θ_y, w_0, $w_{0,x}$, and the displacements of the face laminate are expressed in terms of these core midplane displacements. Figure 4 shows the cross sectional details of a sandwich composite beam. The strain-displacement relations of each sub-laminate are defined as follows:

Face Laminate (t):

$$u_t(x,z,t)=u_0(x,t)+d_t\,\theta_y(x,t)-Z_t\frac{\partial w_0}{\partial x}(x,t)\,, \tag{26}$$

$$w(x,z,t)=w_0(x,t)\,.$$

The corresponding strain components are,

$$\varepsilon_{xt}^{m}=\frac{\partial u_0}{\partial x};\ \varepsilon_{xt}^{gb}=d_t\frac{\partial \theta_y}{\partial x};\ \varepsilon_{xt}^{lb}=-Z_t\frac{\partial^2 w_0}{\partial x^2}. \tag{27}$$

Face Laminate (b):

$$u_b(x,z,t)=u_0(x,t)-d_b\,\theta_y(x,t)-Z_b\frac{\partial w_0}{\partial x}(x,t)\,, \tag{28}$$

$$w(x,z,t)=w_0(x,t)\,.$$

And the strain components are,

$$\varepsilon_{xb}^{m}=\frac{\partial u_0}{\partial x};\ \varepsilon_{xb}^{gb}=-d_b\frac{\partial \theta_y}{\partial x};\ \varepsilon_{xb}^{lb}=-Z_b\frac{\partial^2 w_0}{\partial x^2}. \tag{29}$$

Core Laminate (c):

$$u_c(x,z,t)=u_0(x,t)+Z\theta_y(x,t), \tag{30}$$

$$w(x,z,t)=w_0(x,t)\,.$$

The strain components are,

$$\varepsilon_{xc}^{m}=\frac{\partial u_0}{\partial x};\ \varepsilon_{xc}^{gb}=Z\frac{\partial \theta_y}{\partial x}, \tag{31}$$

and $$\gamma^{s}_{xzc} = \theta_y + \frac{\partial w_0}{\partial x}.$$

Note that in the above equations, m – membrane, gb – global bending, lb – local bending, and s – transverse shear.

2.4.2 *Smart Sandwich Beam Finite Element*

A two noded sandwich beam element is developed with four mechanical and three electrical degrees of freedom per node. Accordingly, the generalized displacements u_0, θ_y and the electric potentials ϕ_{1t}, ϕ_{1b}, ϕ_{1c} are C^0 interpolated over an element as

$$u_0(x,t) = \sum_{i=1}^{2} u_{0i}(t) N_i(x),$$

$$\theta_y(x,t) = \sum_{i=1}^{2} \theta_{yi}(t) N_i(x), \tag{32}$$

$$\phi_{1j}(x,t) = \sum_{i=1}^{2} \phi_{1i} N_i(x),\ j = t,\ b,\ c$$

where N_i are the linear interpolation functions.

Since the face laminate is modelled using classical beam theory, C^1 interpolation is done for w_0.

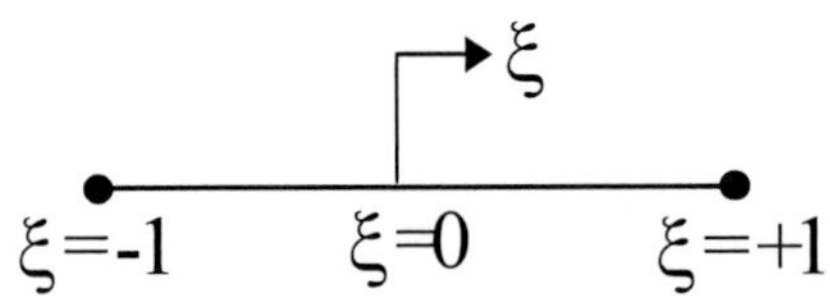

Figure 5 Two-node smart beam element

$$w_0(x,t) = \sum_{i=1}^{2} [w_{0i}(t) \tilde{N}_i(x) + \frac{\partial w_{0i}}{\partial x}(t) \bar{N}_i(x)], \tag{33}$$

where $\tilde{N}_i, \bar{N}_i$ are Hermite cubic interpolation functions,

$$\tilde{N}_1 = (2 - 3\xi + \xi^3)/4$$
$$\bar{N}_1 = (1 - \xi - \xi^2 + \xi^3)/4$$
$$\tilde{N}_2 = (2 + 3\xi - \xi^3)/4$$
$$\bar{N}_2 = (-1 - \xi + \xi^2 + \xi^3)/4$$

Substituting equations (32), (33) in the field equations and transforming into global coordinate system yields,

$$\{\varepsilon^{(i)}\} = \sum_{j=1}^{2} [J]^{-1} [B_u^{(i)}]_j \{\bar{u}\}_j, \quad \text{where } i = t, b, c$$
$$\text{and } \{\bar{u}\} = \{u_0 \; \theta_y \; w_0 \; \partial w_0/\partial x\}^T. \tag{34}$$

$$\{E^{(i)}\} = \sum_{j=1}^{2} [J]^{-1} [B_\phi^{(i)}]_j \{\bar{\phi}\}_j,$$
$$\text{where } \{\bar{\phi}\} = \{\bar{\phi}_{1t} \; \bar{\phi}_{1b} \; \bar{\phi}_{1c}\}^T. \tag{35}$$

In equations (34) and (35), [J] is a Jacobian matrix and B_u, B_ϕ are the shape function derivative matrices in local coordinates. Using these equations together with the material constitutive equations, a stationary value of the energy equation can be sought, which results in:

$$\sum_{i=t,b,c} ([M_{uu}^{(i)}]\{\ddot{\bar{u}}\} + [K_{uu}^{(i)}]\{\bar{u}\} + [K_{u\phi}^{(i)}]\{\bar{\phi}\}) - \{f_m^{(i)}\} = 0$$
$$\sum_{i=t,b,c} ([K_{\phi u}^{(i)}]\{\bar{u}\} - [K_{\phi\phi}^{(i)}]\{\bar{\phi}\}) - \{f_{el}^{(i)}\} = 0 \tag{36}$$

where

$$[M_{uu}^{(i)}] = \int [N_u^{(i)}]^T [\bar{\rho}^{(i)}][N_u^{(i)}]|J| d\xi \;\; ; \;\; [K_{uu}^{(i)}] = \int [B_u^{(i)}]^T [\bar{C}][B_u^{(i)}]|J| d\xi \; ;$$

$$[K_{u\phi}^{(i)}] = \int [B_u^{(i)}]^T [\bar{d}^{(i)}][B_\phi^{(i)}]|J| d\xi \;\; ; \;\; [K_{\phi\phi}^{(i)}] = \int [B_\phi^{(i)}]^T [\bar{\kappa}^{(i)}][B_\phi^{(i)}]|J| d\xi \; ;$$

$$\{f_{el}^{(i)}\} = \frac{\kappa_{33}^{(i)} b}{h_k^{(i)} - h_{k-1}^{(i)}} \int [N_\phi^{(i)}]^T \varphi^{(i)} |J| d\xi \;\; ; \quad \left[K_{\phi u}^{(i)}\right] = [K_{u\phi}^{(i)}]^T.$$

Here φ is a specified voltage, which can be imposed as electric boundary condition while solving actuator equation and $[\bar{c}]$, $[\bar{d}]$, $[\bar{\kappa}]$, are the material constitutive matrices.

They are defined as

$$[\bar{c}] = \begin{bmatrix} A & B & 0 \\ B & D & 0 \\ 0 & 0 & A_s \end{bmatrix}, [\bar{d}] = \begin{bmatrix} A_{PE} \\ B_{PE} \\ A_{PE}^{s} \end{bmatrix}, [\bar{\kappa}] = \begin{bmatrix} A_{DE} \\ D_{DE} \\ A_{DE}^{S} \end{bmatrix}$$

and $\bar{\rho}$ is the mass property matrix,

$$[\bar{\rho}]^{(i)} = \begin{bmatrix} I_n & 0 & 0 & 0 \\ 0 & I_r & 0 & 0 \\ 0 & 0 & I_n & 0 \\ 0 & 0 & 0 & I_r \end{bmatrix}, \quad (I_n, I_r) = \sum_{k=1}^{n} \int_{z_{k-1}}^{z_k} \rho_k (1, z^2)\, dz. \tag{37}$$

3. Bending and Free Vibration Behaviour of Smart Sandwich Beams

The developed beam element has been used to model and analyse the coupled behaviour of piezoelectric sandwich beam structures. For the shear actuation, an axially polarized PZT-5H core is sandwiched between the aluminium layers and for the extension actuation model, two transversely polarized PZT-5H patches are surface bonded onto an aluminium substrate. The material data used in the analysis is given in table 1. The dimension of the sandwich beams are shown in figure 6. To evaluate the piezoelastically developed displacement, $\pm 10\ V$ is applied to the surface bonded actuators, and $-20V$ is applied to the shear actuator (piezoelectric core). The transverse deflections are estimated along the length of beams with actuators size L_a = 100 mm. The estimated displacements are depicted in figure 7 along with those reported by Benjeddou *et al.* (1999) and ABAQUS analysis.

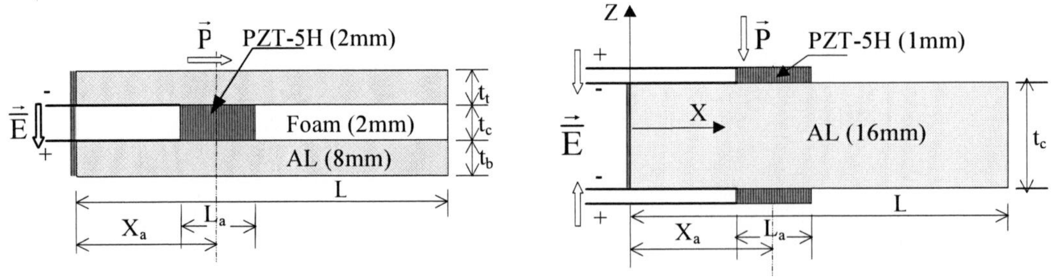

Figure 6. Smart sandwich beams with EAM and SAM

The piezoelectric analysis is done in ABAQUS using the solid element (C3D8E), which has the electro-mechanical coupling capability. A good correlation is observed for both EAM and SAM modelling, which shows the potential of the present element to analyse the behaviour of hybrid actuation. In the second part, the piezoelectric sandwich beams are studied to see how the sensors respond to the mechanical deformations. A tip shear load is applied to obtain the same deflection (w_{max}) as in the case of converse piezoelectric study.

The beams are modelled with 5 elements and each element is assumed as electrode sensor patches.

Table 1. Material Data of the Aluminium Sandwich Beams

Properties	Aluminum	Foam	PZT-5H
E (GPa)	**70.3**	35.3	-
ν	0.343	0.383	-
C_{11}, C_{22}, C_{33} (GPa)	-	-	126.0
C_{44}, C_{55}, C_{66} (GPa)	-	-	23.3
C_{12}, C_{21} (GPa)	-	-	79.5
C_{31}, C_{32} (GPa)	-	-	84.1
d_{31} (C/m^2)	-	-	-6.5
d_{33} (C/m^2)	-	-	23.3
d_{15} (C/m^2)	-	-	17.0
κ_{11}, κ_{22} (F/m)	-	-	1.503×10^{-8}
κ_{33} (F/m)	-	-	1.3×10^{-8}
ρ (Kg/m^3)	2690.0	32.0	7730.0
Geometric Properties: Length = 0.1m, Width = 0.01m			

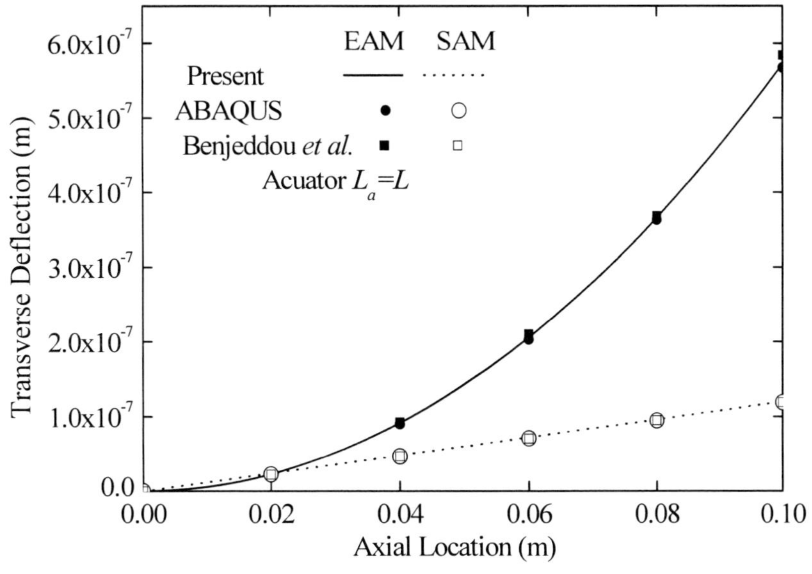

Figure 7. Bending behaviour of C-F sandwich beams

It is observed that while all the 5 shear sensors developed the same voltage due to a constant cross sectional rotation along the beam's length, the extension sensors predicted varying trend as expected (see figure 8). In a distributed parameter system control, the actuator and sensor locations play an important role. However, the direct piezoelectric coupling study shows that the sensor location is not a significant parameter in the shear-actuated cantilever beams.

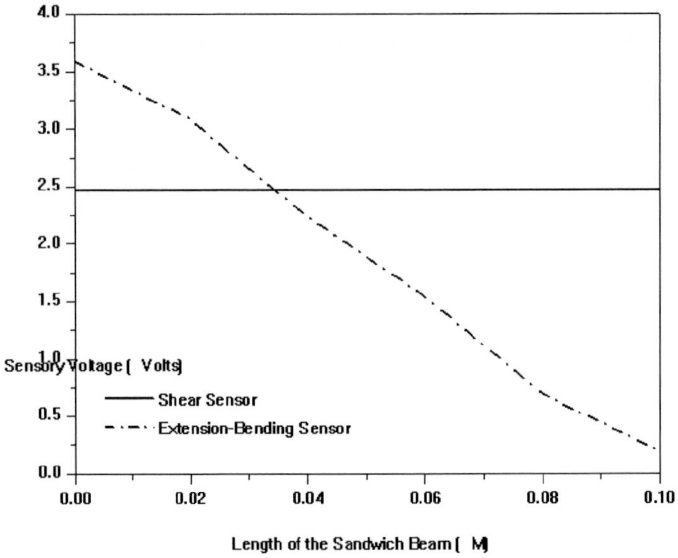

Figure 8. Sensor characteristics of C-F sandwich beams

The developed beam element is employed to carry out free vibration analysis. The same beam configurations are taken as model problems with clamped-free and clamped-clamped boundary conditions. Free vibration frequencies are obtained for beams, having full length actuators as in the static analysis. The obtained natural frequencies are compared with those predicted by ABAQUS (C3D8 element) in tables 2 and 3.

Table 2. Natural Frequencies of PZT_{EAM}/AL/PZT_{EAM} Beams

Mode No	Clamped-Free Frequency (Hz)		Clamped-Clamped Frequency (Hz)	
	Present	ABAQUS	Present	ABAQUS
1	1389.6	1398.1	7529.8	7547.6
2	7604.2	7601.3	17506.4	17434.0
3	11873.2	11936.0	23749.3	23976.0
4	18328.8	18237.0	29423.8	29190.0
5	30742.3	30462.0	42272.0	41805.0

Table 3. Natural Frequencies of AL/PZT_{SAM}/AL Beams

Mode No	Clamped-Free Frequency (Hz)		Clamped-Clamped Frequency (Hz)	
	Present	ABAQUS	Present	ABAQUS
1	1322.4	1333.1	7255.0	7335.9
2	7373.4	7406.7	17101.3	17217.0
3	11873.2	11933.0	23749.3	23955.0
4	18023.3	18052.0	28998.3	29105.0
5	30563.2	30533.0	41933.8	41988.0

3.1 Bending Behaviour of Hybrid Actuated Sandwich Beams

In the second analysis, a hybrid actuated cantilevered sandwich beam is constructed with collocated shear and extension actuators and the beam configuration is presented in figure 9. The two types of actuators are activated individually, as well as simultaneously to see the bending behaviour of sandwich beam. It is seen that a collective actuation effect made the beam to have more active stiffening effect in the clamped-free case (figure 10). The full-length extension type actuators can generate only blocking forces, if the edges are constrained, therefore the segmented piezoelectric wafers are considered to analyse the deflection behaviour of hybrid actuated sandwich beams for various boundary conditions. Figure 11 shows one such analysis as an example with clamped-clamped conditions.

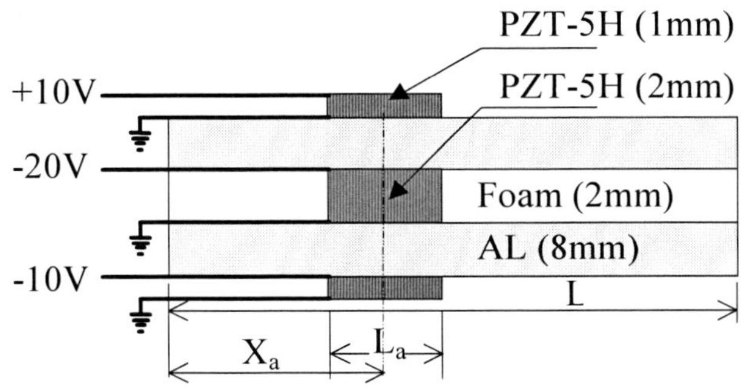

Figure 9. Hybrid actuated sandwich beam

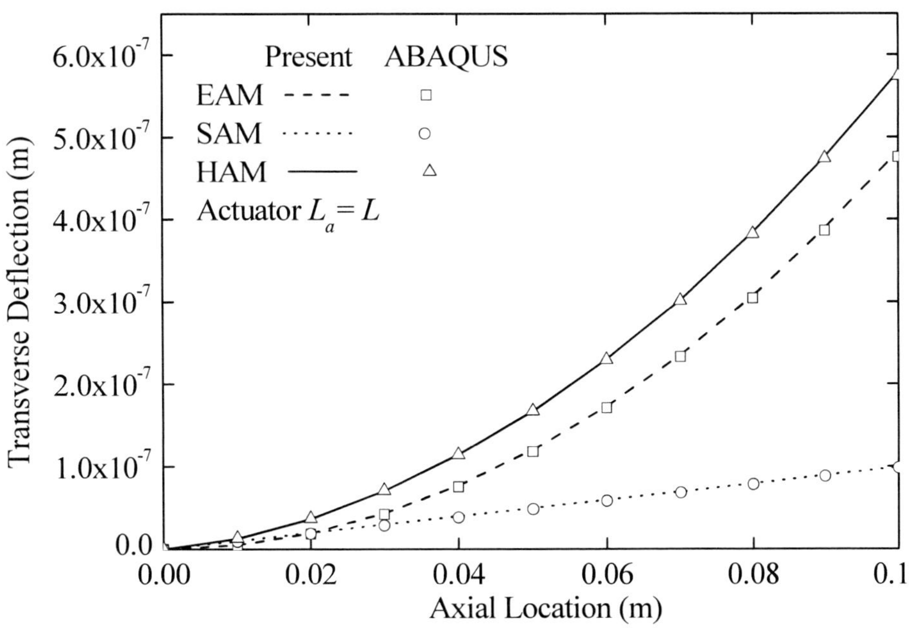

Figure 10. Hybrid actuated sandwich C-F beam

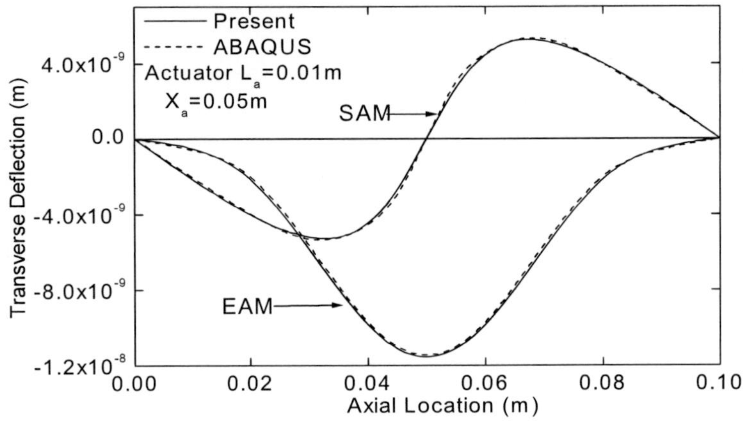

Figure 11. Hybrid actuated C-C sandwich beam with segmented actuators

3.1.1 With Collocated Actuators

The segmented collocated actuators (EAM and SAM) are placed at different locations along the length of beam to see their efficiency in deflecting the sandwich beams. The Clamped-Free (C-F), Clamped-Clamped (C-C), Clamped-Hinged (C-H), and Hinged-Hinged (H-H) boundary effects are considered in the analysis. The actuators are activated individually and simultaneously by applying voltages ±10 to EAM and -20 to SAM. The electro-mechanically induced transverse displacements are estimated and the results are presented in figures 12 to 14.

It is seen that the boundary effects play a significant role on the piezoelectric actuation behaviour. The HAM captures the combined actuation effect (stiffening) of EAM and SAM in the case C-F due to the same deflection patterns. However, this trend is not followed in other cases. Interesting deflection patterns are observed with different boundary conditions; which may be exploited for mode shape control applications. Further, the combination of EAM and SAM also can control more than one elastic mode by appropriately activating them (refer to deflection pattern). In general both actuators are efficient at locations, where the strains seem to be significant.

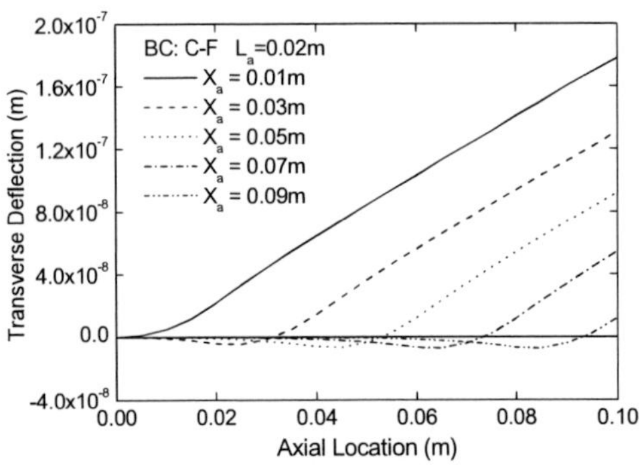

12 (a)

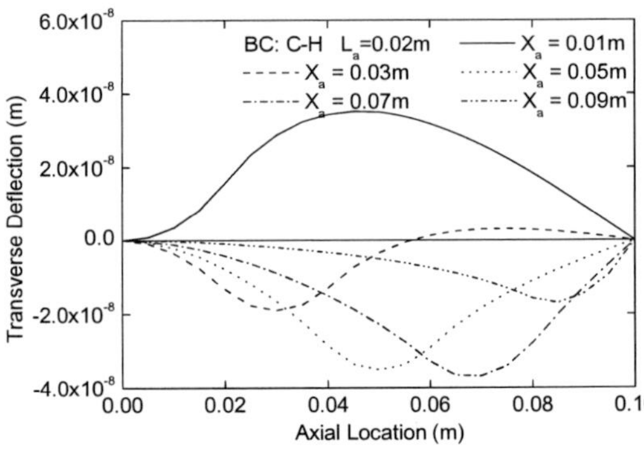

12 (b)

Figure 12. (Continued on next page.)

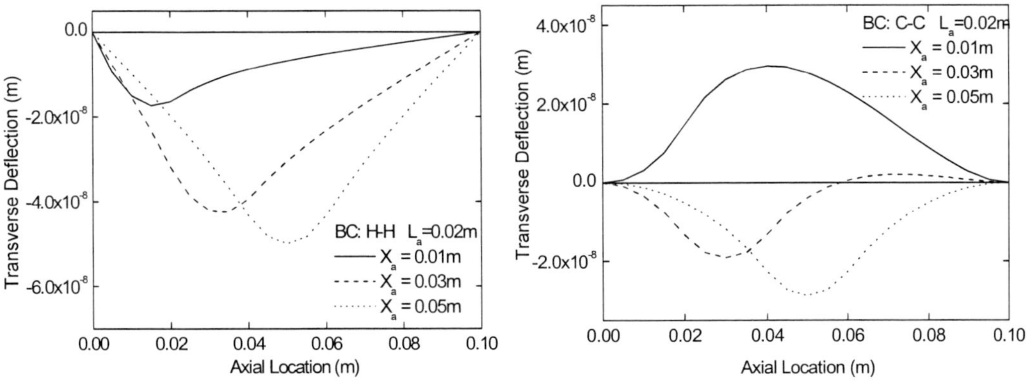

12 (c) 12 (d)

Figure 12. Bending behaviour of sandwich beams with collocated EAM/SAM -EAM active

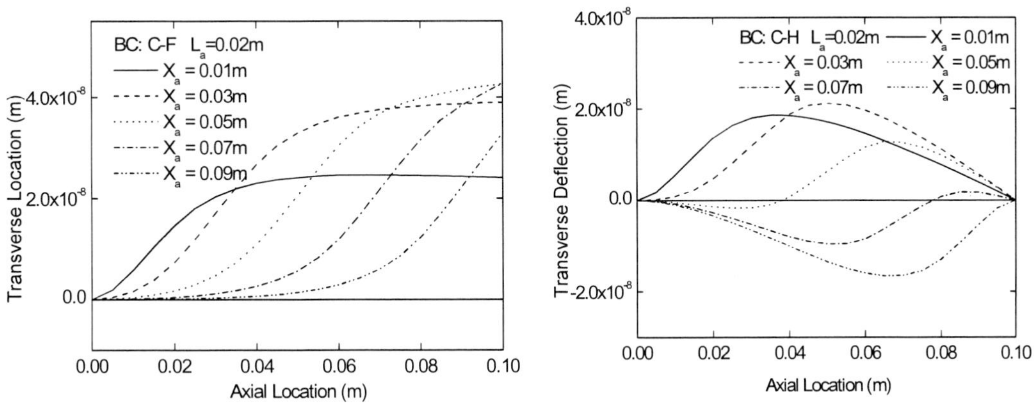

13 (a) 13 (b)

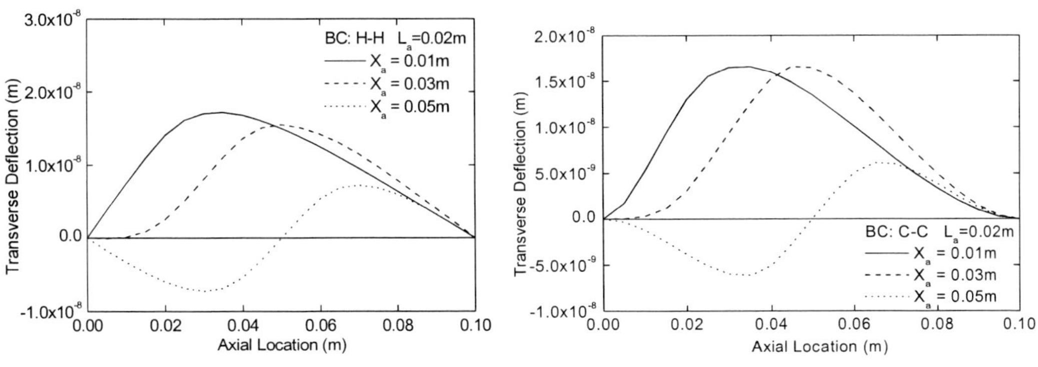

13 (c) 13 (d)

Figure 13. Bending behaviour of sandwich beams with collocated EAM/SAM -SAM active

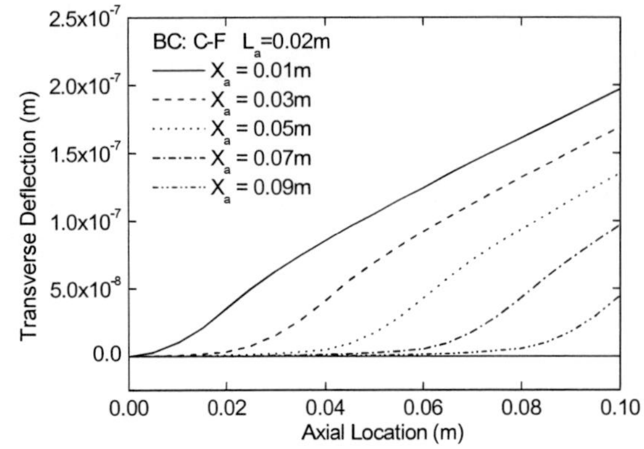

14 (a)

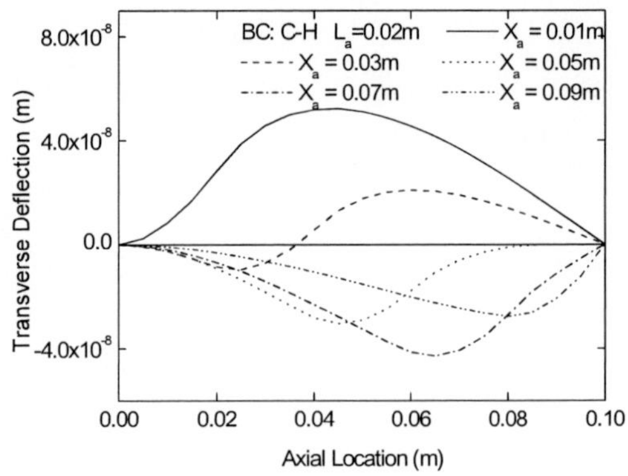

14 (b)

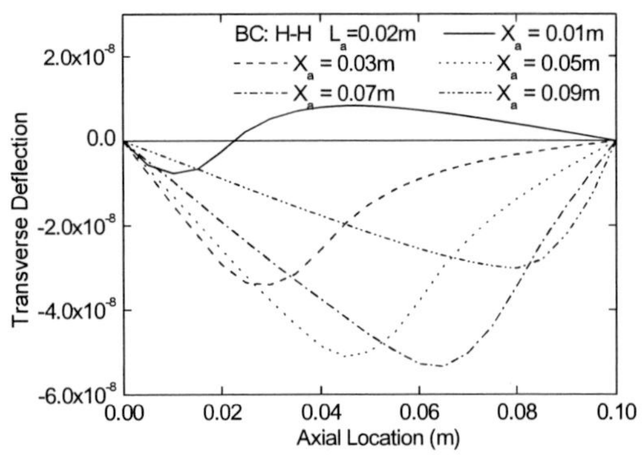

14 (c)

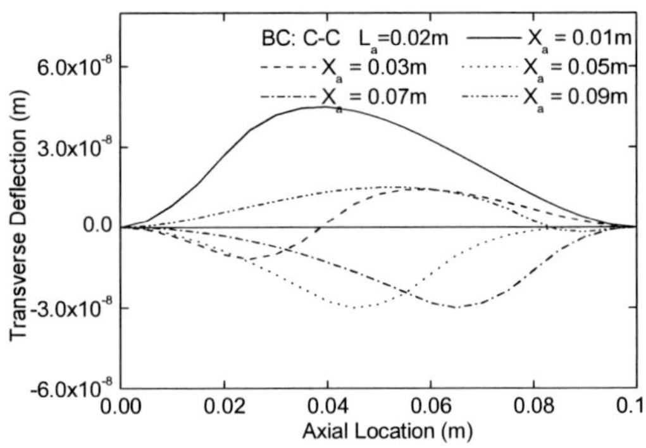

14 (d)

Figure 14. Bending behaviour of sandwich beams with collocated EAM/SAM- both active

3.1.2 With Non-collocated Actuators

The active stiffening effect with hybrid actuation using non-collocated EAM and SAM has been studied (figure 15). Unlike in collocated EAM/SAM, the non-collocated configuration has generated more deflection not only in the C-F case, but in the C-C and C-H cases also (figure 16). However in the H-H case, the SAM influence appears to be predominant. Therefore to achieve a cumulative actuation effort of EAM/SAM, they can be placed in a non-collocated fashion for better control action.

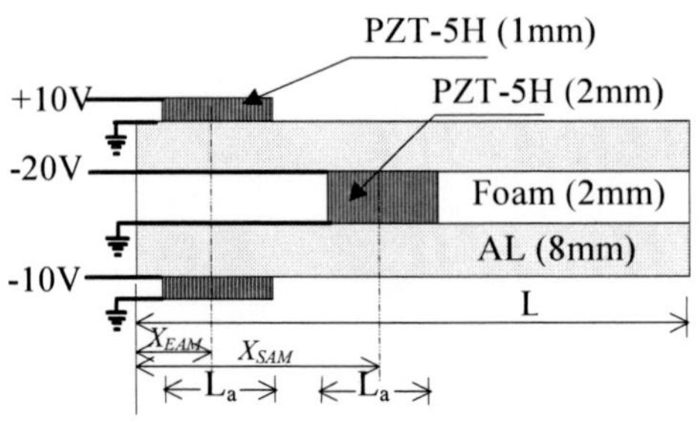

Figure 15. Hybrid actuated sandwich beams with non-collocated actuators

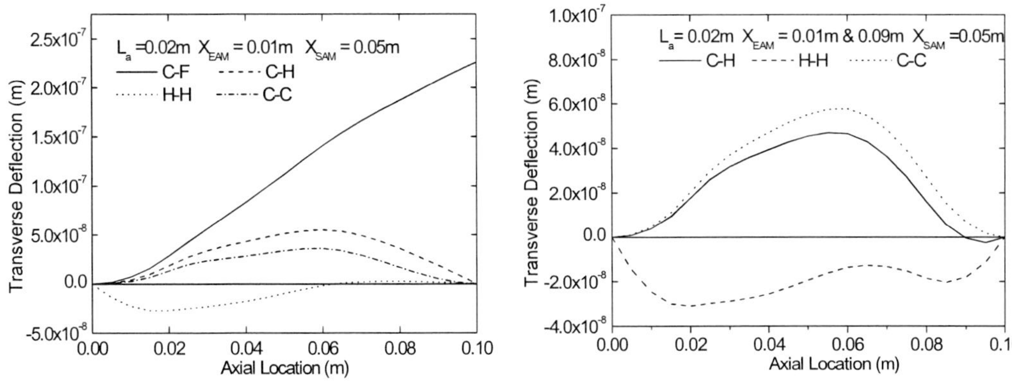

16 (a) 16 (b)

Figure 16. Bending behaviour of sandwich beams with non-collocated EAM/SAM -both active

4. DISTRIBUTED ACTIVE VIBRATION CONTROL

The flexible structural system is a distributed parameter system that has a large number of degrees of freedom (Eigen modes). Also the dynamic behaviour of a structural system is a function of spatial and time variables. To control such a flexible system, it is necessary to use a number of actuators and sensors, spatially distributed so that they are sensitive to spatially distributed structural behaviour. Numerical models (Finite Element Method: FEM) are mostly employed to estimate the structural parameters, namely, stiffness, mass and damping (Rayleigh's proportional damping) for the prediction of system response. The finite element model represents a structure in the form of a multi-degree of freedom system. Further it generates a set of second order simultaneous equations (Equation 36) that will define the dynamic equilibrium of the structural system. In order to design a viable control law for the control of elastic modes of the structure, the discrete numerical system model (FEM) must be condensed or reduced to a reasonable size.

Control design methods normally assume that a full state vector is available. However, in reality to construct the system state completely, a large number of sensors are needed; otherwise an estimator must be designed to build the system state. In addition, when a large structural system is reduced (condensed) for use in control design, the un-modelled modes may some time destabilise the closed loop system. Therefore, in order to tolerate the model errors, it is desirable to optimise the robustness of controller i.e., tolerance to model errors and disturbance. The output feedback control is designed by taking a few selected modes only. Due to incomplete modelling of the structural system, when such a control is applied on real time structure experimentally, the actual damping ratio and free vibration frequency of each mode may not be the same as in the simulated model. Therefore, additional sensors and actuators are needed to increase the degrees of freedom of the control system to provide the necessary stability. Otherwise, a state observer or estimator is required to get the additional system states.

Independent Modal Space Control (IMSC) is a modal filter approach to transfer the responses of multi-degrees of freedom system into independent modal coordinates (single input and single output). Using the IMSC concept, the feedback control force is made independent (as a function of modal coordinates) completely to decouple the structural modes in a feedback control environment. Since the complete control energy is driven to concentrate on a particular mode, spillover problem (exciting uncontrolled modes statically: residual energy) may be avoided. A structural system is controllable, if the installed actuators excite all the possible modes to be controlled. Similarly, the system is observable, if the installed sensors detect the motions of all the possible modes to be controlled.

The feedback control can compensate the external disturbances only in a limited frequency band that is known as control bandwidth. Thus, the control bandwidth is normally limited by the accuracy of the model. There is always some destabilisation of the flexible modes outside the control bandwidth i.e., the disturbance is actually amplified by the control system. Therefore, the control bandwidth must be sufficient enough to ensure better closed loop performance with the designed actuators and sensors configuration. Also, out of bandwidth correction (DC compliance) may be included to avoid influence of higher uncontrolled modes.

4.1 Active Control Strategies

The multifunctional piezoelectric lamina as an actuator (fully active actuator) is able to supply mechanical power to the structural system to modify its dynamic response. The actuator lamina can be used to generate a secondary vibration response, which can reduce the total system response by destructive interference with the original response of the system due to primary source of vibration. Feedback controller requires actually no knowledge of the incoming disturbance to the system and acts to change the system response by changing the system resonance and damping. In this control, both the primary source and the secondary actuators (piezoelectric) influence the sensor signal.

4.1.1 Displacement Feedback (Proportional Control)

The sensor signal (charge) is used directly in a closed loop feedback system, which is proportional to the strain developed (due to deflection) in a piezoelectric lamina. The feedback voltage is obtained by amplifying the sensor signal by a feedback gain.

The displacement gain is $F_d = \left(\dfrac{\phi_a}{\phi_s}\right) = \dfrac{(amplitude\ of\ feedback\ voltage)}{(amplitude\ of\ sensor\ voltage)}.$

In the displacement control, the feedback gain represents a true amplification ratio. Here the vibration suppression is achieved by generating actuation signal 180^0 out of phase with respect to disturbance signal. Actually, the anti-phase actuation signal modifies the strain energy and actively stiffens the vibrating elastic system (see figure 17 a).

4.1.2 Velocity Feedback (Derivative Control)

The sensor signal is differentiated (charge rate), conditioned and then fedback into actuator lamina. The signal used in the velocity feedback is actually strain/unit-time (i.e., strain rate). Since the elastic strain field is expressed in terms of displacements, the time derivative of the displacement is velocity. The electric charge (q) denotes the system displacement, and therefore the charge rate (dq/dt = i: sensor current) is the system velocity.

$$\text{The velocity gain is } F_v = \left(\frac{\phi_a}{\dot{\phi}_s}\right) = \frac{(amplitude\ of\ feedback\ voltage)}{(amplitude\ of\ sensor\ voltage\ rate)}.$$

In the velocity feedback control, the actuation signal is in-phase with the system velocity (90^0 phase lead with system displacement) so as to introduce a dissipative force (active damping) into the system (refer to figure 17 b).

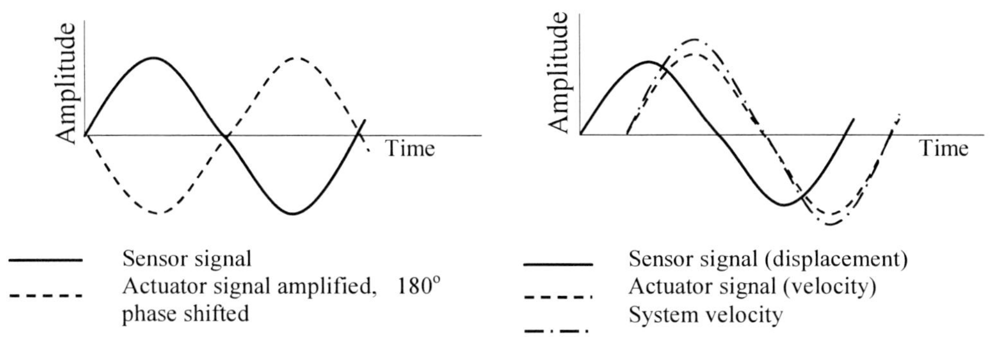

(a) Displacement feedback

(b) Velocity feedback

Figure 17. Active control concepts using feedback controller

4.2 State-Space Formulation

4.2.1 State-Space Models in Physical Domain

The dynamic equation of a smart structural system (Linear Time Invariant) is presented with feedback control as,

$$M_{uu}\,\ddot{\bar{u}} + C_{uu}\,\dot{\bar{u}} + K_{uu}\,\bar{u} = b_u f_m + b_0\,\phi_a \tag{38}$$

where

M_{uu} (n x n) is the mass matrix,

C_{uu} (n x n) is the damping matrix,

K_{uu} (n x n) is the stiffness matrix,

$b_0\,(n \times a) = K_{u\phi_a}$, is an actuator influence matrix with ‘a’ as number of actuators,

b_u is an (n x 1) matrix of influence functions associated with mechanical force vector f_m,

and

$\bar{u}$ is the system displacement, $\ddot{\bar{u}}$ is the system acceleration,

$\dot{\bar{u}}$ is the system velocity, $\phi_a\,(a \times 1)$ is the control input vector.

The block diagram (figure 18) shows the signal flow of an output feedback control system.

The sensor output can be related to system states (displacement, velocity) as follows:

$$\phi_s = c_d \bar{u},$$
$$\dot{\phi}_s = c_v \dot{\bar{u}},$$
$$\text{where } \phi_s, \dot{\phi}_s \text{ are the voltage and voltage rate, respectively,} \tag{39}$$
$$\text{and}$$
$$c_d, c_v = K_{\phi\phi s}^{-1} K_{\phi us},$$

$K_{\phi\phi s}$ is the capacitance of piezoelectric sensor lamina,

$K_{\phi us}$ is the charge sensitivity matrix of sensor lamina.

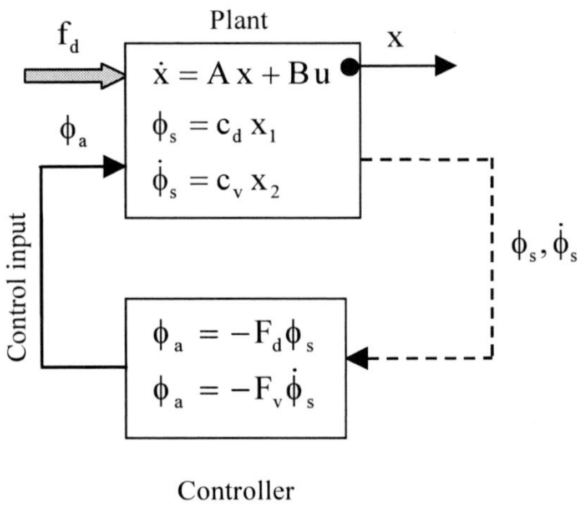

Figure 18. Block diagram of control system (Physical domain)

It may be observed that the sensor patch senses the charge induced due to the nodal displacements associated with it and outputs as sensor voltage. The velocity information of the sensor patch further can be obtained by differentiating the charge signal using a differentiator (analog or digital). This can be realised by properly selecting the circuit capacitance (C_f) and resistance (R_f) for the charge amplifier such that the time period $\tau = C_f R_f$. The equivalent circuit makes the sensor output independent of crystal capacitance and resistance.

A state vector x is defined with two new states (x_1: displacement, x_2: velocity) as follows:

$$x = \begin{Bmatrix} \bar{u} \\ \dot{\bar{u}} \end{Bmatrix} = \begin{Bmatrix} x_1 \\ x_2 \end{Bmatrix}; \; \ddot{\bar{u}} = \dot{x}_2 \text{ and } \dot{x}_1 = \dot{\bar{u}} = x_2. \tag{40}$$

The second order dynamic equation (38) is reduced to two first order equations using equation (40) as,

$$\begin{aligned} \dot{x}_1 &= x_2, \\ \dot{x}_2 &= -M_{uu}^{-1}C_{uu}x_2 - M_{uu}^{-1}K_{uu}x_1 + M_{uu}^{-1}b_0\phi_a + M_{uu}^{-1}b_u f_m. \end{aligned} \tag{41}$$

The state-space equation (41) can be expressed in matrix form as,

$$\begin{aligned} \dot{x}(t) &= \begin{bmatrix} 0 & I \\ -M_{uu}^{-1}K_{uu} & -M_{uu}^{-1}C_{uu} \end{bmatrix} x(t) + \begin{bmatrix} 0 & 0 \\ M_{uu}^{-1}(b_u) & M_{uu}^{-1}b_0 \end{bmatrix} u(t), \\ x(t) &= \begin{bmatrix} x_1(t) \\ x_2(t) \end{bmatrix}, \; u(t) = \begin{bmatrix} f_d(t) \\ \phi_a(t) \end{bmatrix}, \end{aligned} \tag{42}$$

where f_d is the disturbance vector due to mechanical forces.

The output vector is given by,

$$y(t) = \{\phi_s \quad \dot{\phi}_s\}^T = [c_d \quad c_v]\, x(t) \tag{43}$$

The equations (42) and (43) can be written in standard state variable form as follows:

$$\begin{aligned} \dot{x}(t) &= A\,x(t) + B\,u(t) \\ y(t) &= \tilde{c}\,x(t) \end{aligned} \tag{44}$$

4.2.2 State-Space Models in Modal Domain

The dynamic equation (38) is reduced to modal form using the following transformation,

$$\bar{u} = \Phi\, r \tag{45}$$

where $\Phi(n \times n)$ is a modal matrix containing the Eigen vectors of the open-loop system, normalised with respect to mass and r is the modal coordinates.

The equation of motion is decoupled using the modal matrix and is presented here in modal form with feedback control as,

$$\ddot{r}_i + \bar{C}\,\dot{r}_i + \bar{K}\,r_i = b_{0m}\,\phi_{ami} + b_m f_{dmi},\; i = 1, \ldots, n \text{ modes}, \tag{46}$$

where

$$\begin{aligned} &\bar{M} = \Phi^T M_{uu} \Phi = \text{Identity matrix}, \; \bar{C} = \Phi^T C_{uu} \Phi = \text{diag}(2\zeta_i \omega_i), \\ &\bar{K} = \Phi^T K_{uu}\, \Phi = \text{diag}(\omega_i^2) \text{ and } b_{0m} = \Phi^T b_0,\; b_m = \Phi^T b_u, \end{aligned}$$

and f_{dmi} is the generalised load vector of mechanical disturbance.

The output vector in modal form is given by,

$$y_m = \{\phi_{sm} \quad \dot{\phi}_{sm}\}^T = [c_{dm} \quad c_{vm}]\begin{Bmatrix} r \\ \dot{r} \end{Bmatrix},$$
$$c_{dm} = \Phi^T c_d, \; c_{vm} = \Phi^T c_v. \tag{47}$$

A state vector x_m containing modal displacement η_1 and modal velocity η_2 is defined as follows:

$$x_m = \begin{Bmatrix} r \\ \dot{r} \end{Bmatrix} = \begin{Bmatrix} \eta_1 \\ \eta_2 \end{Bmatrix}; \; \ddot{r} = \dot{\eta}_2 \text{ and } \dot{\eta}_1 = \dot{r} = \eta_2. \tag{48}$$

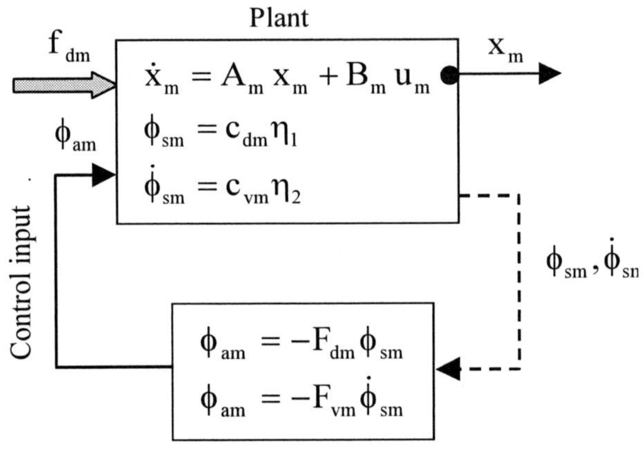

Figure 19. Shows the signal flow in modal output feedback control system.

The modal state-space equation can be expressed in matrix form as,

$$\dot{x}_{mi}(t) = \begin{bmatrix} 0 & 1 \\ -\omega_i^2 & -2\zeta_i\omega_i \end{bmatrix} x_{mi}(t) + \begin{bmatrix} 0 & 0 \\ b_m(i) & b_{0m}(i,a) \end{bmatrix} u_{mi}(t),$$
$$x_{mi}(t) = \begin{bmatrix} \eta_1(t) \\ \eta_2(t) \end{bmatrix}, \; u_{mi}(t) = \begin{bmatrix} f_{dmi}(t) \\ \phi_{ami}(t) \end{bmatrix}. \tag{49}$$

where f_{dmi} is the disturbance vector due to mechanical force.

The modal output vector is given by,

$$y_{mi}(t) = \{\phi_{smi} \quad \dot{\phi}_{smi}\}^T = [c_{dm}(s,i) \quad c_{vm}(s,i)]\, x_m(t) \tag{50}$$

The equations (49) and (50) can be written in standard state variable form as follows:

$$x_{mi}(t) = A_{mi}\, x_{mi}(t) + B_{mi}\, u_{mi}(t),$$
$$y_{mi}(t) = \tilde{c}_{mi}\; x_{mi}(t), \tag{51}$$

where the matrices A_{mi} is of size (2 x 2), B_{mi} is of size (2 x a), $\tilde{c}_{mi}$ is of size (s x 2) for the ith mode, and s is the number of sensors.

4.3 Modal Control

A distributed parameter system can be idealised as many single degree of freedom systems (modes) in modal domain. A modal controller can be further designed to stabilise and control these individual modes independently.

4.3.1 State Feedback

The following linear control law is adopted in the vibration control:

$$\phi_{ami} = -F_{dmi}\eta_1$$
$$\phi_{ami} = -F_{vmi}\eta_2 \tag{52}$$

where F_{dmi}, F_{vmi}, are the optimal modal gains. One may note that the feedback voltage is proportional to modal displacement and modal velocity.

The optimal modal feedback gain is estimated such that the feedback control law (equation 52) minimises the modal performance index,

$$J_i = \int_0^{\infty}(x_{mi}^T Q_i x_{mi}) + (\phi_{ami}^T R_i \phi_{ami})dt \rightarrow \min, \tag{53}$$

where Q_i is the state penalty matrix and R_i is the control penalty matrix.

The matrices Q and R will be selected such that they are non-negative and positive definite. This can be accomplished by picking either a positive real value or zero as the diagonal elements for Q and R. In addition, $|R| \neq 0$; otherwise large components may appear in the control gains and that will saturate the actuator devices.

The modal steady state solution S_i is next obtained by solving the following Arithmetic Riccati Equation (ARE):

$$0 = S_i A_{mi} + A_{mi}^T S_i - S_i B_{mi} R_i^{-1} B_{mi}^T S_i + Q_i \ . \tag{54}$$

The optimum modal gain (F_{mi}: F_{dmi}, F_{vmi}) in steady state can be expressed as,

$$F_{mi} = R_i^{-1} B_{mi}^T S_i \tag{55}$$

Equations (54) and (55) can be solved using MATLAB function (LQR).

4.3.2 Output Feedback

The modal sensor output (voltage, voltage rate) is related to the modal state vector x_m in equation 48. The linear output feedback control law adopted in the present investigation is,

$$\begin{aligned} \phi_{ai} &= -F_{dmi}\,\phi_{smi} \\ \phi_{ai} &= -F_{vmi}\,\dot{\phi}_{smi} \end{aligned} \qquad (56)$$

The optimal modal output feedback gain is determined for individual mode using the modal steady solution such that the feedback control law (equation 56) minimizes the following performance index,

$$J = \int_0^{\infty} (x_{mi}^T Q_i\, x_{mi}) + (\phi_{ami}^T R_i\, \phi_{ami})\, dt \rightarrow \min, \qquad (57)$$

where $Q_i = [c_{mi}]^T [c_{mi}]$ is the state penalty matrix and R_i is the control penalty matrix of ith mode.

The actuator voltages of the ith mode are proportional to the ith mode modal displacement and modal velocity and are defined as

$$\begin{aligned} \phi_{ami} &= -F_{dmi}\, c_{dm}(s,i)\,\eta_1, \\ \phi_{ami} &= -F_{vmi}\, c_{vm}(s,i)\,\eta_2, \end{aligned} \qquad (58)$$

where F_{dmi}, F_{vmi} are the diagonal gains corresponding to displacement and velocity, respectively.

4.4 Condensation of System Matrices

Control system design will be computationally expensive and practically difficult for flexible modes control, as the system size is very large. A model reduction scheme can be adopted to reduce the degrees of freedoms (dof's) by representing the condensed dof's in terms of master dof's that are to be retained in the system matrices. There are several reduction schemes reported in the literature, such as, static and dynamic condensation techniques and System Equivalent Reduction and Expansion Process (SEREP). A simplified procedure is outlined in this section to implement the condensation schemes for the structural control problem [Raja *et al.* 2003].

Let $\overline{u}_r$ are the master dof's that can be measured and the remaining dof's $\overline{u}_c$ are considered as slave dof's. Thus, the total dof's of the system are separated as,

$$\{\overline{u}(t)\}^T = [\{\overline{u}_r(t) \;\; \{\overline{u}_c(t)\}]. \qquad (59)$$

Accordingly, the mass, stiffness, damping and modal matrices also get partitioned as,

$$M_{uu} = \begin{bmatrix} M_{rr} & M_{rc} \\ M_{cr} & M_{cc} \end{bmatrix}, K_{uu} = \begin{bmatrix} K_{rr} & K_{rc} \\ K_{cr} & K_{cc} \end{bmatrix}, \ C_{uu} = \begin{bmatrix} C_{rr} & C_{rc} \\ C_{cr} & C_{cc} \end{bmatrix}, \ [\Phi] = \begin{bmatrix} \Phi_r \\ \Phi_c \end{bmatrix}. \tag{60}$$

The important task is to introduce a transformation matrix $\tilde{W}$ of size (n x n_r) that relates the (nx1) total dof's with (n_rx1) master dof's.

$$\{\bar{u}(t)\} = [\tilde{W}]\{\bar{u}_r(t)\} \tag{61}$$

The transformation matrix can be defined, if one adopts static condensation technique as follows:

$$\left[\tilde{W}\right] = \begin{bmatrix} I \\ -K_{cc}^{-1} K_{cr} \end{bmatrix} \tag{62}$$

Similarly, if one adopts the dynamic condensation technique for system reduction then

$$\left[\tilde{W}\right] = \begin{bmatrix} I \\ -\bar{D}_{cc}^{-1} \bar{D}_{cr} \end{bmatrix}, \tag{63}$$

where $\bar{D}$ is the dynamic stiffness matrix given by $\bar{D} = \left[K_{uu} - \omega^2 M_{uu}\right]$ and ω is the frequency at which the reduction is made.

Finally, if one adopts the SEREP reduction technique, the transformation matrix reads

$$[\tilde{W}] = \begin{pmatrix} \Phi_r \\ \Phi_c \end{pmatrix} [\Phi_r]^+ \tag{64}$$

where $\left[\Phi_r\right]^+$ denotes pseudo-inverse of $\left[\Phi_r\right]$ and is given by

$$\left[\Phi_r\right]^+ = \left[\Phi_r^T \Phi_r\right]^{-1} \Phi_r^T .$$

The displacement vector is partitioned into master and slave dof's first and then put in the following form using modal matrix,

$$\begin{aligned} \{\bar{u}_r(t)\} &= [\Phi_r]\{r(t)\}, \\ \{\bar{u}_c(t)\} &= [\Phi_c]\{r(t)\}. \end{aligned} \tag{65}$$

This leads to an expression for the generalized coordinate vector $\{r(t)\} = [\Phi_r]^+ \{\bar{u}_r\}$.

Using the transformation matrix $\tilde{W}$, the reduced mass, stiffness, damping and piezoelectric coupling matrices for the beam can be obtained as,

$$[M_r] = [\tilde{W}]^T [M_{uu}] [\tilde{W}],\ [K_r] = [\tilde{W}]^T [K_{uu}] [\tilde{W}],$$
$$[C_r] = [\tilde{W}]^T [C_{uu}] [\tilde{W}],\ [K_{u\phi r}] = [\tilde{W}]^T [K_{u\phi}]. \tag{66}$$

Note that the reduced structural matrices, being function of $\tilde{W}$, are non-linearly dependent on structural parameters.

4.5 Pole Placement and Mode Shape Control

The objectives of vibration control are frequency control or pole placement, i.e., alter the Eigenvalue by maintaining the system stability, and amplitude/mode shape control. Thus, the controller design has to be such that the amplitude of corresponding elements of closed loop system Eigenvectors are to be reduced significantly as compared to open loop ones. To achieve these targets, a set of desired Eigenvalues Λ^d and their associated Eigenvectors $\bar{\Psi}^d$ will be chosen in such a way that their norms remain the same as those of open loop system but with a set of desired amplitude constraints.

The system response $\bar{u}(t)$ can be represented as a linear combination of the set of desired Eigenvectors $\bar{\Psi}^d$. Using equations (40) and (44), the closed loop dynamic equation with feedback control can be derived as,

$$[(A + BF\tilde{c}) - \Lambda_i^d I]\{\bar{\Psi}_i^d\} = 0, \quad i=1,\ldots, p \text{ modes}. \tag{67}$$

Now, the task is to determine the feedback gain matrix [F], such that the Eigenvalues/Eigenvectors of the closed loop system are as close as possible to the desired values.

In the problem of Pole Allocation and Eigenstructure Assignment, it is assumed that the desired closed loop Eigenvectors can be a linear combination of ‘p’ open loop Eigenvectors, if ‘p’ modes are to be controlled.

The open loop Eigenvectors can be arranged in a modal matrix form as,

$$[\Phi^o] = [\bar{\Psi}_1 \ \ \bar{\Psi}_2 \ \ \bar{\Psi}_3 \ \ldots \bar{\Psi}_p]. \tag{68}$$

Then the desired closed loop eigenvectors may be estimated as follows:

$$\{\bar{\Psi}_i^d\} = [\Gamma][\Phi_i^o], \quad i=1,\ldots, p \text{ modes}. \tag{69}$$

where $[\Gamma]$ is a matrix of appropriate size with 1 (specified) or 0 (non-specified) as elements.

The mode shape constraints can be applied on the specified elements to achieve a desired shape for a particular mode.

Let m_i be a modal control vector and is defined by,

$$m_i = F\tilde{c}\bar{\Psi}_i^d, \quad i=1,\dots, p \text{ modes.} \tag{70}$$

Equation (67) reads

$$\bar{\Psi}_i^d = [\Lambda_i^d I - A]^{-1} B m_i, \quad i=1,\dots, p \text{ modes.} \tag{71}$$

The implication of equation (71) is important because the ith Eigenvector corresponding to Λ_i^d must lie in the space spanned by the column of $[\Lambda_i^d I - A]^{-1} B$.

The vector m_i has a particular role in the controller design and is used as a design quantity in active shape control analysis.

From equation (71),

$$m_i = (L_i^T L_i)^{-1} L_i^T \bar{\Psi}_i^d, \text{ where } L_i = [\Lambda_i^d I - A]^{-1} B. \tag{72}$$

The achievable Eigenvector $\bar{\Psi}_i^a$ can be obtained using the modal control vector m_i as

$$\bar{\Psi}_i^a = L_i m_i, \quad i=1,\dots, p \text{ modes.} \tag{73}$$

The achievable eigenvectors of 'p' modes are then arranged in matrix form.

$$\bar{\Psi}^a = [\bar{\Psi}_1 \ \bar{\Psi}_2 \ \dots\dots \bar{\Psi}_p] \tag{74}$$

Finally the feedback gain is estimated, employing the achievable Eigenvector as follows:

$$F = (B^T B)^{-1} B^T (\bar{\Psi}^a \Lambda^d - A\bar{\Psi}^a)(\tilde{c}\bar{\Psi}^a)^{-1}. \tag{75}$$

4.6 Active Vibration Control and Modal Response Analysis

The efficiency of shear and extension actuators in controlling the bending vibration is studied on the smart composite sandwich beams using modal control analysis (LQR/IMSC). Figure 20 shows the configuration of the beams and the material data is given in table 4. The free vibration frequencies obtained from FE analysis are presented in table 5. For shear actuation, foam is introduced as a core along with PZT to simulate an equivalent sandwich model. It is observed that the higher mode frequencies of the shear-actuated beam are reduced significantly because of foam material. Therefore, for the purpose of comparison, only first three bending modes are considered in the vibration control analysis to evaluate the performance of the actuators. The proposed modal control scheme (refer to section 4.3) is applied to estimate the active modal damping introduced by shear as well as extension actuators in each mode.

The control gains of each mode are obtained taking same R_i for both actuation mechanisms, i.e., same control effort. This is done in order to qualitatively compare the actuator efficiency (extension & shear actuators) in vibration control application. Subsequently, a modal response study is carried out for the open-loop system with 1% viscous damping (proportional, $\zeta = 0.01$) using Newmark time integration approach and a comparison is made with the response of the closed-loop system. The sum of active damping and the 1% open loop damping is presented as the closed-loop system damping. The response analyses are carried out with a sinusoidal input disturbance force (1 N force amplitude) at resonant frequencies.

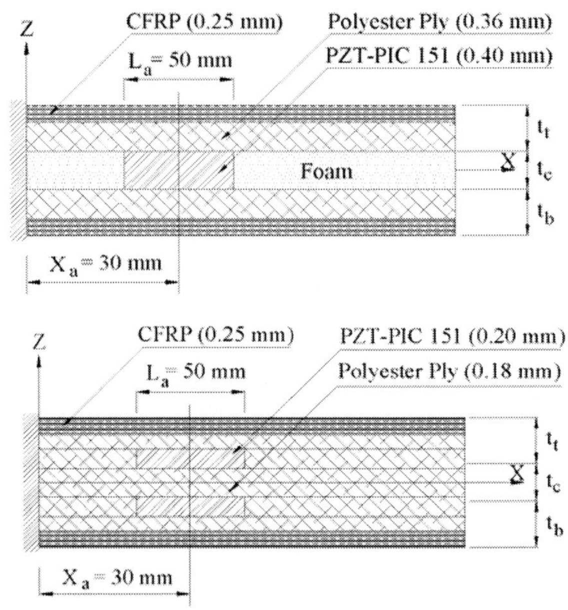

Figure 20. Composite sandwich active/sensory beams

Table 4. Material Data of the Composite Sandwich Beams

Properties	CFRP	PZT (PIC 151)	Polyester Ply	Foam
E (GPa)	65.0	60.0	3.15	35.3
G (GPa)	30.06	23.08	1.17	12.76
ν	0.081	0.3	0.344	0.383
ρ (Kg/m^3)	1770.0	7800.0	1380.0	32.0 kg/m^3
d_{31}(C/m^2)	-	12.6	-	-
d_{15}(C/m^2)	-	13.38	-	-
κ_{11}, κ_{33} (F/m)	-	18.5x10^{-9}	-	-
Geometric Properties: Length = 180 mm, Width = 40 mm.				

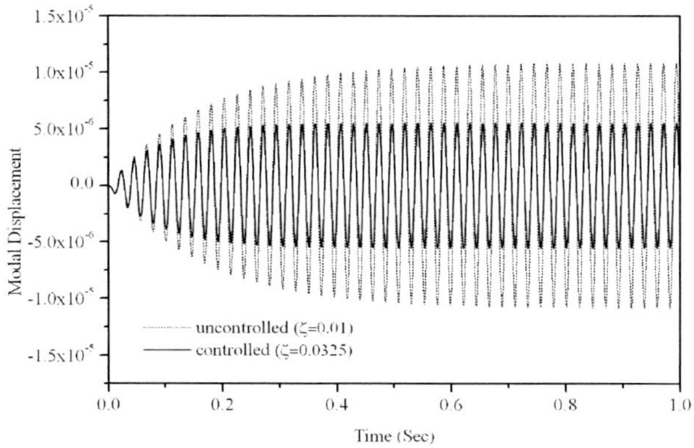

Figure 21. First mode control by extension actuators

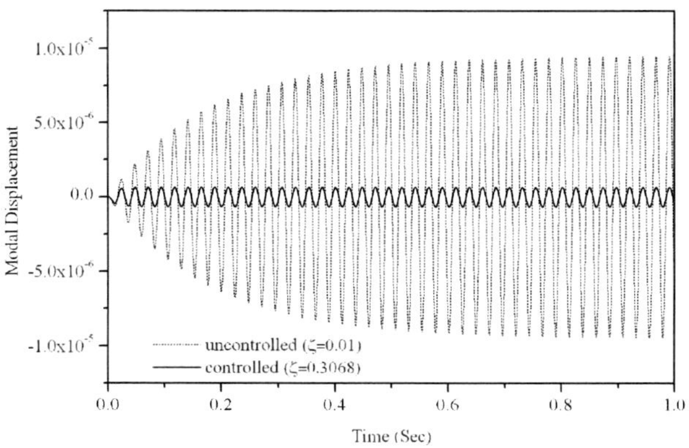

Figure 22. First mode control by shear actuator

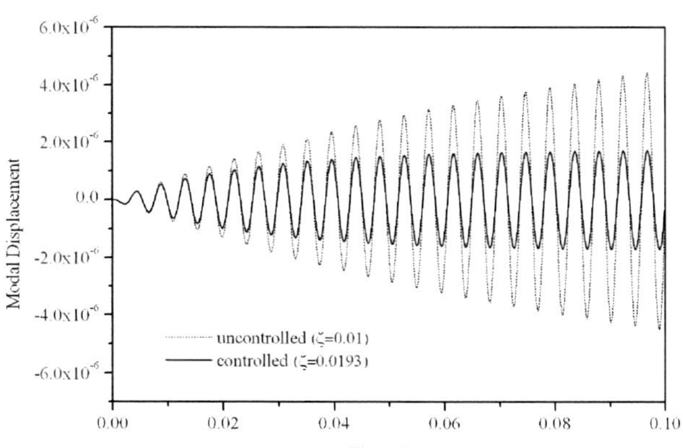

Figure 23. Second mode control by extension actuators

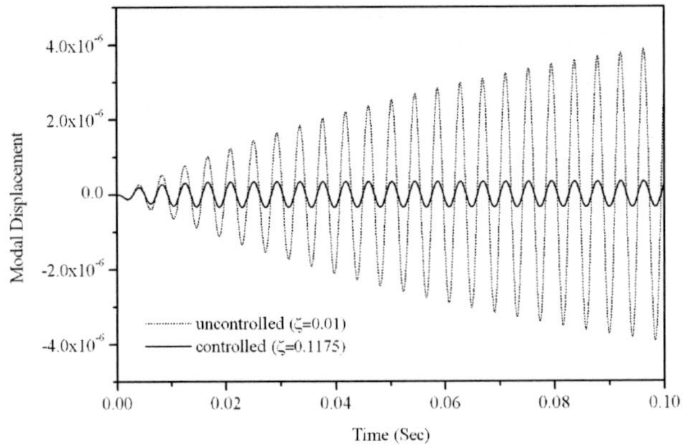

Figure 24. Second mode control by shear actuator

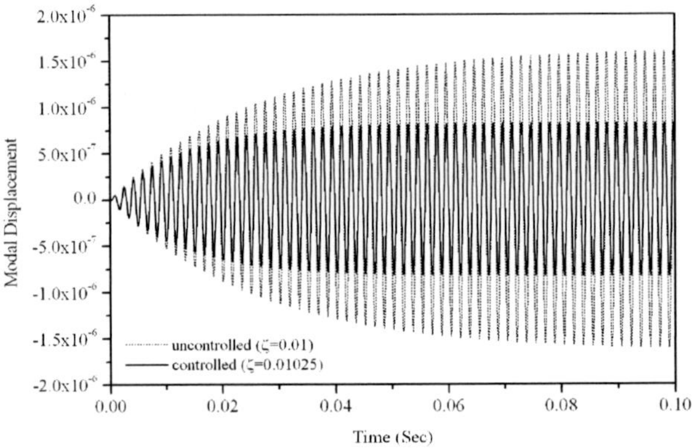

Figure 25. Third mode control by extension actuators

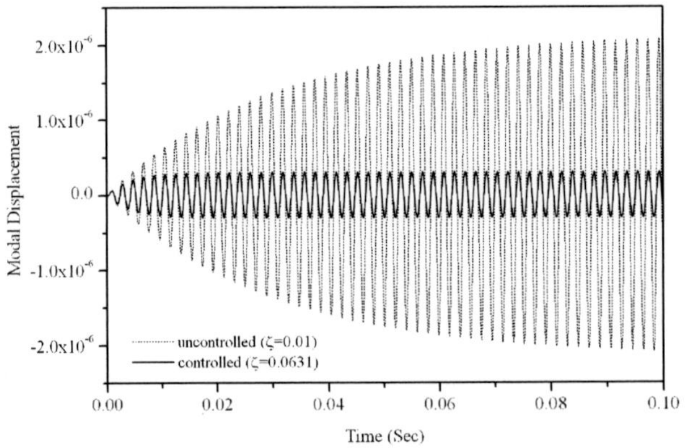

Figure 26. Third mode control by shear actuator

Table 5. Free Vibration Frequencies of Composite Sandwich Beams

Mode No	Extension Actuation Model Experimental Frequency in Hz (Raja, 1999)	Extension Actuation Model (present FEM)		Shear Actuation Model (present FEM)	
		10 el.	20 el.	10 el.	20 el.
1	42.90	44.28	44.26	42.38	42.23
2	211.60	227.46	226.66	238.96	237.37
3	545.60	596.23	588.19	518.59	513.81
4	1090.00	1219.2	1182.54	892.28	886.73
5	1800.00	2046.3	1962.41	1357.9	1345.18

From the modal control analysis, the following observations are made (see figures 21 to 26).

The shear actuator is found more efficient in controlling the first three elastic modes compared to the extension-bending actuators. This is due to the reason that θ_y is directly coupled with electric potential in shear actuation.

For the same control effort, the shear actuator introduces more active damping than the extension-bending actuators; around ten times in the first mode control and six times in the second and third mode controls.

4.7 Pole Placement and Mode Shape Control

A cantilevered steel beam with full-length active lamina (EAM) on top and bottom is considered to study the pole placement technique and mode shape control (refer to section 4.5). The beam geometry is shown in figure 27 and the material data is taken in the analysis as in table 6.

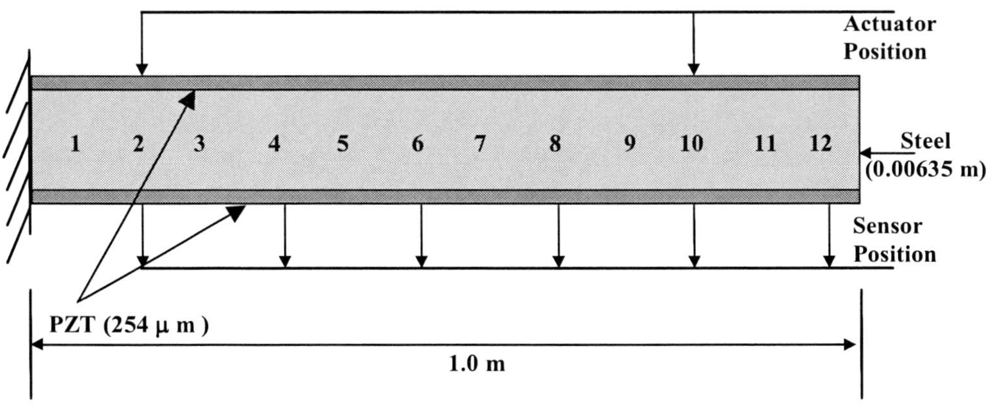

Figure 27. Cantilever PZT/Steel/PZT Beam

The beam is discretized using 12 elements and dynamic analysis is carried out. The main objectives of active vibration control are to place the desired Eigenvalues (for first three modes) and control the tip motion of the cantilever beam to achieve desired mode shapes. In the present control study model, two actuators and six sensors are employed. The following shape constraints are imposed: Deflection at the free end is to be $1/10^{th}$ of the open loop system and the deflection at grid point 4 as in the open loop system to maintain a smooth deflected shape. The optimal positions of the actuators are selected by minimising the number of unstable modes.

The following states are assumed to be measured: displacements at grids 2, 6, 10 and velocities at grids 4, 8, 12. The measured states are taken as specified elements in equation 69.

The desired Eigenvalues of the closed loop system is chosen as

$$\Lambda^d = \Lambda_R \pm j\Lambda_I \quad (76)$$

where $\Lambda_I = \omega_d$ is the damped natural frequency (rad/sec) and $\Lambda_R = \zeta\omega_n$, ω_n is the undamped natural frequency (rad/sec).

Table 6. Material Data of the PZT/Steel/PZT Beam

Properties	Steel	PZT
E (GPa)	210.0	63.0
G (GPa)	80.8	24.2
μ	0.3	0.3
ρ (Kg/m^3)	7750.0	7600.0
d_{31}(C/m^2)	-	11.28
κ_{11}, κ_{33} (F/m)	-	16.5x10^{-9}
Geometric Properties: Length = 1000 mm, Width = 50.8 mm.		

And the damping factor is,

$$\zeta = -\Lambda_R / \sqrt{(\Lambda_R^2 + \Lambda_I^2)}\,. \qquad (77)$$

The closed loop desired Eigenvalues are given below along with the open loop eigen values in parentheses.

$$\begin{aligned} \Lambda_{1,2}^d &= -2.0 \pm j30.0 \ (0 \pm j34.19) \\ \Lambda_{3,4}^d &= -2.5 \pm j200.0 \ (0 \pm j215.18) \\ \Lambda_{5,6}^d &= -6.5 \pm j600.0 \ (0 \pm j622.25) \end{aligned} \qquad (78)$$

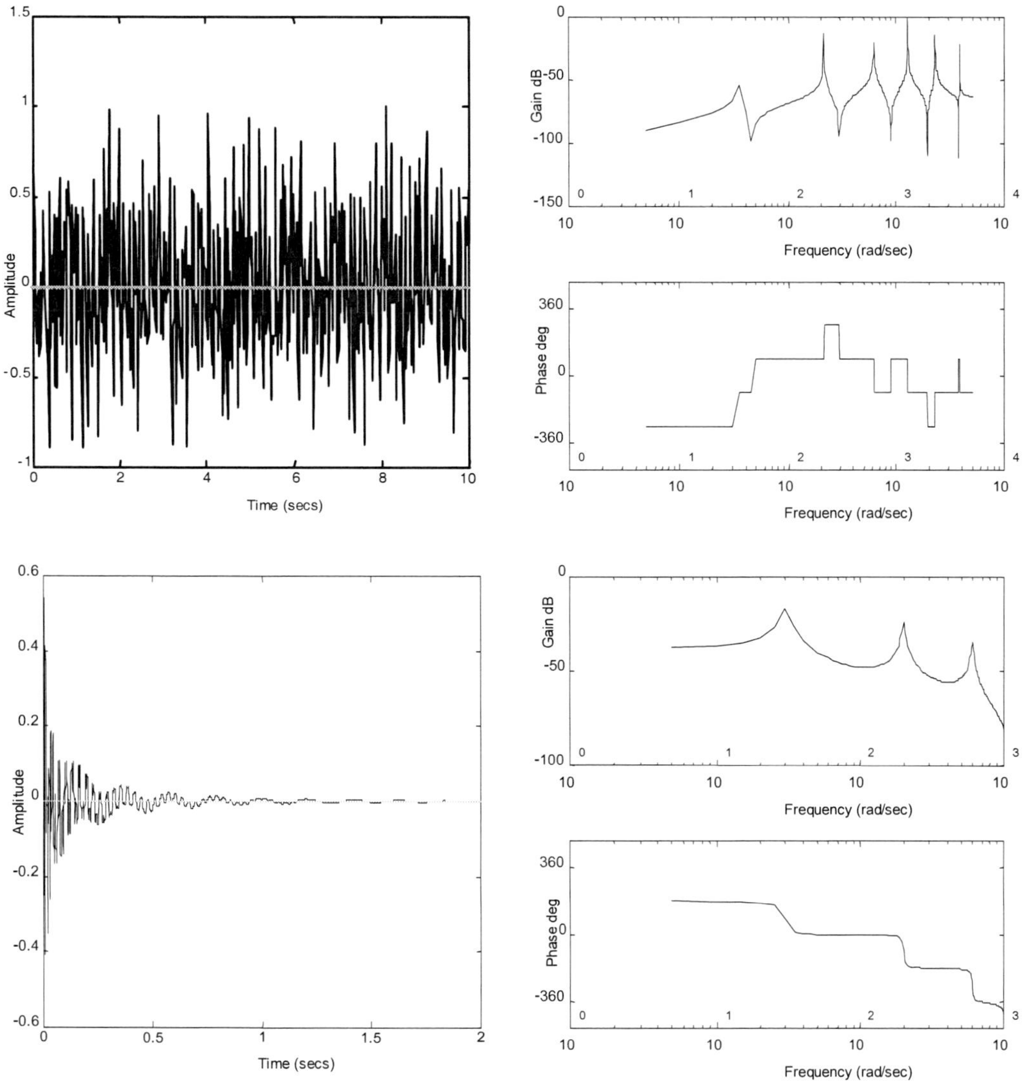

Figure 28. Time and frequency responses of open loop and closed loop systems.

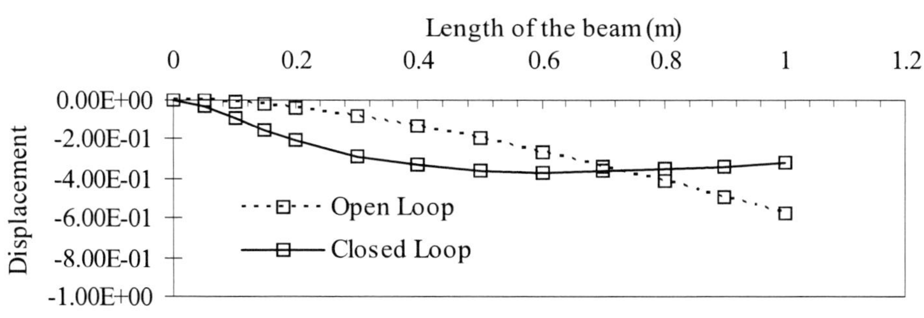

Figure 29. First mode shape of the smart cantilever beam

The real parts in these Eigenvalues are arrived at optimal values after considering the facts that the uncontrolled modes are stable and the developed control scheme could be successfully applied to place the poles. Figure 28 shows the time and frequency response analyses of open loop and closed loop systems (impulse function in MATLAB is used for this purpose). It may be observed that vibration levels of the controlled modes are significantly reduced. From table 7, it can be noticed how exactly the first three Eigenvalues are altered without affecting much of the other modes. The uncontrolled and controlled mode shapes of the first three elastic modes are depicted in figures 29 to 31. The controlled mode shapes are observed to be quite different from uncontrolled ones and this is achieved through the application of shape constraints.

It is worth noticing that the damping factors, actuators and sensors locations are found to be significant factors in order to arrive at a constant gain matrix (controller), which not only controls few elastic modes, but also keeps the system stable.

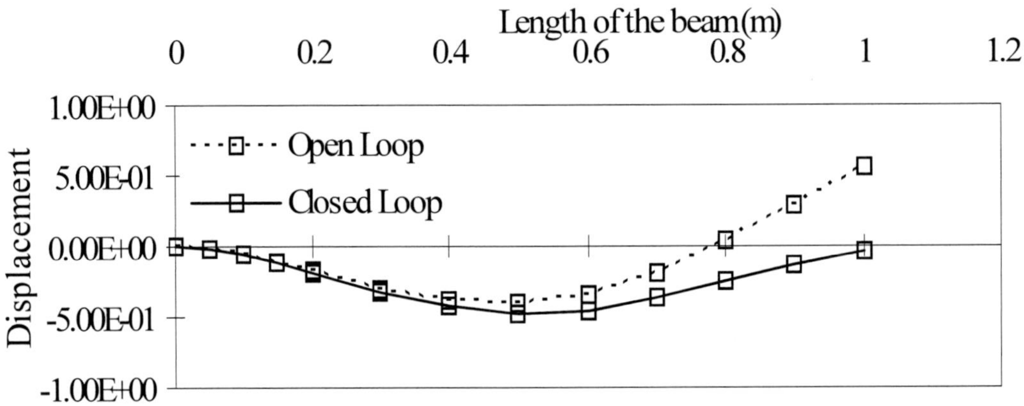

Figure 30. Second mode shape of the smart cantilever beam

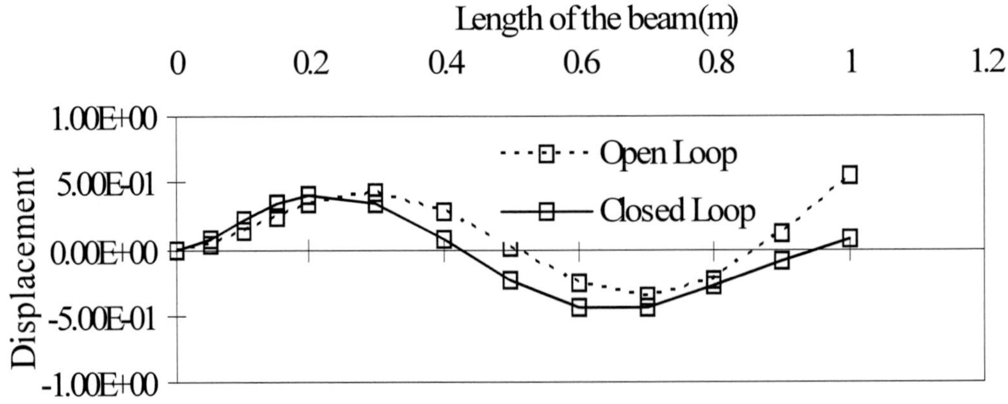

Figure 31. Third mode shape of the smart cantilever beam

Table 7. Dynamic Characteristics of PZT/Steel/PZT Beam

Mode	Open loop		Closed loop	
	Damping (%)	Frequency (rad/sec)	Damping (%)	Frequency (rad/sec)
1,2	0	± 34.19	6.65	± 30.06
3,4	0	± 215.18	1.25	± 200.02
5,6	0	± 622.25	1.08	± 600.04
7,8	0	± 1289.51	2.22	± 1214.9
9,10	0	± 2302.89	0.90	± 2243.4
11,12	0	± 3824.29	0.95	± 3847.0
13,14	0	± 6127.07	0.51	± 6113.9
15,16	0	± 9494.63	0.45	± 9528.0
17,18	0	± 14,847.95	0.52	± 14,850.0
19,20	0	± 25,649.66	0.02	± 25,653.0
21,22	0	± 43,355.17	0.06	± 43,358.0
23,24	0	±130,891.75	0.15	± 130,888.0

5. Conclusions

A hybrid piezoelectric actuation concept is developed using shear and extension actuators. Finite element procedures are derived to characterise the hybrid actuation mode by combining shear and extension actuation constitutive models. Further, a two node sandwich beam element is developed using three layer theory and sub-laminate concept to analyse smart sandwich structures. Numerical results are presented to qualify the element to model shear actuation, extension actuation independently and simultaneously. Interesting results have been obtained to demonstrate the influence of active stiffening and active damping effects of shear actuators and extension actuators in the structural control of sandwich beams. Finally, a shape control concept is presented with an example to show the capability of distributed PZT actuators to place desired poles and modify their associated mode shapes.

References

Abramovich, H., H.-R.Meyer-Piening, 2001, "Actuation and Sensing of Soft Core Sandwich Plates with a Built-In Adaptive Layer," *Journal of Sandwich Structures and Materials*, Vol.3, pp.75-86.

Allik, H and T.Hughes, 1970, "Finite Element Method for Piezoelectric Vibration," *International Journal for Numerical Methods in Engineering*, Vol. 2, pp.151-168.

Aldraihem, O.J., and A.A.Khdeir, 2000, "Smart Beams with Extension and Thickness – Shear Piezoelectric Actuators," *Smart Materials and Structures*, Vol.9, pp.1-7.

Barrett, R., 1992, "Active Plate and Wing Research Using EDAP Elements," *Smart Materials and Structures*, Vol.1, pp.214-226.

Baruch, P., and H.Abramovich, 1997, "Consistent Methodology for the Modelling of Piezolaminated Shells," *AIAA J.*, Vol.35 (8), pp.1316-1326.

Baz, A., and Poh, S., 1996, "Optimal Vibration Control with Modal Positive Position Feedback," *Optimal Control Application and Methods*, Vol.117, pp.141-149.

Benjeddou, A., M.A.Trindade and R.Ohayon, 1997, "A Unified Beam Finite Element Model for Extension and Shear Piezoelectric Actuation Mechanisms," *Journal of Intelligent Material Systems and Structures*, Vol.8, pp.1012-1025.

Benjeddou, A., M.A.Trindade and R.Ohayon, 1999, "New Shear Actuated Smart Structure Beam Finite Element," *AIAA Journal*, Vol.37 (3), pp.378-383.

Benjeddou, A., and J.-F.Deue, 2001-a, "Piezoelectric Transverse Shear Actuation and Sensing of Plates, Part 1: Three-Dimensional Mixed State Space Formulation," *Journal of Intelligent Material Systems and Structures*, Vol.12, pp.435-449.

Chang-Ho Hong and I.Chopra, 1999, "Modelling and Validation of Induced Strain Actuation of Composite Plates," *AIAA Journal*, Vol.37 (3), pp.372-377.

Chopra I., "Review of state of art of smart structures and integrated systems", *AIAA Journal* 2002, 40(11): 2145-2188.

Ha, S.K., C.Keilers and F.K.Chang, 1992, "Finite Element Analysis of Composite Structures Containing Distributed Piezoceramic Sensors and Actuators," *AIAA Journal*, Vol.30 (3), pp.772-779.

Heyliger, P.R., G.Ramirez and D.A.Saravanos, 1994, "Coupled Discrete Layer Finite Elements for Laminated Piezoelectric Plates," *Communications in Numerical Methods in Engineering*, Vol.10, pp.971-981.

Hwang, W.S., and H.C.Park, 1993, "Finite Element Modelling of Piezoelectric Sensors and Actuators," *AIAA Journal*, Vol.31 (5), pp.930-937.

Koconis, D.B., Lazlo P.Kollar and George S.Springer, 1994, "Shape Control of Composite Plates and Shells with Embedded Actuators I. Desired Voltages Specified," *Journal of Composite Materials,* Vol.28 (5), pp.415-457.

Lazarus, K.B., and Crawley, E.F., 1992, "Multivariable High Authority Control of Plate-Like Active Structures," 33rd AIAA-ASME-ASCE-AHS-ASC Structures, *Structural Dynamics and Materials Conference*, Dallas TX, pp.931-945.

Nailon, M., H.Coursant and F.Besnier, 1983, "Analysis of Piezoelectric Structures by Finite Element Method," *ACTA Electronica*, Vol.25 (4), pp.341-362.

Park, Young-Kun, and S.-B.Choi, 1999, "Vibration Control of a Cantilevered Beam via Hybridisation of Electro-Rheological Fluids and Piezoelectric Films," *Journal of Sound and Vibration*, Vol.225 (2), pp.391-398.

Raja S, M. Rose, K. Rohwer, "Piezothermoelastic Modeling and Active Vibration Control of Laminated Piezoelectric Composite Beam", *J. of Intelligent Material Systems and Structures* 1999, 10(11): 890-899.

Raja S., P. K. Sinha, G. Prathap, "Active Vibration Control of Composite Sandwich Beams with Distributed Piezoelectric Extension-Bending and Shear Actuators", *Smart Materials and Structures* 2002, 11 (1):63-71.

Raja S., P. K. Sinha, G. Prathap, "Active Stiffening and Active Damping Effects on Closed Loop Vibration Control of Composite Beams and Plates", *J. of Reinforced Plastics and Composites* 2003, 22 (12):1101-1121.

Raja S., Sreedeep R., G. Prathap, "Bending Behavior of Hybrid-Actuated Piezoelectric Sandwich Beams", *J. of Intelligent Material Systems and Structures* 2004, 15(8): 611-619.

Raja S., S. Rajappa, R. Sreedeep, "Active Vibration Control of Shear Actuated Piezoelectric Sandwich Beam – Experiment", 3rd International Conference on Theoretical, Applied, Computational and Experimental Mechanics December 2004, IIT Kharagpur, India.

Saravanos, D.A, and P.R.Heyliger, 1995, "Coupled Layerwise Analysis of Composite Beams with Embedded Piezoelectric Sensors and Actuators," *Journal of Intelligent Material Systems and Structures*, Vol.6 (3), pp.350-363.

Saravanos, D.A, and P.R.Heyliger and D.A.Hopkins, 1997, "Layerwise Mechanics and Finite Element for Dynamic Analysis of Piezoelectric Composite Plates," *International Journal of Solids and Structures*, Vol.34 (3), pp.359-378.

Senthil S. Vel, R. C. Batra, 2001, "Exact Solution for Rectangular Sandwich Plates with Embedded Piezoelectric Shear Actuators", *AIAA Journal*, Vol. 39(7), pp. 1363 – 1373.

Tzou, H.S., and C.I.Tseng, 1990, "Distributed Piezoelectric Sensor/Actuator Design for Dynamic Measurement/Control of Distributed Parameter Systems – A Piezoelectric Finite Element Approach," *Journal of Sound and Vibration*, Vol.138 (1), pp.17-34.

Tzou, H.S., and C.I.Tseng, 1991, "Distributed Vibration Control and Identification of Coupled Elastic Piezoelectric Systems: Finite Element Formulation and Applications," *Mechanical Systems and Signal Processing*, Vol.5 (3), pp.215-231.

Tzou, H.S., and J.J.Hollkamp, 1994, "Collocated Independent Modal Control with Self-Sensing Orthogonal Piezoelectric Actuators," *AIAA*-94-1737CP, pp.57-64.

Zhang, X.D., and C.T.Sun, 1996, Formulation of an Adaptive Sandwich Beam," *Smart Materials and Structures*, Vol.5, pp.814-823.

Zhang, X.D., and C.T.Sun, 1999, "Analysis of a Sandwich Plate Containing a Piezoelectric Core," *Smart Materials and Structures*, Vol.8, pp.31-40.

In: Research in Hybrid Materials
Editors: Simon J. Brunner and Julian W. Egger
ISBN 978-1-60456-718-2

Chapter 7

STUDIES ON DUMBBELL-SHAPED BISFULLERENE AND CONJUGATED OLIGOMER HYBRIDS

Ningzhang Zhou, Li Wang, Mark Tobin, David W. Thompson[2] and Yuming Zhao[1]
Department of Chemistry, Memorial University of Newfoundland, St. John's, Newfoundland, Canada A1B 3X7

ABSTRACT

Hybrid molecules comprised of [60]fullerene and various π-conjugated oligomers have gained much attention owing to their promising potentials in molecular optoelectronics and photovoltaics. In this article, we intend to describe our recent study on a series of dumbbell-shaped C_{60}-oligomer-C_{60} triad molecules, primarily aimed at better understanding the roles of conjugation motif and dimensionality of the oligomer bridge in affecting their electronic and photophysical properties. Two types of conjugated oligo(phenylene ethynylene)/oligo (phenylene vinylene) (OPE/OPV) hybrid oligomers with linear or cruciform molecular topologies were designed and prepared via sequential Pd-catalyzed cross-coupling and Wittig-Horner reactions. These oligomers were readily endcapped with fullerenyl groups via an *in situ* ethynylation protocol, affording the desired C_{60}-oligomer-C_{60} triads. Electrochemical and photophysical properties of these novel bis(fullerene)-OPE/OPV hybrids were characterized by cyclic voltammetry, UV-vis and emission spectroscopic methods. Our results clearly indicate that properties of these triad molecules are subject to the influence of the π-bridge to varied degrees at the ground and the excited states. Moreover, evidence for rapid photoinduced electron and/or energy transfer was found in our excited state characterization. The photophysical behavior revealed suggests that these C_{60}-oligomer-C_{60} molecular dumbbells offer not only a new model system for in-depth understanding of the structure-property relationship but also a class of potentially useful candidates for advanced C_{60} based optoelectronic materials.

[2] E-mail address: dthompso@mun.ca; yuming@mun.ca (Corresponding authors.)

1. INTRODUCTION

Buckminsterfullerene (C_{60}) or *Bucky ball* is an aesthetic 3-dimensional allotropic form of carbon that has captivated immense research interest among the scientific community since its first discovery in the early 1980s [1]. Over the past two decades, a tremendous amount of efforts of fullerene research have been invested on functionalization of C_{60} with numerous complex molecular structures, including organic chromophores, organometallics, dendrimers, biomolecules, conjugated oligomers/polymers, and to name a few [1,2]. These studies have greatly enhanced the scope of our understanding in respect to the structure-dependent molecular behavior of various fullerene derivatives, which in turn has prompted continued exploration on the potential applications for fullerene-based materials in the area of advanced molecular devices [3]. The most prominent application of fullerene derivatives is to fabricate the so-called "bulk-heterojunction" organic photovoltaic cells using materials constituted by fullerene and various organic semiconductors [3-6]. In a general sense, C_{60} features excellent electron accepting properties and remarkable excited state dynamics [7]. Thus, it can be used as both an efficient chromophore and electron acceptor. Association of fullerene moieties with various photoactive electron donor groups, in either a covalent or a non-covalent manner, has provided a rational and widely applied strategy for the design of new fullerene-based optoelectronic materials [8,9,10e]. Nevertheless, there are still many factors and parameters that need to be elucidated and refined before realistic device applications, such as organic solar cells, can be eventually realized.

Among the numerously designed fullerene-donor derivatives in the recent literature, covalently linking C_{60} cages with monodisperse π-conjugated oligomers has been an attractive approach since it can produce ideal systems for probing the relationship between molecular structure and essential photophysical behavior in photo-electron conversion where photoinduced energy/electron transfer and charge separation are normally the major events. A straightforward and intuitive design motif for fullerene-oligomer hybrid based optoelectronic materials is to construct the molecules in a linear dyad molecular architecture, *i.e.* C_{60}-π, which is generally viewed as a "molecular heterojunction" [10]. It has been reported that such C_{60}-π hybrid molecules could potentially maximize the efficiency of organic solar cells to a significant extent in comparison to non-covalent fullerene-polymer blends [11]. In addition to the relatively simple dyad structure, triad fullerene derivatives characterized by a symmetric dumbbell-shaped molecular structure, namely C_{60}-π-C_{60}, have grown rapidly in number in recent publications [12]. So far, a plethora of π-conjugated oligomers with defined and monodisperse structures and bandgaps have been employed as the bridging unit in the C_{60}-π-C_{60} system, for instance, linearly conjugated oligo(thiophenes), oligo(phenylene arylene)s, oligo(phenylene ethynylene)s [12]. These conjugated π-units were typically bonded to two fullerenyl groups at their termini via a variety of covalent linkages to function as electroactive bridges. In comparison to the relatively simple C_{60}-π dyad system, the dumbbell-shaped triads are actually not so demanding in respect to synthetic efforts, since the symmetrical structure requires no selective protection at one terminus of the oligomer. Moreover, as Martín and co-workers pointed out in two recent review articles [9,12a], the electronic interactions between the two C_{60} cages in either the ground state or the excited state varied significantly depending on the detailed structures and electronic nature of the bridging π-units incorporated. The presence of a second fullerenyl unit into the molecular system has been reported to significantly influence the photophysical consequences, the origin of which was presumably ascribed to the unique stabilizing effect on the photogenerated radical ion pair by the additional C_{60} cage [12a,b].

In our work, we are interested in understanding the effects on the photophysical properties of the fullerene derivatives brought about by the monodisperse conjugated oligomer π-bridge in a C_{60}-π-C_{60} triad system [13]. A new model system of dumbbell-shaped C_{60}-π-C_{60} triads was designed and synthesized, where the oligomer unit takes either a linear or a cruciform hybrid OPE/OPV structure (Schemes 1 to 4). The following will elaborate the synthetic methods, ground-state properties, and excited state properties for these novel C_{60}-π-C_{60} triad systems. Overall, access to these C_{60}-oligomer hybrid compounds has allowed the roles of the oligomer bridge to be better understood, and lent insight into such molecular factors as electron delocalization, conjugation length and dimensionality, and electronic coupling. The intention here is to illustrate that the molecular factors given above deserve serious and judicious considerations in the design of new functional C_{60}-oligomer materials.

2. Results and Discussion

2.1. Synthetic Methods

The synthetic methods towards the new C_{60}-π-C_{60} hybrids are briefly summarized in Schemes 1 to 4. The detailed experimental procedures will be reported in a subsequent manuscript [14]. The first series of conjugated hybrid oligomers are linear OPE/OPV hybrid oligomers 6-9, in which the central OPV units are flanked by phenylacetylene groups on each side. An iterative synthesis of these oligomers is described in Scheme 1. Two essential building blocks, compounds 3 and 4, were first obtained via Pd-catalyzed cross-coupling reactions. Compound 4 then underwent a Wittig-Horner reaction with the ylide species generated by treating diphosphonate 5 with lithium hexamethyldisilazide (LHMDS) to afford the first OPE/OPV hybrid oligomer 6. Oligomer 6 was then desilyated with K_2CO_3 and cross-coupled with phenylacetylene iodide 3 under Pd/Cu catalysis to yield a longer linear OPE/OPV oligomer, 8. Another step of protiodesilylation using K_2CO_3 converted 8 into 9 in a decent yield.

The second series of OPE/OPV hybrid oligomers (12-13 and 15-16) are in the cruciform shape, which are the constitutional isomers of the linear oligomers 6-9. As outlined in Scheme 2, the synthesis of these cruciform-shaped oligomers began with a diiodoarylene 10, which was prepared *via* an established synthetic method reported by Bunz and co-workers [15-17]. Sonogashira coupling of 10 and trimenthylsilylacetylene (TMSA) under the catalysis of Pd/Cu led to the formation of ethynylated arylene 11 in a good yield. Compound 11 then underwent protiodesilylation and Sonogashira coupling reaction with 3 to form the first cruciform oligomer 12. The TMS endcapping groups on 12 could be easily removed with K_2CO_3 yielding oligomer 13 in a high yield. In a similar manner, a longer version of cruciform oligomer, 15, was obtained *via* the Sonogarhsira coupling of 10 and 14. Finally, oligomer 15 was deprotected using tetrabutylammonium fluoride (TBAF) in THF.

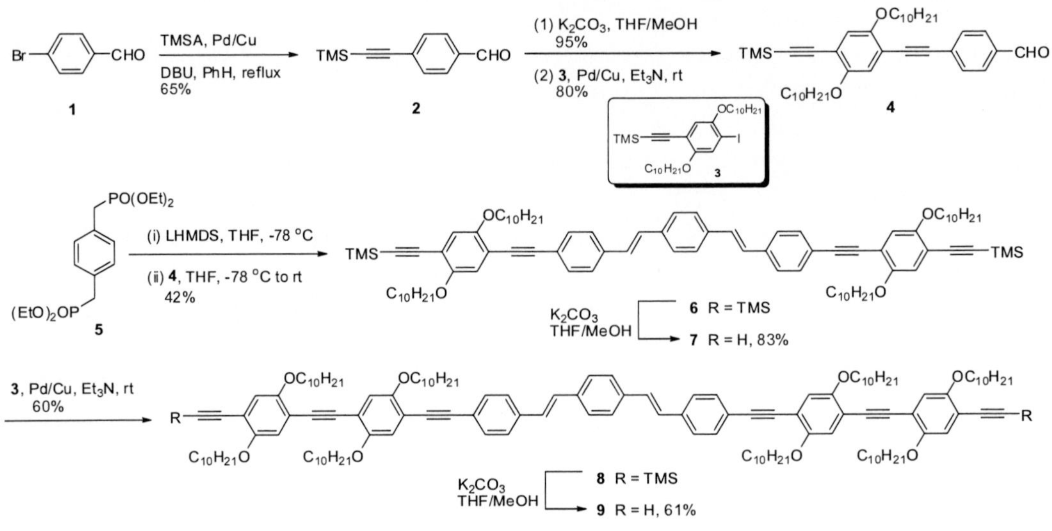

Scheme 1. Synthesis of linear OPE/OPV hybrid oligomers.

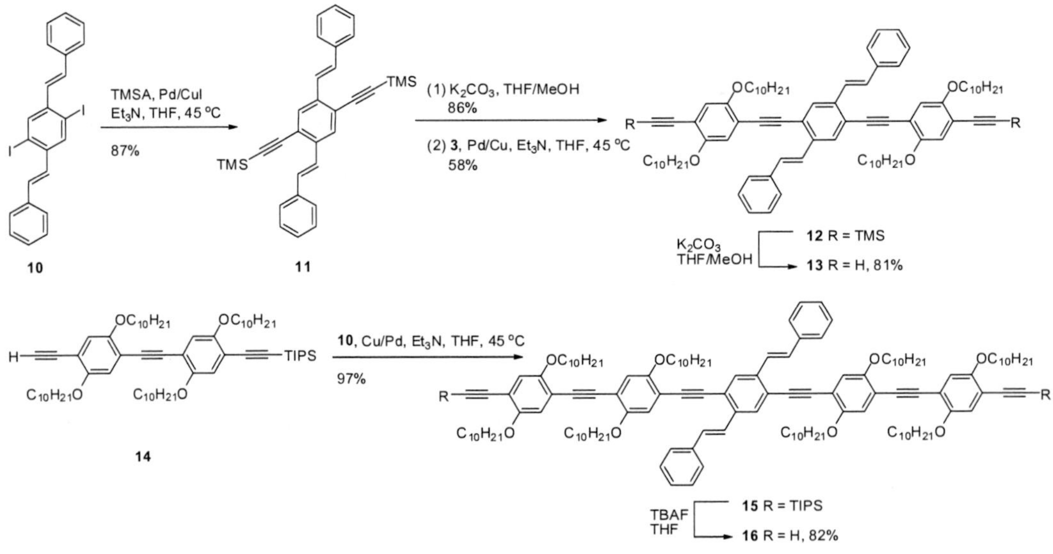

Scheme 2. Synthesis of cruciform OPE/OPV oligomers.

Compared with the linear OPE/OPV oligomer series, the molecular skeleton of the cruciform oligomer is more flexible and accessible in terms of adding more functionalities into the oligomer backbone. This feature should be advantageous in further molecular design and tailoring because it enables more opportunities to fine-tune the electronic and optical properties at the molecular level. To explore this potential, a strong electron donating group, *N,N*-diphenylamino group, was selected to be attached to the termini of the OPV axial in the cruciform OPE/OPV architecture (Scheme 3). From aryldiiodie 17, amino group-functionalized cruciform oligomers 19-20 and 22-23 could be synthesized through the same strategy as used in the unfunctionalized cruciform oligomers shown in Scheme 2.

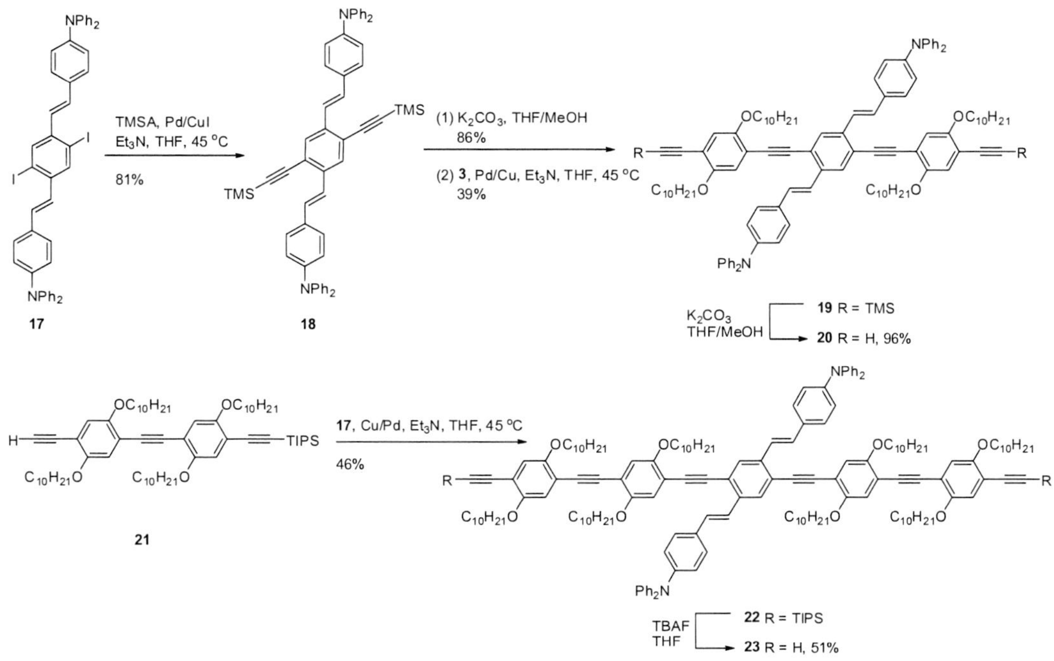

Scheme 3. Synthesis of donor-attached cruciform OPE/OPV hybrid oligomers.

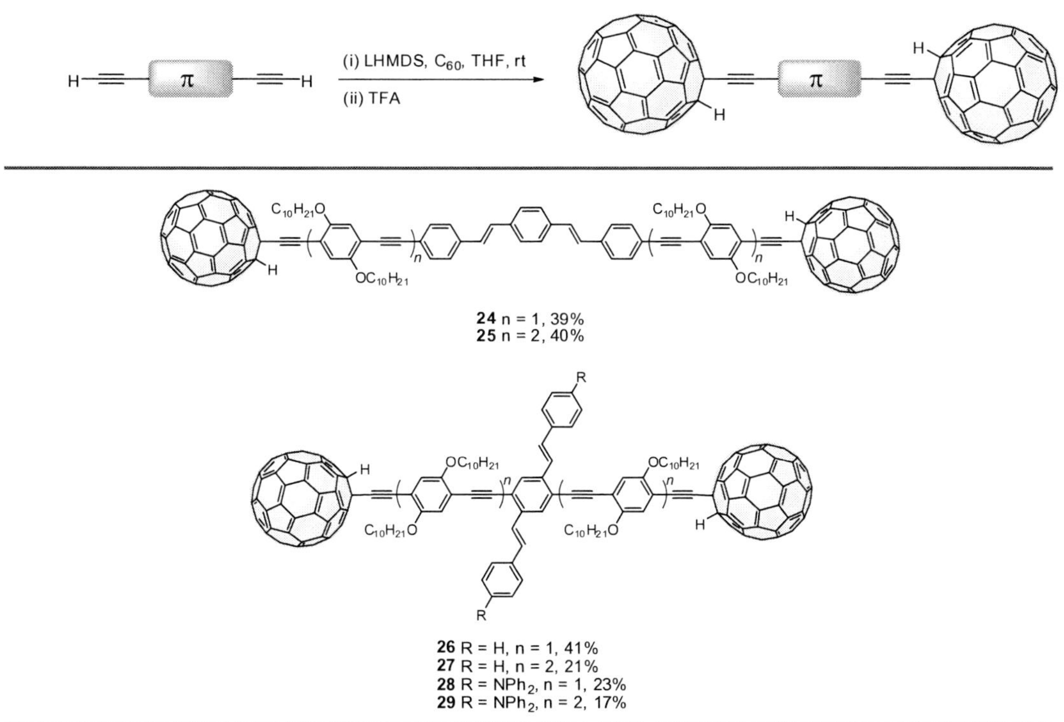

Scheme 4. Synthesis of dumbbell shaped C_{60}-π-C_{60} triads via the *in situ* ethynlation reaction.

With all these oligomers successfully prepared, the synthesis of the targeted C_{60}-oligomer-C_{60} triads was straightforward. In our work, we employed a well-established *in situ* ethynylation protocol, which has been found particularly reliable and efficient in forming multifullerene-oligomer hybrids [18,19]. As outlined in Scheme 4, oligomers with terminal akynyl groups can react with excess C_{60} through a nucleophilic addition mechanism under the condition of slowly adding a strong base–lithium hexamethyldisilazide (LHMDS)–in THF solution. The resulting fullerides can be then easily quenched with trifluoroacetic acid (TFA) to afford the desired C_{60}-π-C_{60} oligomers 24-29 in reasonable yields. It is noteworthy that all these bisfullerene compounds, despite their extremely large molecular weights (2,000–6,000 Da), can be satisfactorily purified by silica flash column chromatography using carefully controlled chromatographic conditions. Very crucial to successful purification was to apply suitably chosen eluent gradient in each of the column chromatographic separations.

2.2. NMR Spectroscopic and Electrochemical Characterization

All the C_{60}-oligomer-C_{60} compounds and their OPE/OPV oligomer precursors have been fully characterized by IR, ^{1}H and ^{13}C NMR spectroscopy and mass spectrometry, which convincingly confirmed their structures and purities [14]. Particularly remarkable is the ^{13}C NMR spectroscopic data for the six dumbbell-shaped bisfullerene compounds 24-29. The

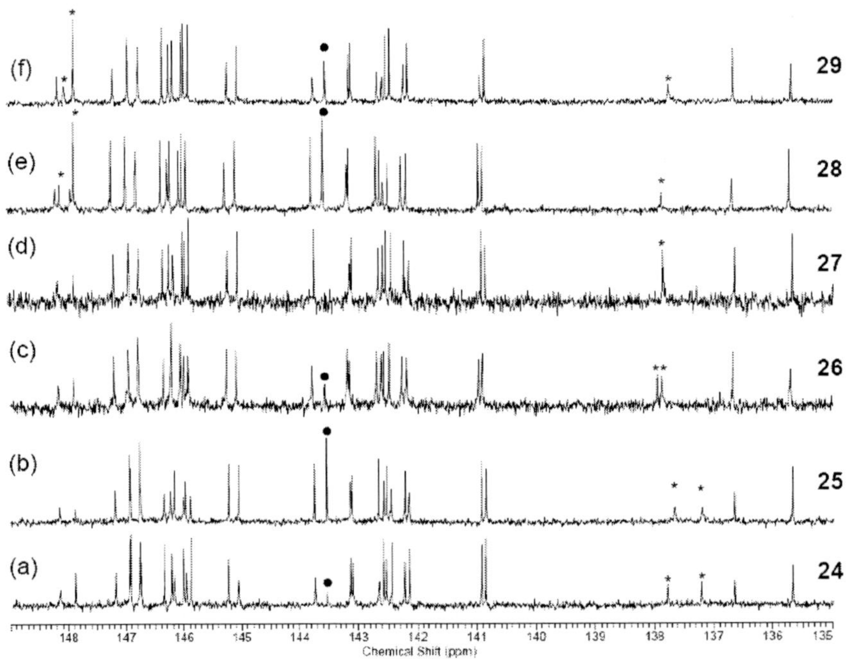

Figure 1. Comparison of ^{13}C NMR spectra for C_{60}-π-C_{60} compounds 24-29 in the region of 135-149 ppm. Signals labeled with * are due to aromatic carbons in the oligomer bridges. Signals labeled with • are due to unreacted C_{60}. (a) Spectrum of 24: 26 out of 30 sp^2 fullereneyl carbon signals were observed; (b) spectrum of 25: 26 out of 30 sp^2 fullereneyl carbon signals were observed; (c) spectrum of 26: 28 out of 30 sp^2 fullereneyl carbon signals were observed; (d) spectrum of 27: 26 out of 30 sp^2 fullereneyl carbon signals were observed; (e) spectrum of 28: 24 out of 30 sp^2 fullereneyl carbon signals were observed; (b) spectrum of 29: 27 out of 30 sp^2 fullereneyl carbon signals were observed.

centrosymmetric molecular structure in each C_{60}-π-C_{60} compound makes the two fullerenyl groups chemically equivalent. In addition, each of the C_{60} cages possesses a mirror symmetry plane bisecting itself. Theoretically 30 distinctive signals should appear in the region of 135 to 149 ppm corresponding to all the sp^2 carbons on the two-fold symmetric C_{60} cage [18]. Experimentally, we observed well-resolved ^{13}C NMR spectra for all the six C_{60}-π-C_{60} compounds in this region as given in Figure 1. From the comparison of the ^{13}C NMR shown in Figure 1, it is obvious that the different electronic properties of the oligomer π-bridges cause no significant changes in the chemical shifts of each type of sp^2 carbons on the C_{60} cage. This observation indicates that electronic effects between the C_{60} cage and the oligomer bridge in the ground state are small, although the two conjugated moieties are closely linked through only one sp^3 carbon center.

Electrochemical redox properties for the six bisfullerene molecular dumbbells were characterized by cyclic voltammetric experiments (CV). The detailed CV data are summarized in Table 1 and cyclic voltammograms are given in Figure 2.

Table 1. Summary of cyclic voltammetric data for compounds 24-29.[a]

Entry	Oxidation		Reduction	
	E_{pa} (V)	E_{pc} (V)	E_{pa} (V)	E_{pc} (V)
24	1.27	1.12	-0.47, -0.83, -1.10, -1.40	-0.59, -0.79, -0.99, -1.24, -1.51, -1.97
25	–	1.10	-0.35, -1.10, -1.34	-0.69, -0.80
26	–	–	-0.38, -0.84, -1.12, -1.27	-0.59, -1.00, -1.24, -1,52, -1.73
27	–	–	-1.03, -1.27	-0.29, -0.60, -0.80, -1.00, -1.51
28	–	–	–	-0.37, -0.84, -1.04, -1.28, -1.74
29	–	–	–	-0.59, -0.85, -0.99, -1.16, -1.42, -1.52

[a]Cyclic voltammetric experiments were performed at room temperature. Data were recorded in solutions of *o*-dichlorobenzene-CH_3CN (4:1, v/v). Bu_4NBF_4 (0.1 M) as the supporting electrolyte, glassy carbon as the working electrode, and Pt wire as the counter electrode. Potentials are given in volts versus a Ag/AgCl reference electrode. Scan rate: 100 mV s^{-1}.

The linear OPE/OPV bridged bisfullerene dumbbells 24 and 25 show notable amphoteric electrochemical behavior in their cyclic voltammograms. For compound 24, the oligomer bridge was oxidized at a formal potential of *ca.* 1.20 V, while for compound 25 with further extended phenylacetylene axial the oxidation peak can only been vaguely seen at relatively higher potential. In the reduction region of 24, four chemically reversible single-electron reducing steps can be clearly seen, which are due to the consecutive reductions on the C_{60} cages compared against the previously reported C_{60}-OPE-C_{60} systems [18-20]. However, no clear evidence for appreciable electron communications between the two C_{60} cages and the bridge can be found in the CV profile of 24, albeit a shoulder cathodic peak is discernible right before the second reduction peak. For compound 25, the reversibility of the redox processes on the C_{60} cage is dramatically lost, although four reduction steps on C_{60} are still distinguishable in the cathodic process.

No obvious oxidation features can be observed in both the cyclic voltammograms of the cruciform oligomer bridged compounds 26 and 27 under 1.5 V, which indicates that the cruciform OPE/OPV oligomers are less prone to be oxidized than their linear cousins. The

CV profiles for these two compounds in the reduction region are, however, similar to the features seen in the cyclic voltammograms of their constitutional isomers 24 and 25. Again, this observation agrees with the insignificant electronic effects between the oligomer bridge and the C_{60} cages in the ground state implied by their ^{13}C NMR data (see Figure 1).

In the cyclic voltammograms of 28 and 29, the presence of two strong electron donating diphenylamino groups in each of the molecules causes distinctive oxidation features in the positive region of their cyclic voltammograms. The oxidation processes are irreversible and no distinctive peak potentials can be unambiguously located. In the reduction region, both compounds show multiple-step reductions ascribed to the C_{60} cages. However, as these processes are all irreversible and their profiles are obscure, it is difficult to make any specific assignments on the redox processes, not to mention drawing clear conclusions as to the electronic effects between the C_{60} cages and the electronic-rich oligomer bridges.

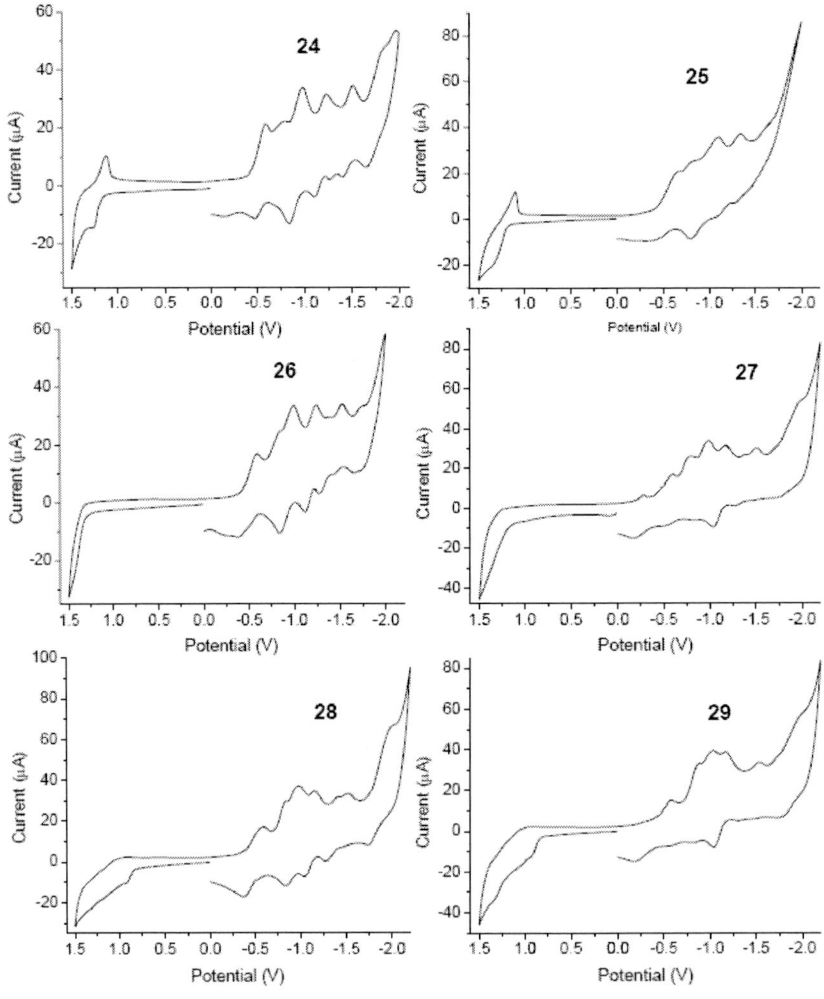

Figure 2. Cyclic voltammograms for C_{60}-π-C_{60} triads 24-29.

Nevertheless, a general pattern can be clearly found in our electrochemical results, that is, the reversibility of the redox waves due to reductions on the C_{60} cage tends to be lost significantly as the conjugation length or the electronic-richness of the oligomer π-bridge increases. This characteristics is particularly distinctive when comparing our CV data with previously reported C_{60}-π-C_{60} analogues with relatively electron-deficient OPE bridges [18-20].

2.3. Photophysical Characterization

UV-visible (UV-vis) and emission spectral data, extinction coefficients, quantum yields, optical energies, and reorganization energies for the OPE/OPV bridges and C_{60}-π-C_{60} triads are summarized in Table 2. The two main structural motifs of bridge for the C_{60}-π-C_{60} triads, linear OPE/OPV and cruciform OPE/OPV, were examined and compared. The length of the linear OPE/OPV and cruciform OPE/OPV bridges are varied by incorporating alkoxy-substituted phenyleneethynylene as "spacers" where the alkoxy substituents are used to improve the solubility of the C_{60}-π-C_{60} triads. A detailed understanding of the excited state dynamics in the π-conjugated oligomeric bridged C_{60} triads requires thorough characterization on each molecular component which comprise the bridges and the C_{60} terminated triads. The results of the absorption and emission spectroscopic investigations are reported below for phenyleneethynylene spacer and its C_{60} adduct, followed by the linear OPE/OPV hybrid oligomeric bridges and their C_{60} terminated adducts and finally the cruciform OPE/OPV molecular bridges and their corresponding C_{60}-π-C_{60} triads.

2.3.1. The Phenyleneethynylene Spacer (PES) and C_{60}-PES

The absorption and emission spectra of PES are shown in Figure 3. The absorption spectrum of PES is dominated by an intense structured band at 335 nm (ε= 8010 $M^{-1}cm^{-1}$, E_{op} = 29850 cm^{-1}) with a well-defined shoulder at 346 nm (ε= 8010 $M^{-1}cm^{-1}$, E_{op} = 29850 cm^{-1}) assigned to a $S_o \rightarrow S_1$ transition. There is a low energy transition that tails out past 400 nm tentatively assigned to an $n \rightarrow \pi^*$ transition associated with the alkoxy substituents. Light excitation at 335 nm into the $S_0 \rightarrow S_1$ transition leads to an intense emission band at 367 nm (ϕ_f = 0.62, E_{em} = 29850 cm^{-1}) reasonably assigned to a $S_1 \rightarrow S_0$ radiative transition. There is a much weaker emission band at 715 nm which probably arises from a $T_1 \rightarrow S_0$ phosphorescence [7].

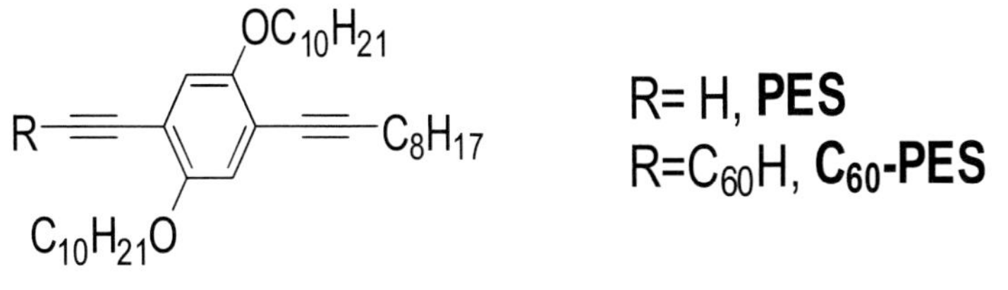

Figure 3.

Table 2. Summary of UV-vis and emission spectral data, extinction coefficients, quantum yields, optical energies, and reorganization energies for the OPE/OPV bridges and C_{60}-π–C_{60} triads in N_2 saturated $CHCl_3$ at 298(± 3) K

Entry	Absorption			Emission			λ_t(cm^{-1})
	λ_{abs} (nm)	ε (×10^4 M^{-1}cm^{-1})	E_{op} (cm^{-1})	λ_{em} (nm)	E_{em} (cm^{-1})	Φ_f	
PES [(a)]	276 335 346	2.6 0.80 0.71	36,230 29,850 28,900	367 715	27,250	0.62	825
C_{60}-PES[(a)]	256 335 346	0.40 1.8 1.4	39,060 29,850 28,900	378 411 730(T)	26,460 24,330 13,700	2.6×10^{-3}	1220
7	293 401	0.25 1.1	34,130 24,940	445 472 510 (sh)	22,470 21,190 19,610	0.63	1233
24	253(sh) 329 404 430(sh)	0.92 1.4	39,530 30,400 24,750 23,260	453 711(T)	22,080 14,060	6.0×10^{-3}	1339
9	308 411	12 41	32,470 24,330	456 486 523(sh)	21,930 20,580 19,120	0.89	1200
25	309[(a)] 330[(a)] 414[(a)] 435(sh) [(a)]	24 22 33 28	32,360 30,300 24,160 22,989	460 488(sh) 710(T)	21,740 20,490 14,080	< 10^{-3}	1208
13	324 360 390(sh)	5.8 7.8 5.4	30,860 27,780 25,320	428 452 (sh)	23,360 22,120	0.96	[(b)]
26	254(sh) 325 360 403(sh)	1.8 0.94 0.87 0.60	38,910 30,770 27,780 24,815	451 469 (sh) 711	22,175 21,320 14,045	~ 10^{-4}	[(b)]

Entry	Absorption			Emission			λ_r(cm^{-1})
	λ_{abs} (nm)	ε ($\times 10^4$ M^{-1}cm^{-1})	E_{op} (cm^{-1})	λ_{em} (nm)	E_{em} (cm^{-1})	Φ_f	
16	312	2.1	32,970	452	22,125	0.96	[b]
	330	2.4	30,300	484 (sh)	20,660		
	372	3.3	26,890				
	410	3.2 , br	24,390				
27	255	26	39,220	[c]			[b]
	315(sh)	15	31,950				
	326	16	30,680			< 0.01	
	370	14	27,030	711	14,045		
	415	13	24,100				
20	309	5.6	32,360	507	19,720	0.80	[b]
	387	7.9	25,840	530(sh)	18,870		
	435(sh)	5.1	22,990				
28	256	16	39,060	509	19,650	1.3×10^{-3}	[b]
	309	9.4	32,360	708(T)	14,120		
	391	8.1	25,580				
	463(sh)	3.2	21,600				
23[d]	309	1.6	32,360	509	19,650	0.46	2490
	406	2.8	24,630				
29	332(sh) [a]	0.36	30,120	520(br)	19,650	$\leq 10^{-3}$	≤3120[d]
	405 [a]	0.30	24,690		18,450		
	434(sh) [a]	0.27	23,040				

[a] in Toluene.

[b] Stokes shift cannot be determined with certainty due to complexity of absorption band envelope.

[c] Obscured by emission from trace amounts of 16.

[d] There is some uncertainty in this value due to the broadness and high signal to noise ratio found in the emission spectrum of 29.

The absorption band for C_{60}-PES is quite distinct from that found for PES. There is an intense band at 335 nm ($\varepsilon = 1.85 \times 10^4$ $M^{-1}cm^{-1}$, $E_{op} = 29850$ cm^{-1}) which tails out past 600 nm. The 335 nm transition envelope is assigned to overlapping $\pi \rightarrow \pi^*$ transitions from the PES and C_{60} moieties based on preliminary spectral modeling analyses. The long tailed absorption feature is associated with the C_{60} and lacks the characteristic weak intensity highly structured transitions found in the 420 to 600 nm range in pristine C_{60} [21]. Excitation at 320 nm gives rise to emission bands at 378 nm ($\phi_f = 2.6 \times 10^{-3}$) and 730 nm. It should be noted that at 320 nm excitation excites both the PES and C_{60} components. The emission at 378 nm which is Stokes shifted by 2440 cm^{-1} is assigned to $S_1 \rightarrow S_0$ transition from PES. The radiative transition at 730 nm ($E_{em} = 13700$ cm^{-1}) has yet to be firmly assigned. It is in a spectral region found for emission from $^1C_{60}$, or $^3C_{60}$ luminescence [7,22]. Time-resolved emission and transient absorption spectroscopic experiments are currently underway to better characterize this emission band and probe the dynamic processes that lead to the 240 fold decrease in for the fluorescence quantum yield from PES [14].

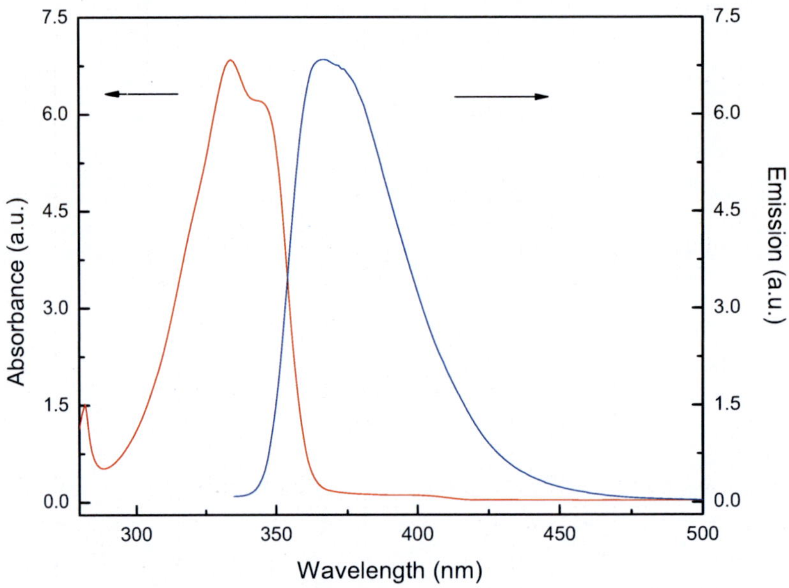

Figure 4. Absorption (red) and emission (blue) spectra for PES in N_2 saturated toluene solution at 298 (± 3) K. The emission spectrum was obtained using 330 nm excitation. The low energy emission at 715 nm is not shown.

2.3.2. Linear OPE/OPV Hybrid Oligomers

Spectroscopic parameters for 7 (Scheme 1) are given in Table 2. The absorption spectrum for 7 is dominated by an intense $\pi \rightarrow \pi^*$ transition at 401 nm with an apparent shoulder ~ 410 nm. Following excitation at 400 nm, a highly structured emission spectrum with the most intense transition at 445 nm with vibronic progressions at 472 and a shoulder at 510 nm was observed. Based on the energy differences between the vibronic progressions, $\hbar\omega \sim 1300$ cm^{-1} which consistent with the quantum spacings found in other aromatic systems [23]. The Stokes shift is related to the total reorganization energy by [23,24]

$$E_{op} - E_{em} = 2\lambda_t \qquad [1]$$

where λ_t is the total reorganization energy between the excited and ground state and is the sum of the vibrational (λ_{vib}) and solvent reorganization energies (λ_o). λ_t is 1230 cm^{-1} based on the optical data given in Table 2.

The absorption and emission data for 24 are given in Table 2 and shown in Figure 5.

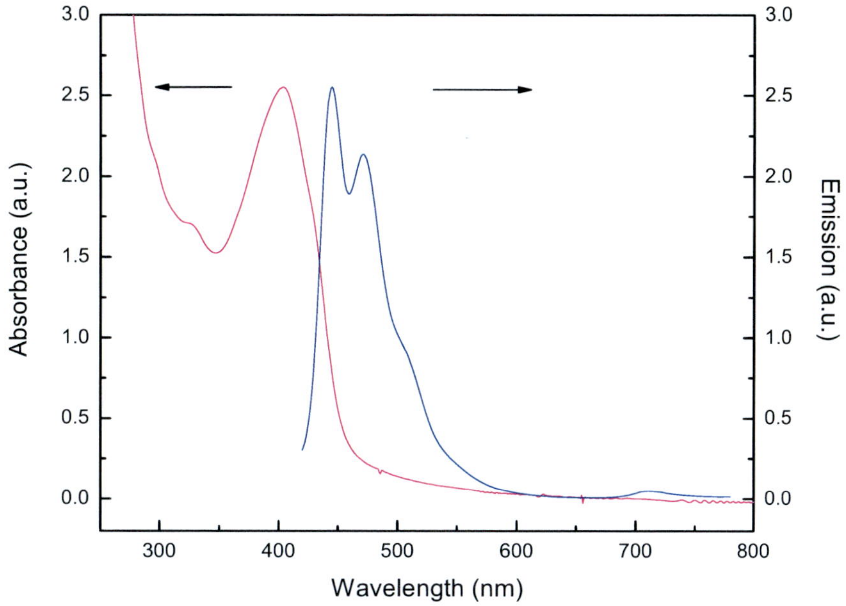

Figure 5. Absorption (red) and emission (blue) spectra for 24 in N_2 saturated $CHCl_3$ solution at 298 (± 3) K. The emission spectrum was obtained using 400 nm excitation.

The absorption spectra for 24 is similar to 7 with an intense $\pi \rightarrow \pi^*$ transition at 404 nm, but with substantially increased absorptivity at 329 nm (ε= 1.4 × 10^4 $M^{-1}cm^{-1}$). The spectral band envelope has contributions from both C_{60} and OPE/OPV bridge in the 404 nm regime, however at higher energies the increased absorptivity is due to the C_{60} termini [13]. The emission spectrum obtained from 404 nm excitation has an emission band at 453 nm (ϕ_f = 6 × 10^{-3}) which is Stokes shifted by 2700 cm^{-1} with no apparent vibronic progressions. The quenched fluorescence is assigned to the π-bridge and has a reorganization energy of ~ 1300 cm^{-1} based on the Stokes shift, eq. [1]. A detailed understanding of the emission spectra awaits the results of the emission spectral fitting analysis developed by Meyer *et al* [24]. The spectral fitting analysis has lead to a quantitative evaluation of the electronic and structural differences between the ground and excited states and has exceedingly successful in understanding the excited state dynamics in d^6 metal polypyridyl complexes. The radiative transition at 710 nm has yet to be firmly assigned, but is likely to be associated with emission from $^1C_{60}$, or $^3C_{60}$ [7].

The absorption and emission spectroscopic properties for 9 (Scheme 1) and 25 given in Table 2, follow the same trends outline above for 7, and 24.

2.3.3. Cruciform OPE/OPV Hybrid Oligomers

Absorption spectra for 13 (Scheme 2) and 26 have a series of overlapping $\pi \rightarrow \pi^*$ transitions localized on bridge in the 300–450 nm regime, Table 2. These $\pi \rightarrow \pi^*$ bands are superimposed transitions localized on the OPV fragment, and the connecting phenyleneethynylene spacers [13]. The absorption spectrum for 13 differs significantly from that observed for 26 with increased absorption below 300 nm and a broad sloping shoulder that extends out past 700 nm. These transitions are associated with the C_{60} termini [13]. The transitions at 325 and 360 nm in 13 are also apparent in 26, albeit the intensities of these absorption bands in 26 are smaller. The origin of the diminished intensity in 26 is not known and awaits a more detailed spectral deconvolution and theoretical analysis.

Absorption spectra for 13 and 16 have a series of overlapping bridge localized $\pi \rightarrow \pi^*$ transitions in the 300–450 nm regime. The ε_{max} values for 13 and 16 range from 2.0×10^4 to 7.8×10^4 M^{-1} cm^{-1} respectively, consistent with assignment of highly allowed $S_0 \rightarrow S_1$ transitions. Generally, the energies of the $\pi \rightarrow \pi^*$ transitions are found at lower energies for 16 *vs.* 13, an expected observation based on the increased conjugation in 16 giving rise to more extended charge delocalization. Comparative spectra for 16 and 27 follow the same trends as those described above for 13 and 26, albeit the absorption spectral band intensities in 27 are higher than 16, a trend that is reversed from observations made for 13 and 26.

The emission spectrum for 13 in room temperature N_2 saturated toluene has an emission band 428 nm (Φ_{em} = 0.96) and possesses a lifetime of 2.8 ns ($k = 3.8 \times 10^8$ s^{-1}) which is reasonably assigned as $S_1 \rightarrow S_0$ radiative transition. The emission spectrum for 26 is more complex than that observed in 13; the most intense emission transition of 26 occurs at 451 nm, some 1185 cm^{-1} lower in energy than 13, with $\Phi_f \sim 10^{-4}$ for 26 *vs.* 0.96 for 13, and a new emission band appears beyond 700 nm. The emission at 711 nm is tentatively assigned to either $^1C_{60}$ or $^3C_{60}$ based emission based on comparison with analogous systems [7].

The molecular assemblies 20 and 28 differ from 13 and 26 due to the diphenyl amino substituent placed on the cruciform OPV bridge fragment. The absorption and emission spectroscopic properties of these assemblies are given in Table 2. The absorption spectra of 20 and 26 possess similar band shape envelopes as those described for 13 and 26 but are shifted to lower energy by approximately 2000 cm^{-1}. The lower energy absorption bands are ascribed to the electron donating amino substituent which in essence decreases the energy difference between the π and π^* levels of the bridge. The emission spectra of 20 and 26 have similar energies (E_{em} ~19700 cm^{-1}) with Φ_f of 0.8 and 1×10^{-3} respectively. A low energy radiative transition at 708 nm (E_{em} ~14125 cm^{-1}) is found in the emission spectrum of 28 and is tentatively assigned to emission from C_{60}.

The absorption and emission spectra for 23 is shown in Figure 6. In contrast to the absorption spectrum of 20, the spectrum found for 23 has two dominant transitions at 309 and 406 nm assigned to $S_o \rightarrow S_2$ and $S_o \rightarrow S_1$ transitions respectively. Apparently the placement of two phenyleneethynylene spacers between the substituted OPV and C_{60} results in a more symmetrical $S_o \rightarrow S_1$ transition. The total reorganization energy for this system is markedly increased with an energy of 2500 cm^{-1} some 1200 cm^{-1} higher in energy than those found in the linear OPE/OPV bridged triads. The microscopic origins of the increased reorganization energies as well as substantial changes in absorption band envelope are not understood at present and therefore more detailed theoretical investigations will be undertaken and described in a future publication. The absorption spectrum of 29 is centered at 405 nm with markedly increased absorptivity at λ < 350 nm which likely due to absorption by the C_{60} termini. The observed emission is broad and weak ($\Phi_f \sim 10^{-3}$).

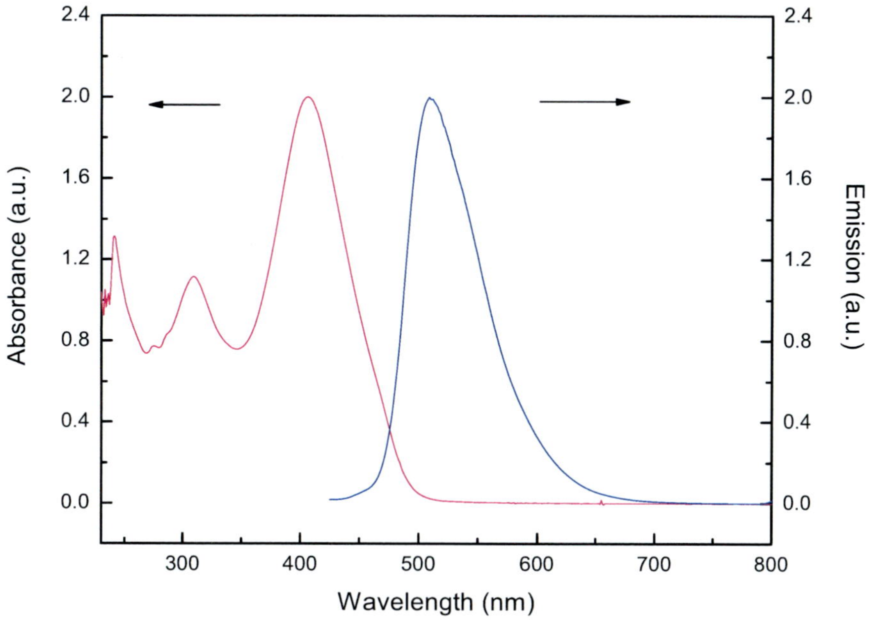

Figure 6. Absorption (red) and emission (blue) spectra for 23 in N_2 saturated $CHCl_3$ solution at 298 (± 3) K. The emission spectrum was obtained using 400 nm excitation.

2.3.4 Excited State Dynamics

The attenuated bridge based emission in all of the C_{60}-π-C_{60} triads relative to the linear and cruciform π-bridges are due to additional non-radiative decay pathway(s), presumably energy transfer from the bridge to C_{60} creating a C_{60} based excited state which may deactivate via radiative and nonradiative pathways to reform the ground state or engage in oxidative excited state electron transfer of the linear and cruciform π-bridges to form a charge separated state in a fashion analogous systems described elsewhere [14]. The exact nature of these decay pathways is under investigation using time resolved emission, nanosecond laser flash photolysis experiments and time-resolved infrared spectroscopic techniques. The results from these experiments are forthcoming [14].

3. CONCLUSION

The synthetic protocols for the two constitutional isomers of π-conjugated OPE/OPV conducting oligomeric bridge, linear and cruciform, have been described and their structure and electrochemical properties have been fully characterized by various spectroscopic methods. Furthermore, it has been demonstrated that the architecture of the cruciform can be altered by placing substituents at the termini of the OPV moiety. This feature is important in the development of new 2-dimensional conducting organic oligomers and lends itself to the synthetic design of more sophisticated 3-dimensional molecular arrays for device applications.

The photophysical data obtained thus far show important and substantial differences depending on the architecture of the OPE/OPV bridging unit. The linear OPE/OPV oligomers display rich spectroscopic signature and remarkably small reorganization energies relative to the donor-attached cruciform OPE/OPV oligomers. The absorption band envelopes found in the cruciform bridged C_{60}-π-C_{60} triads and their π-conjugated precursors display complexity and consist of a number of yet to be identified transitions. It is anticipated that Frank-Condon line shape analysis of the absorption and emission spectra as well as density functional theory (DFT) calculations will be exceedingly useful in the elucidation of the electronic structure of these intriguing molecules. In the C_{60} adducts, the electronic withdrawing C_{60} greatly alters the shape of the absorption band envelop. Therefore, it appears at least for this family of compounds significant electronic coupling exists in both the ground and the exited states. All of the C_{60}-π-C_{60} triads display significantly quenched emission reasonably assigned to non-radiative energy/electron transfer quenching dynamics described in detail in other analogous systems. Transient spectroscopic investigations are now underway to clarify the mechanistic details regarding both energy and electron transfers.

ACKNOWLEDGEMENT

The authors acknowledge NSERC Canada, the Canadian Foundation for Innovation, Industrial Research Innovation Fund (IRIF), and Memorial University of Newfoundland for funding. Li Wang thanks Dr. C. R. Jablonski for funding. Dane Sheppard and Paul Inder are acknowledged for preliminary investigations.

REFERENCES

[1] Guldi, D. M.; Rahman, G. M. A.; Sgobba, V.; Ehli, C. *Chem. Soc. Rev.* 2006, 35, 471-487.

[2] (a) Hirsch, A.; Brettreich, M. *Fullerene: Chemistry and Reactions*; Weily-VCH: Weinheim, 2005. (b) Koruga, D., Hameroff, S.; Withers, J.; Sundareshan, M. *Fullerene C60: History, Physics, Nanobiology, Nanotechnology,* North-Holland: Amsterdam; New York, 1993.

[3] Roncali, J. *Chem. Soc. Rev.* 2005, 34, 483-495.

[4] Shaheen, S. E.; Brabec, C. J.; Sariciftci, N. S.; Padinger, F.; Fromherz, T.; Hummelen, J. C. *Appl. Phys. Lett.* 2001, 78, 841-843.

[5] Bendikov, M.; Wudl, F.; Perepichka, D. F. *Chem. Rev.* 2004, 104, 4891-4946.

[6] Brabec, C. J.; Sariciftci, N. S.; Hummelen, J. C. *Adv. Funct. Mater.* 2001, 11, 15-26.

[7] Guldi, D. M.; Prato, M. *Acc. Chem. Res.* 2000, 33, 695-703.

[8] Nierengarten, J.-F. *New. J. Chem.* 2004, 28, 1177-1191.

[9] Segura, J. L.; Martín, N.; Guldi, D. M. *Chem. Soc. Rev.* 2005, 34, 31-47.

[10] (a) Possamai, G.; Maggini, M.; Menna, E.; Scorrano, G.; Franco, L.; Ruzzi, M.; Corvaja, C.; Ridolfi, G.; Samorì, P.; Geri, A.; Camaioni, N. *Appl. Phys. A* 2004, 79, 51-

58. (b) Martín, N.; Giacalone, F.; Segura, J. L.; Guldi, D. M. *Synth. Met.* 2004, 147, 57-61. (c) Gu, T.; Tsamouras, D.; Melzer, C.; Krasnikov, V.; Gisselbrecht, J.-P.; Gross, M.; Hadziioannou, G.; Nierengarten, J.-F. *ChemPhysChem* 2002,124-127. (d) Kanato, H.; Takimiya, K.; Otsubo, T.; Aso, Y.; Nakamura, T.; Araki, Y.; Ito, O. *J. Org. Chem.* 2004, 69, 7183-7189. (e) Ostubo, T.; Aso, Y.; Takimiya, K. *J. Mater. Chem.* 2002, 12, 2565-2575. (f) Guldi, D. M.; Luo, C.; Swartz, A.; Gómez, R.; Segura, J. L.; Martín, N.; Brabec, C.; Sariciftci, N. S. *J. Org. Chem.* 2002, 67, 1141-1152. (g) Guldi, D. M.; Giacalone, F.; de la Torre, G.; Segura, J. L.; Martín, N. *Chem. Eur. J.* 2005, 11, 7199-7210.

[11] Dhanabalan, A.; van Duren, J. K. J.; van Hal, P. A.; van Dongen, J. L. J.; Janssen, R. A. J. *Adv. Funct. Mater.* 2001, 11, 255-262.

[12] (a) Sánchez, L.; Herranz, M. Á.; Martín, N. *J. Mater. Chem.* 2005, 15, 1409-1421. (b) Sánchez, L.; Sierra, M.; Martín, N.; Guldi, D. M.; Wienk, M. W.; Janssen, R. A. J. *Org. Lett.* 2005, 7, 1691-1694. (c) Guldi, D. M.; Luo, C.; Swartz, A.; Gómez, R.; Segura, J. L.; Martín, N. *J. Phys. Chem. A* 2004, 108, 455-467. (d) Segura, J. L.; Gómez, R.; Martín, N.; Luo, C.; Guldi, D. M. *Synth. Met.* 2001, 119, 65-66. (e) Langa, F.; Gomez-Escalonilla, M. J.; Rueff, J.-M.; Duarte, T. M. F.; Nierengarten, J.-F.; Palermo, V.; Samorì, P.; Rio, Y. Accorsi, G.; Armaroli, N. *Chem. Eur. J.* 2005, 11, 4405-4415.

[13] (a) Zhou, N.; Wang, L.; Thompson, D. W.; Zhao, Y. *Tetrahedron Lett.* 2007, 48, 3563-3567. (b) Chen, G.; Zhao, Y. *Tetrahedron Lett.* 2006, 47, 5069-5073.

[14] Zhou, N.; Wang, L.; Tobin, M.; Thompson, D. W.; Zhao, Y. *manuscript in preparation.*

[15] Wilson, J. N.; Joscowicz, M.; Wang, Y.; Bunz, U. H. F. *Chem. Commun.* 2003, 2962-2963.

[16] Wilson, J. N.; Smith, M. D.; Enkelmann, V.; Bunz, U. H. F. *Chem. Commun.* 2004, 1700-1701.

[17] Wilson, J. N.; Bunz, U. H. F. *J. Am. Chem. Soc.* 2005, 127, 4124-4125.

[18] Zhao, Y.; Shirai, Y.; Slepkov, A. D.; Cheng, L.; Almeny, L. B.; Sasaki, T.; Hegmann, F. A.; Tour, J. M. *Chem. Eur. J.* 2005, 11, 3643-3658.

[19] Shirai, Y.; Zhao, Y.; Cheng, L.; Tour, J. M. *Org. Lett.* 2004, 4, 2129-2132.

[20] Atienza, C.; Insuasty, B.; Seoane, C.; Martín, N.; Ramey, J.; Aminur Rahman, G. M.; Guldi, D. M. *J. Mater. Chem.* 2005, 15, 124-32.

[21] Gallagher, S. H.; Armstrong, R. S.; Lay, P. A.; Reed, C. A. *J. Phys. Chem.* 1995, 99, 5817-5825.

[22] Sandanayaka, A. S. D.; Taguri, Y.; Araki, Y.; Ishi-i, T.; Mataka, S.; Ito, O. *J. Phys. Chem. B* 2005, 109, 22502-22512.

[23] Turro, N. J. *Modern Molecular Photochemistry*; University Science Books: Sausalito, CA, 1991.

[24] Chen, P.; Meyer, T. J. *Chem. Rev.* 1998, 98, 1439-1478.

215In: Research in Hybrid Materials
Editors: Simon J. Brunner and Julian W. Egger
ISBN 978-1-60456-718-2

Chapter 8

INFLUENCE OF EPOXY-AMINE RATIO ON DEGRADATION OF EPOXY-SILICA HYBRID MATERIALS

Jelena Macan and Marica Ivanković
Faculty of Chemical Engineering and Technology,
University of Zagreb, HR-10001 Zagreb,
Marulicev trg 19, p.p. 177, Croatia

ABSTRACT

Organic-inorganic hybrid materials are nanostructured materials whose inorganic phase is synthesised by sol-gel process. Polymer based hybrid materials contain inorganic phase dispersed in polymer matrix on a nanometric level. Sol-gel process is particularly suitable for preparation of these materials since the inorganic phase is formed at low temperatures, preserving polymeric matrix from degradation during synthesis. Similarly to other polymeric nanocomposites, organic-inorganic hybrid materials show improved thermal stability compared to unmodified polymeric materials. This can be ascribed to poorer diffusion of degradation products through the stable inorganic phase, and also to stabilising effect of silicon.

Organic-inorganic hybrid materials based on diglycidyl ether of bisphenol A (DGEBA) and 3-glycidyloxypropyltrimethoxysilane (GLYMO) were prepared, with a poly(oxypropylene) diamine, Jeffamine D230, as a crosslinking agent. Previous research on these hybrid materials has shown that epoxy groups of GLYMO are able to react only with primary hydrogen atoms in amine groups of D230. Therefore two sets of hybrids were prepared, one with stoichiometric quantity of amine (1 mole of amine groups per 2 moles of epoxy groups) and the other with doubled quantity of amine in relation to epoxy groups of GLYMO. Thermal degradation of all prepared hybrid materials, as well as unmodified epoxy resin, was investigated by combined differential thermal analysis and thermogravimetric analysis (DTA-TGA) in inert atmosphere of nitrogen and oxidative atmosphere of air. It was determined that although increased content of amine improves conversion of epoxy groups and therefore degree of organic crosslinking in the hybrid materials, their overall thermal stability is decreased. Hybrid materials with increased amine content have lower degradation onset temperatures and lower residue at 1000 °C.

This can be attributed to unreacted amine groups that remain in the material, and to lower density of inorganic network. Atmosphere has very little influence on the main degradation step, which probably proceeds in self-generated atmosphere.

INTRODUCTION

Research of polymer nanocomposites, materials in which inorganic phase is dispersed on nanometric level within polymer matrix, increased greatly in recent years. Improved interaction between phases in such materials is expected to result in superior properties compared to classical composites. Organic-inorganic hybrid materials can be regarded as a class of nanocomposites where the inorganic phase is formed either in parallel with the polymer matrix or directly within it, instead of by mixing preformed inorganic nanofiller with monomers or polymer. This manner of preparation ensures dispersion of inorganic phase within the polymer matrix at a molecular level, which leads to much more homogeneous material than would be possible to obtain with hard-to-disperse nanofiller. In consequence, basically the whole volume acts as an interphase [1]. Organic-inorganic hybrid materials unite qualities of both organic and inorganic phase in synergistic manner and are expected to possess superior properties [2].

Sol-gel process is often used to prepare organic-inorganic hybrid materials with polymer matrix because of its flexibility and low reaction temperatures, which prevent possible degradation of the polymer matrix during the preparation of hybrid materials [3]. Inorganic phase is formed by hydrolysis and condensation of metal (usually silicon) alkoxides, as shown in following equations (R represents an alkyl group) [4]:

$$\equiv\text{Si-OR} + H_2O \Leftrightarrow \equiv\text{Si-OH} + \text{ROH} \tag{1}$$

$$\equiv\text{Si-OR} + \equiv\text{Si-OH} \rightarrow \equiv\text{Si-O-Si}\equiv + \text{ROH} \tag{2}$$

$$\equiv\text{Si-OH} + \equiv\text{Si-OH} \rightarrow \equiv\text{Si-O-Si}\equiv + H_2O \tag{3}$$

Reactions of hydrolysis and condensation can be either acid or base catalysed.

In order to achieve better properties of hybrid materials, interphase interaction can be improved by formation of hydrogen or covalent bonds between the phases. For this purpose organically modified silicon alkoxides, $R'\text{-Si(OR)}_3$, are used, where R' is an organic chain containing a reactive group. The reactive group can react either with monomers or with polymer matrix while remaining linked to the inorganic phase through hydrolysis-resistant Si-C covalent bond [1,3].

Similarly to nanocomposites, organic-inorganic hybrid materials have noticeably improved thermal stability in comparison with unmodified polymer material [1,5]. Formation of stable char is one of desirable mechanisms of flame retardation in polymers, since the char layer acts both as a thermal insulator and as a barrier to oxygen diffusion. A good way of augmenting char formation is the addition of inorganic fillers, since the presence of inherently stable inorganic phase produces more voluminous and mechanically stronger char, consequently improving its insulating and barrier properties [6].

In this chapter, thermal degradation kinetics of epoxy-silica organic-inorganic hybrids, with 3-glycidyloxypropyltrimethoxysilane (GLYMO) as the inorganic precursor, is studied. GLYMO is an organically modified silicon alkoxide, which can undergo both sol-gel condensation of the alkoxy groups and curing of the epoxy functionality to form a hybrid network containing covalent bonds between organic and inorganic phase. Several workers prepared hybrid materials based on epoxy resin using sol-gel process [5, 7-26], but there was little research on hybrid materials with GLYMO as the main inorganic precursor [27-33]. Presence of inorganic phase of nanometric dimensions is expected to enhance thermal properties of the epoxy matrix, both by acting as flame retardant and by enhancing char formation [34,35].

Degradation of Epoxy Resins

Epoxy resins are engineering polymers with a wide area of application, either unmodified or as matrices for composite materials. Consequently, degradation of epoxy materials has been widely studied. Choice of curing agent greatly influences the stability of epoxy resins, amine-cured resins being less stable than those cured with anhydrides because of strong nucleophilic nature of amine groups within the network. Since an amine crosslinking agent was used in research presented in this chapter, focus will be on the crosslinking and degradation of epoxy-amine resins.

Amine crosslinking agent can react with epoxy group of resin with one or both of its hydrogen atoms, as shown in equations (4) and (5):

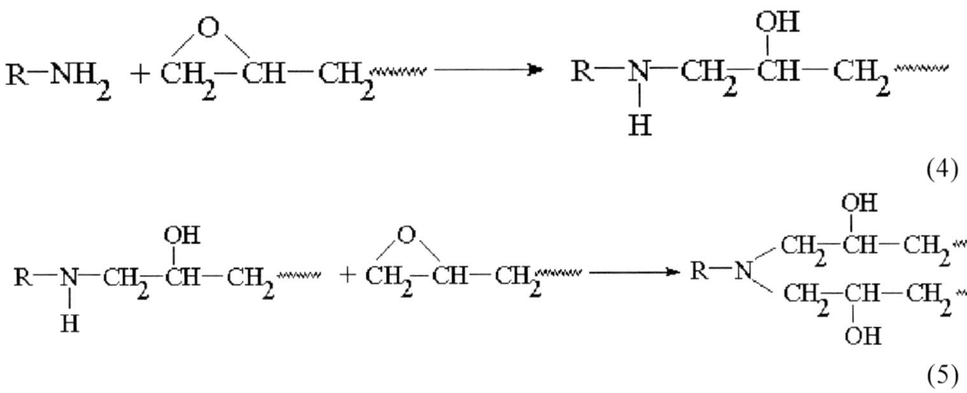

(4)

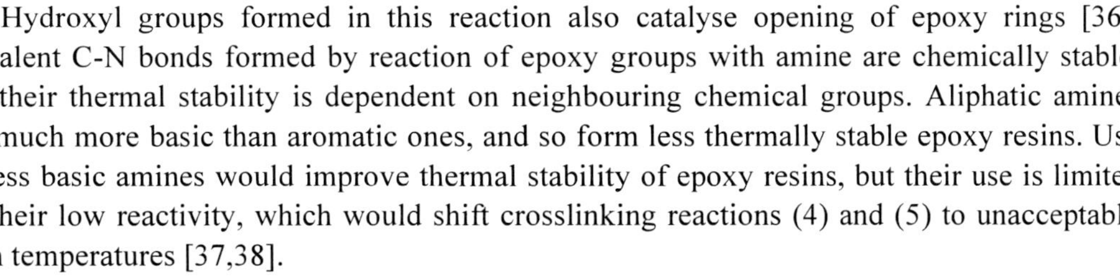

(5)

Hydroxyl groups formed in this reaction also catalyse opening of epoxy rings [36]. Covalent C-N bonds formed by reaction of epoxy groups with amine are chemically stable, but their thermal stability is dependent on neighbouring chemical groups. Aliphatic amines are much more basic than aromatic ones, and so form less thermally stable epoxy resins. Use of less basic amines would improve thermal stability of epoxy resins, but their use is limited by their low reactivity, which would shift crosslinking reactions (4) and (5) to unacceptably high temperatures [37,38].

Degradation mechanism of epoxy resins is very complex, and consists of numerous parallel reactions in which various gaseous and solid products are formed. Degradation at low temperatures proceeds through homolytic scission of covalent bonds. Although not resulting in mass loss of the samples, bond scission leads to changes in the chemical composition of the epoxy resin and its properties, noticeable already at 240 °C and quite advanced by 290 °C. In this temperature interval recombination of radicals formed by degradation also takes place, leading to fresh crosslinking of the resin but also to randomisation of its structure [38,39]. Low mass loss below 300 °C is ascribed to evaporation of residual solvent and adsorbed humidity [40].

The first major mass loss at 300-350 °C in both oxidative and nonoxidative degradation of amine cured epoxies is caused by dehydration due to elimination of water molecule from oxypropylene group, $-CH_2-CH(OH)-$, and subsequent formation of double bonds. This dehydration is concurrent with network breakdown [38-40]. Epoxy-amine resins contain more hydroxyl groups then ones cured with anhydrides, which facilitates dehydration and is probably an additional cause of their lower thermal stability [41].

Scission of $CH-CH_2$ and $O-CH_2$ bonds proceeds in parallel and subsequently with the dehydration [39,40]. The products form unsaturated phenolic compounds by Claisen rearrangement, which in turn either polymerise to form char or are released as gaseous products [41]. The final degradation step consists of isomerization, intramolecular cyclization, chain transfer and other reactions of the radicals formed in the initial stages of degradation [42]. Beside water and phenolic compounds, which are the main products of epoxy-amine resin degradation, other possible products are benzofuranes, anilines and trace aliphatic compounds, and there is evidence of long-lived free radicals within the char [38,40].

Incomplete cure of epoxy resin and resulting free functional groups have, as a rule, detrimental influence on thermal stability since they act as centres of polymer chain scission [42]. Resins with increased amine content are less thermaly stable than those with stoichiometric quantity of amine, and thermal stability improves as degree of crosslinking increases. Excess of epoxy groups does not necessarily have a detrimental influence on thermal stability, since they homopolymerise at higher temperatures according to equation (6) and in this way even contribute to the stability of the resin [43]. Resulting exothermic effect was also ascribed to isomerisation of epoxy groups into aldehydes, which was not experimentally confirmed [39,41]. Homopolymerisation is confirmed by significant release of water during degradation of uncrosslinked DGEBA at 350 °C, probably as a consequence of dehydration of ether linkages formed during previous homopolymerisation of DGEBA [41].

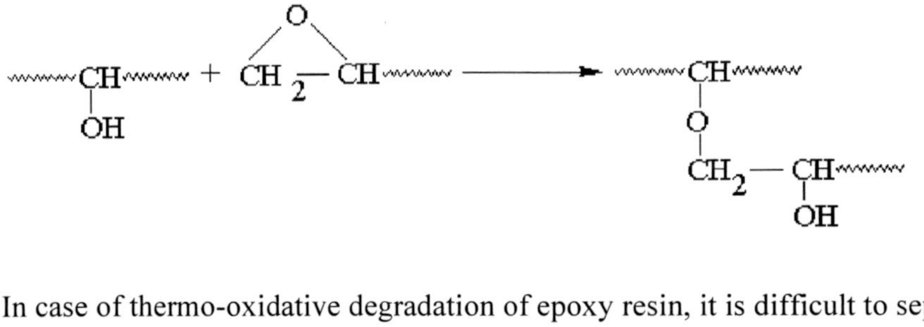

(6)

In case of thermo-oxidative degradation of epoxy resin, it is difficult to separate oxidation from nonoxidative degradation, especially if the resin was previously cured in air, which can cause its partial oxidation. Dehydration is again the primary degradation reaction [44], followed by homolytic scission of covalent C-O-C and C-N bonds as indicated by formation of phenolic products [45]. Another cause of scission in oxidative atmosphere can be hyperoxide degradation, which leads to formation of carbonyl and amide groups [46]. Apparently, at temperatures under 310 °C degradation mechanism is not atmosphere dependent [44], although there is some indication of early reaction with oxygen to form peroxides at lower temperature range 160-250 °C [39]. The mass loss in 250-500 °C range in oxidative atmosphere is lower in comparison with nitrogen atmosphere, which also points to possible bonding of oxygen to resin and temporary improvement of its thermal stability during thermo-oxidative degradation [44]. Endothermic reactions above 360 °C are ascribed to evaporation of heavier products of degradation [40].

New legislation mandates that traditional toxic halogen-based additives for decreasing polymer flammability be replaced with new, ecologically acceptable alternatives. Most often these are phosphorus and silicon compounds, which shift degradation to higher temperatures. During degradation silicon migrates to the surface of the material and creates a protective layer, which prevents thermo-oxidative degradation of epoxy resin [47].

EXPERIMENTAL

An epoxy resin, diglycidyl ether of bisphenol A (DGEBA, Epikote 828 EL, Shell Chemicals) with the epoxy equivalent weight of 190 g/mol, and 3-glycidyloxypropyltrimethoxysilane (98%, Aldrich Chemicals) were used to synthesize organic-inorganic hybrid materials, with poly(oxypropylene) diamine (Jeffamine D230, Huntsman Corporation) acting both as curing agent for epoxy and as basic catalyst for GLYMO hydrolysis. Formulas of all starting materials are given in Figure 1. The materials were used as received.

Previous research on these hybrid materials has shown that epoxy groups of GLYMO are able to react only with primary hydrogen atoms in amine groups of D230 [31]. Therefore two sets of hybrids were prepared, one with stoichiometric quantity of amine (1 mole of amine groups per 2 moles of epoxy groups, Series 1) and the other with doubled quantity of amine in relation to epoxy groups of GLYMO (1 mole of amine groups per 1 mole of GLYMO epoxy groups, Series 2). To prepare hybrid materials, DGEBA and GLYMO were blended in mass ratios of 1:2; 1:1; and 2:1 at room temperature in a closed vessel for 90 min. After addition of D230, the mixtures were stirred for another 60 min. Pure DGEBA and pure GLYMO where also mixed with appropriate amount of D230 for 90 min. Samples were designated according to the mass ratio of GLYMO to DGEBA and amount of D230. Designations and composition of all prepared materials are given in Table 1. Curing was effected at room temperature for 30 days, making use of air humidity for GLYMO hydrolysis. The materials were then postcured for 24 h at 120 °C in an oven.

Mass loss of fully cured samples was measured by combined differential thermal analysis and thermogravimetric analysis (DTA-TGA), using a Seiko SII Exstar 6000. Samples weighted at 14-19 mg with accuracy of ± 0.01 mg were heated in open platinum crucibles from room temperature to 1000 °C at heating rate of 10 °C/min in either a nitrogen gas flow or air flow of 250 cm^3/min.

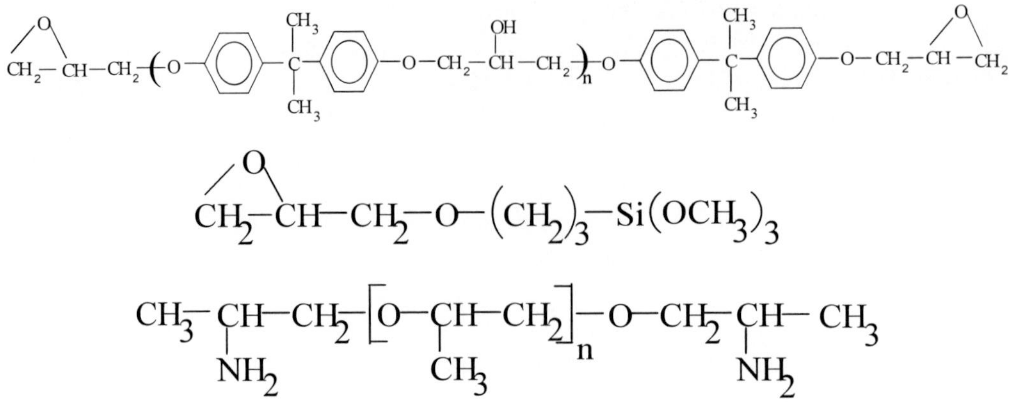

Figure 1. Formulas of DGEBA, GLYMO and D230.

Table 1. Designations and composition of prepared hybrid materials

Designation	Mass ratio GLYMO: DGEBA	Amount of D230 added
Epoxy	0 : 1	0.30 · *m* (DGEBA)
G1E2	1 : 2	0.30 · *m* (DGEBA) + 0.24 · *m* (GLYMO)
G1E2J2		0.30 · *m* (DGEBA) + 0.48 · *m* (GLYMO)
G1E1	1 : 1	0.30 · *m* (DGEBA) + 0.24 · *m* (GLYMO)
G1E1J2		0.30 · *m* (DGEBA) + 0.48 · *m* (GLYMO)
G2E1	2 : 1	0.30 · *m* (DGEBA) + 0.24 · *m* (GLYMO)
G2E1J2		0.30 · *m* (DGEBA) + 0.48 · *m* (GLYMO)
G1E0	1 : 0	0.24 · *m* (GLYMO)
G1E0J2		0.48 · *m* (GLYMO)

Degradation in Nitrogen Atmosphere

Thermogravimetric curves showing degradation of unmodified epoxy resin and hybrid materials are given in Figure 2. The residue at 1000 °C increases with content of GLYMO in the hybrid material. Degradation onset also shifts to higher temperatures, but only for Series 1 hybrids (Table 2), while in Series 2 hybrids the degradation has the same onset temperature as the unmodified epoxy resin. Lowering of degradation onset is probably due to presence of free amine groups and increased concentration of poly(oxypropylene) chains which are more sensitive to degradation than epoxide segments of the material [48].

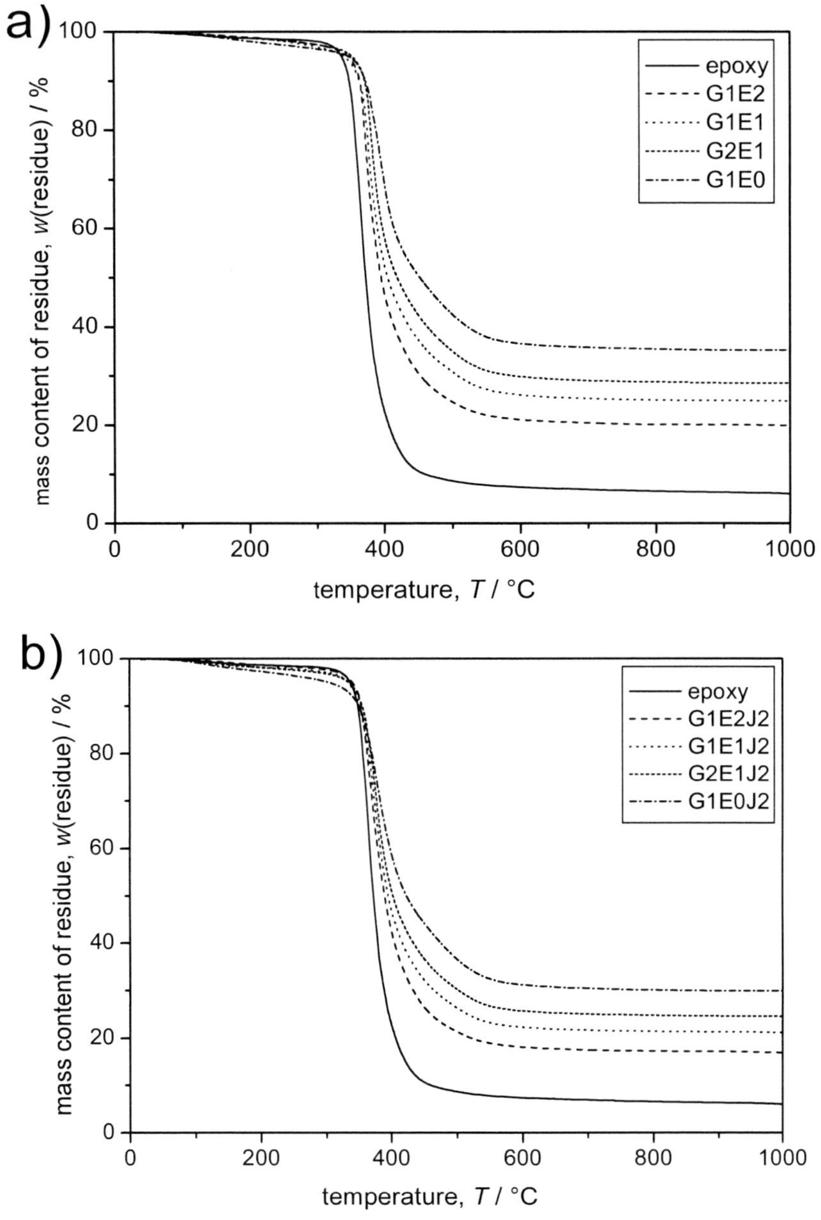

Figure 2. Thermogravimetric curves of investigated samples in nitrogen atmosphere: a) Series 1 hybrids and b) Series 2 hybrids, in comparison with unmodified epoxy resin

The inorganic phase is less dense in Series 2 hybrids due to relatively lower GLYMO content, which facilitates the diffusion of gaseous products of degradation from within the material. This results in decreased formation of carbon char, which is calculated as difference between the total residue at 1000 °C and theoretical SiO_2 content in hybrid material, according to the following equation:

$$w(\mathrm{C}) = w(\mathrm{residue}) - w(\mathrm{SiO_2}) \tag{7}$$

Table 2. Onset temperature of degradation, T_0, mass content of residue at 1000 °C as determined by thermogravimetric analysis, w(residue), theoretical content of SiO_2, $w(SiO_2)$, content of carbon char calculated according to equation (7), w(C), and difference of carbon char content in Series 1 and Series 2 hybrids with same GLYMO:DGEBA mass ratio, Δw(C)

Material	T_0, °C	w(residue), %	$w(SiO_2)$, %	w(C), %	Δw(C), %
Epoxy	313	6.04	0.00	6.04	--
G1E2	323	19.89	6.72	13.17	2.92
G1E2J2	316	16.90	6.65	10.25	
G1E1	318	24.88	10.14	14.74	3.35
G1E1J2	312	21.15	9.76	11.39	
G2E1	330	28.47	13.59	14.88	3.08
G2E1J2	315	24.55	12.75	11.80	
G1E0	336	35.31	20.54	14.77	3.16
G1E0J2	314	29.97	18.36	11.61	

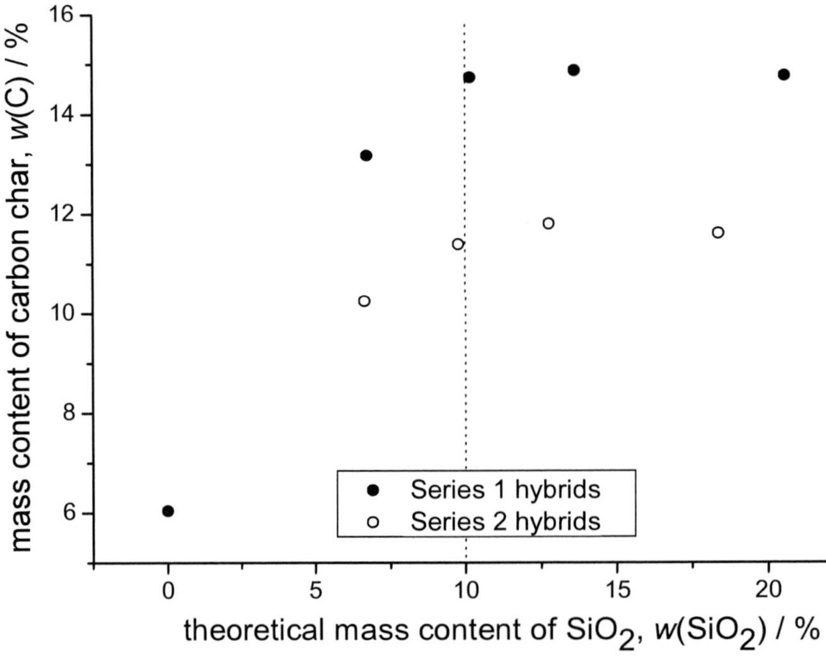

Figure 3. Dependence of carbon char content on theoretical SiO_2 content and amount of amine added.

Carbon char content and the difference of char content for Series 1 and Series 2 hybrids with the same GLYMO:DGEBA ratio is also given in Table 2. The content of carbon char is approximately 3 % lower for Series 2 hybrids, and Figure 3 shows dependence of the carbon char content on the content of inorganic phase. It can be seen that increasing content of inorganic phase favours creation of additional carbon char up to ~10 % SiO_2 content. Since the char further prevents diffusion of degradation products, decrease of total residue is a sign of decreased thermal stability of hybrid materials. Combined with decreased temperature of degradation onset it indicates that increased amine content annuls positive influence of inorganic phase on thermal stability of hybrid materials.

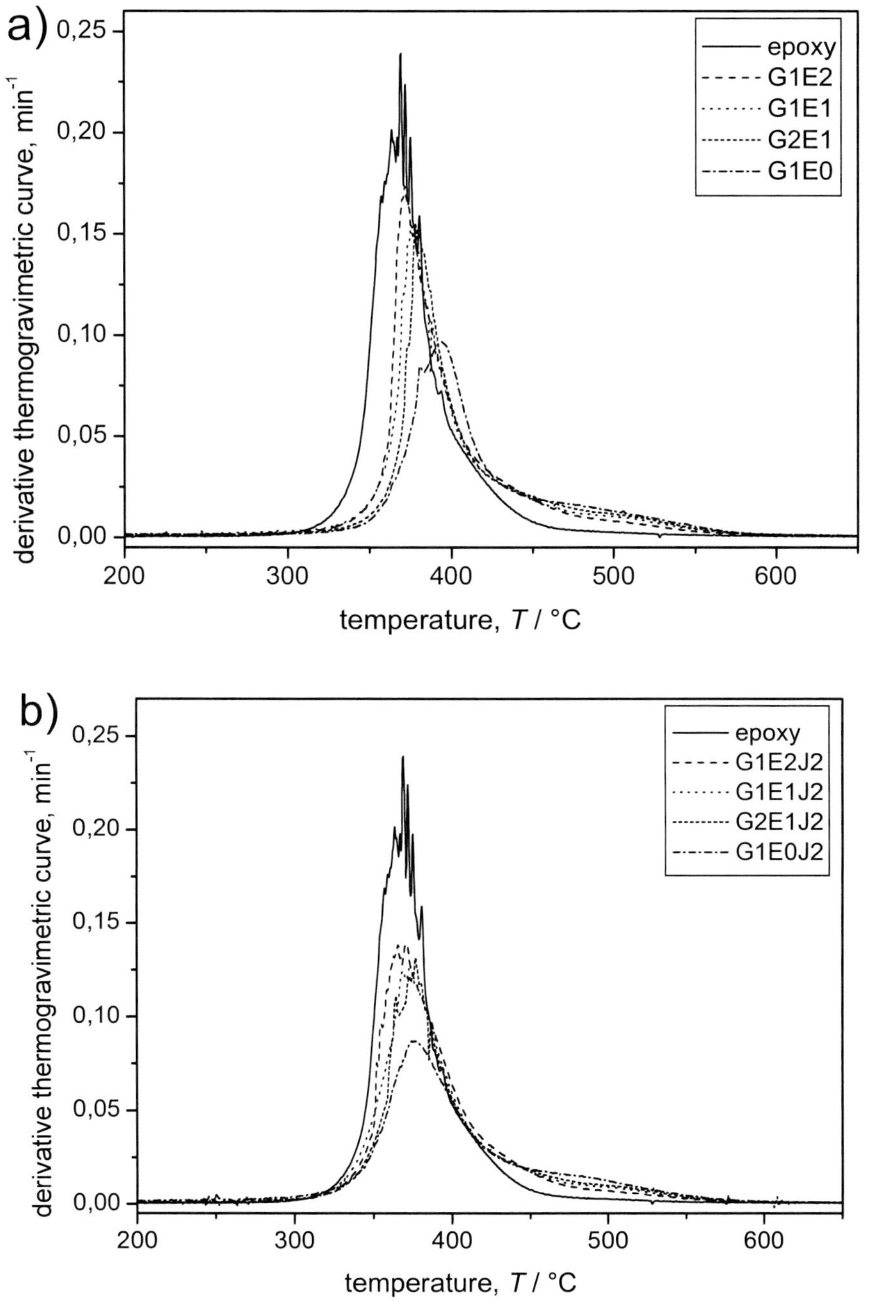

Figure 4. Derivative thermogravimetric curves in nitrogen atmosphere: a) Series 1 hybrids and b) Series 2 hybrids, in comparison with unmodified epoxy.

Basic information on mechanism of degradation can be obtained from derivative thermogravimetric curves, DTG (Figure 4). While unmodified epoxy resin degrades in a single step, hybrid materials show an additional degradation step at higher temperatures (400-600 °C). Mass loss at these temperatures is relatively low, ~1 % (Table 3). The effect of increased amine content is also more clearly visible from derivative mass loss curves. Series 1 hybrids show a visible shift of degradation to higher temperatures, while for Series 2 hybrids the main mass loss maximum mostly overlaps that of unmodified epoxy resin.

Table 3. Temperature of maximal rate of degradation for the first and the second step as determined from derivative thermogravimetric curves, T_{max}, and accompanying relative mass losses, w(loss)

Material	First step		Second step	
	T_{max}, °C	w(loss), %	T_{max}, °C	w(loss), %
epoxy	369	93.96	--	--
G1E2	372	79.11	440	1.00
G1E1	377	74.01	464	1.11
G2E1	378	70.27	452	1.26
G1E0	394	63.32	464	1.37
G1E2J2	366	82.27	442	0.83
G1E1J2	373	77.93	452	0.92
G2E1J2	377	74.43	450	1.02
G1E0J2	374	68.80	462	1.23

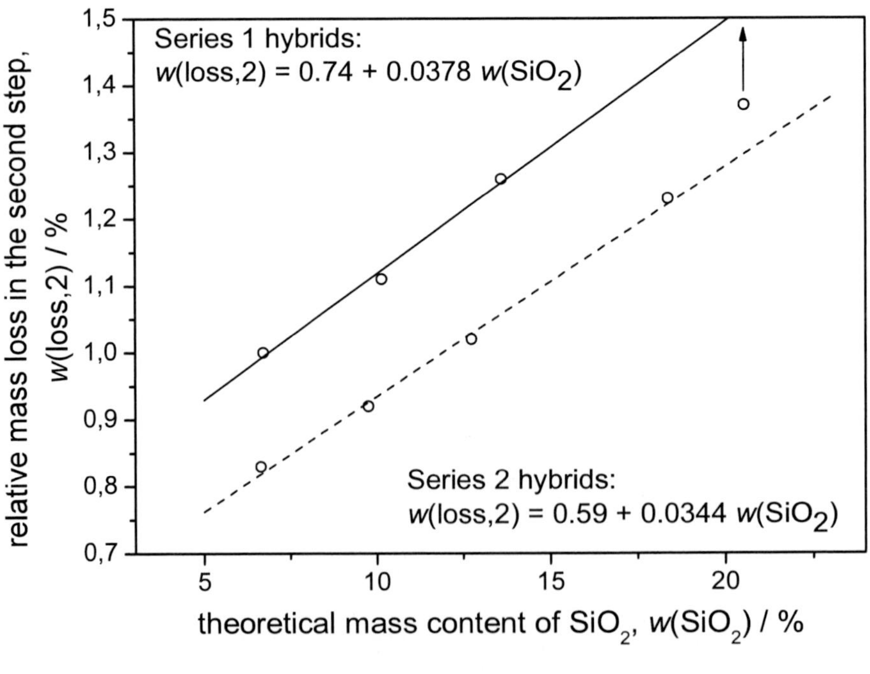

Figure 5. Dependence of mass loss in the second degradation step on theoretical SiO_2 content, for Series 1 (full line) and Series 2 (dashed line) hybrids.

The mass loss in the second step is linearly dependent on theoretical content of SiO_2, as shown in Figure 5. For Series 1 hybrids the mass loss in the first step is smaller, and this difference also directly depends on the theoretical SiO_2 content. It may be supposed that the organic residue which degrades in the second step is linked to the inorganic phase, most probably by the covalent Si-C bond. The residue is more stable in Series 1 hybrids due to decreased diffusion of degradation products through denser inorganic phase. This increases the amount of residue after the first step, which would degrade in the second one. A noticeable offset of G1E0 hybrid from the linear dependence is also a consequence of its particularly dense inorganic phase which prevents degradation of char.

Except for a shift towards lower temperatures, addition of amine does not cause obvious changes in appearance of DTG maxima, as shown in Figure 6 for hybrid materials G1E2 and G1E2J2, as well as G1E0 and G1E0J2. It is therefore likely that amount of amine used in preparation of hybrid materials does not particularly influence their degradation mechanism.

Heat of degradation of the organic phase can be estimated from differential thermal analysis (DTA) curves (Figure 7). It can be seen that the degradation reaction in inert atmosphere is exothermal. The exothermal maxima correspond quite well with mass loss steps, as illustrated in example of hybrids G1E0 and G1E0J2 (Figure 8). Only unmodified epoxy resin shows a complicated set of endothermal reactions between the two exothermal steps.

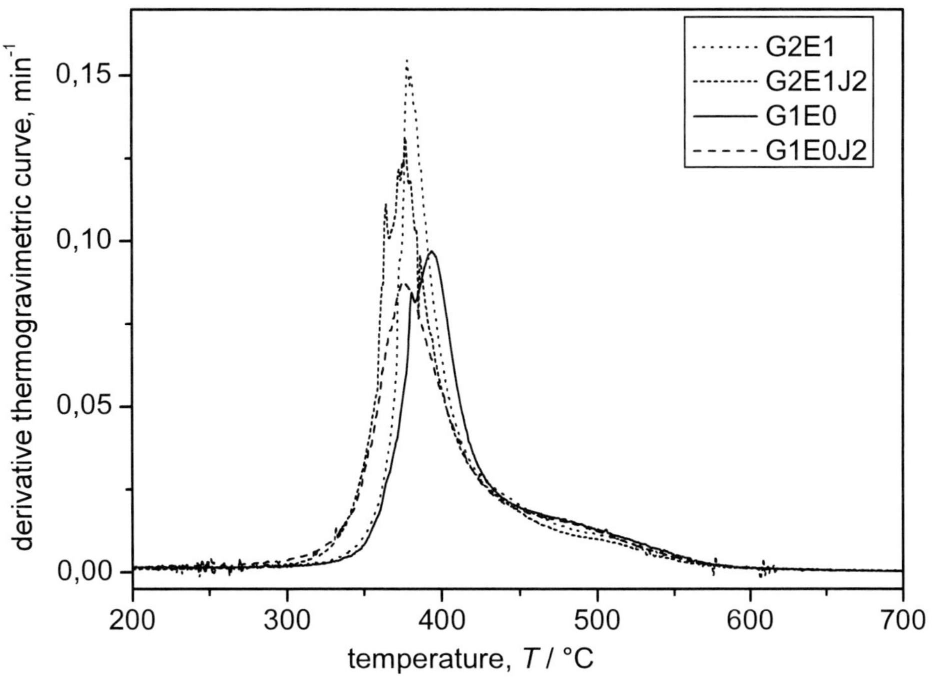

Figure 6. Comparison of derivative thermogravimetric curves for listed hybrids from Series 1 and 2, in nitrogen atmosphere.

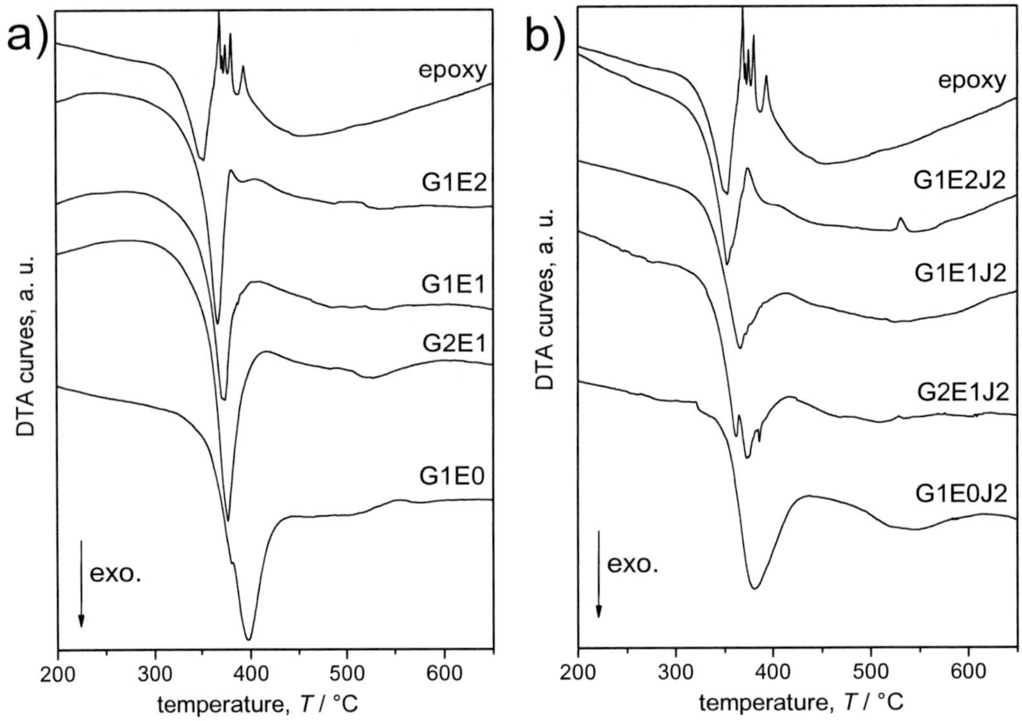

Figure 7. Differential thermal analysis curves in nitrogen atmosphere for a) Series 1 hybrids and b) Series 2 hybrids, in comparison with unmodified epoxy.

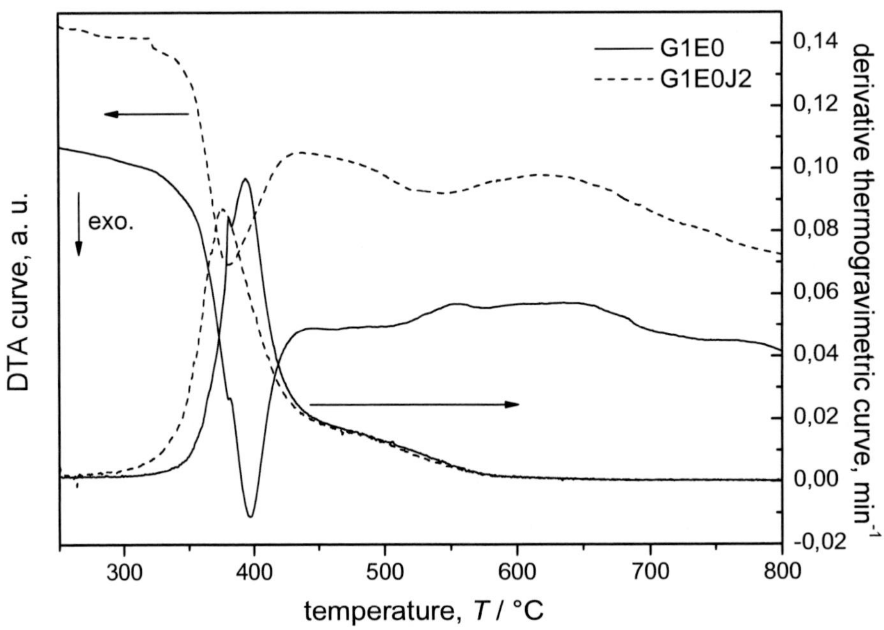

Figure 8. Comparison of differential thermal analysis and derivative thermogravimetric curves for listed hybrids in nitrogen atmosphere.

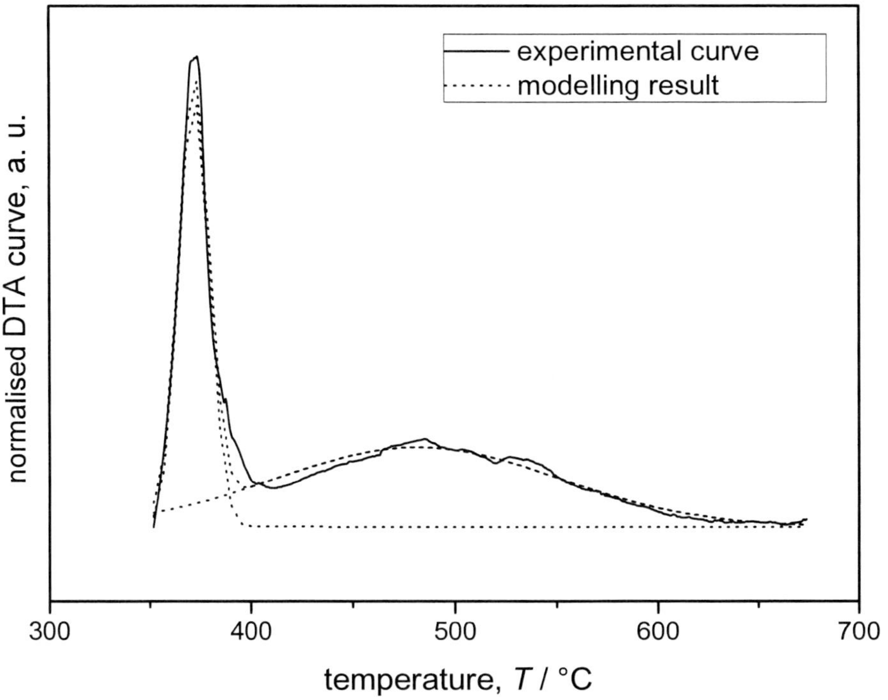

Figure 9. Separation of a normalised differential thermal analysis curve for G1E1 hybrid into two maxima, under assumption that they can be described by Gauss equation.

Table 4. Area under DTA curve, A_{DTA}, temperatures of maximum degradation rate for the first and the second step of degradation as determined from DTA curves, T_{max}, and relative area, A_{rel}, under the corresponding maximum

Material	A_{DTA}, a.u./g	First step		Second step	
		T_{max}, °C	A_{rel}, %	T_{max}, °C	A_{rel}, %
epoxy	224.7	353	13.77	455	86.23
G1E2	193.8	367	19.16	458	80.84
G1E1	207.8	374	21.90	462	78.10
G2E1	237.7	377	27.15	458	72.85
G1E0	245.4	397	35.61	466	64.39
G1E2J2	248.0	352	8.80	475	91.20
G1E1J2	233.9	367	18.70	482	81.30
G2E1J2	212.1	373	23.76	470	76.24
G1E0J2	203.9	381	38.76	489	61.24

Area under DTA maximum is proportional to the heat released in that step of degradation. Normalised DTA curves are approximated with a sum of Gauss maxima in an attempt to determine relative heat of reaction for each step, as illustrated in Figure 9 for hybrid G1E1, and the results are given in Table 4. Total heat of reaction per gram of sample does not depend on the organic content. It also increases in Series 1 hybrids and decreases in Series 2 hybrids. This irregular behaviour is probably caused by overlapping of exothermal reactions with endothermal ones, such as chain scission and evaporation of gaseous products. Consequently, it is difficult to draw any conclusions on the mechanism of degradation without further research, for example by analysis of gaseous products and isothermal measurements at suitable temperatures.

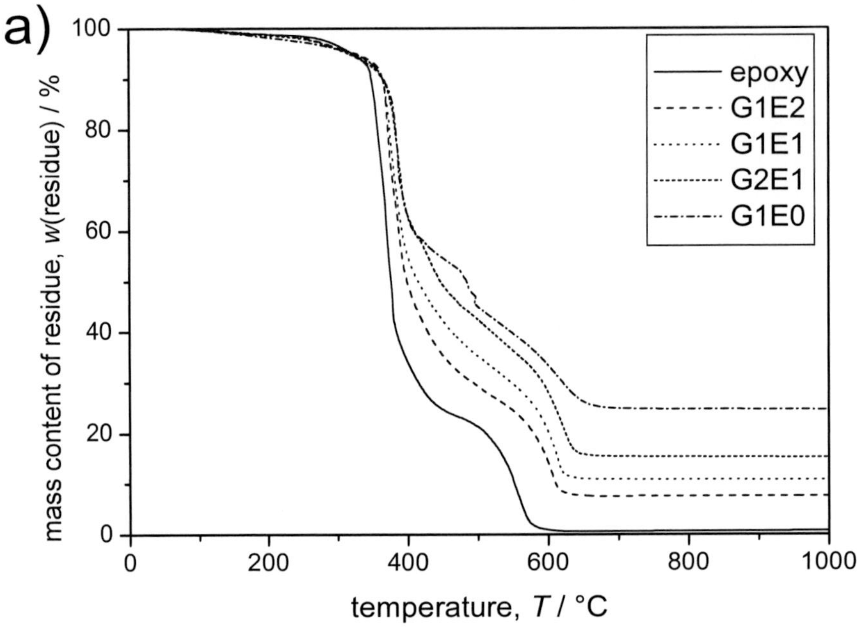

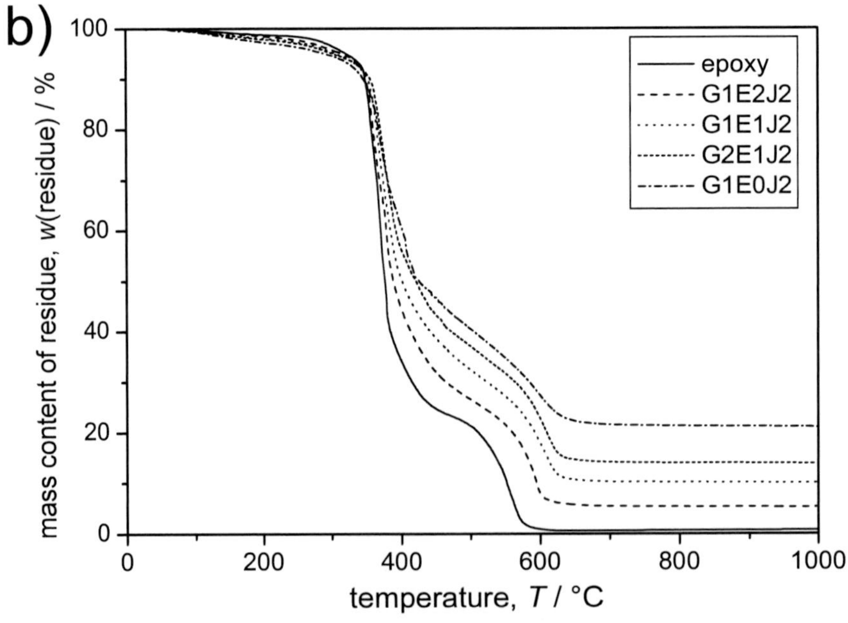

Figure 10. Thermogravimetric curves of investigated samples in atmosphere of air: a) Series 1 hybrids and b) Series 2 hybrids, in comparison with unmodified epoxy.

Table 5. Mass content of residue at 1000 °C, *w*(residue), and after the first degradation step, *w*(residue,1), determined by thermogravimetric analysis; onset temperature, T_o, and maximal degradation rate temperature, T_{max}, for the first and the second degradation step, as determined from derivative thermogravimetric curves

Material	*w*(residue), %	$T_o(1)$, °C	$T_{max}(1)$, °C	*w*(residue,1), %	$T_o(2)$, °C	$T_{max}(2)$, °C
Epoxy	0.73	268	366	23.86	462	543
G1E2	7.52	257	371	27.22	522	594
G1E1	10.75	252	376	33.44	516	599
G2E1	15.20	256	381	41.73	506	608
G1E0	24.65	278	382	44.40	505	604
G1E2J2	5.30	260	364	26.55	500	585
G1E1J2	10.05	258	370	29.99	524	598
G2E1J2	13.88	260	375	34.00	528	598
G1E0J2	21.03	259	366	41.91	487	585

Degradation in Oxidative Atmosphere

Degradation mechanism of epoxy resin and hybrid materials in oxidative atmosphere of air is quite complex, as can be seen from Figures 10 and 11. It can be simplified into two steps: primary degradation, and combustion of char formed in the first step in air. The mass loss, as well as the onset temperature and the temperature of maximal rate of degradation is listed in Table 5 for all investigated materials. At first it seems that the first step does not correspond to the degradation in inert atmosphere, since the onset temperature is over 50 °C lower, and the mass residue after the first step is noticeably higher in oxidative atmosphere (compare Table 2). But the comparison of DTG curves taken in inert and oxidative atmosphere (Figure 12) shows that the most pronounced mass loss in inert atmosphere completely overlaps the first larger mass loss in oxidative atmosphere. Their position and size are identical for each investigated material. It is therefore justified to assume that the mechanism of degradation in this step does not change regardless of the atmosphere. The same behaviour was seen in epoxy resins [44], and was also confirmed for organic-inorganic hybrids prepared from GLYMO and aminosilane [49]. This is ascribed to degradation proceeding in self-generated atmosphere within the material, due to insufficient diffusion of oxygen from without the material [50].

The degradation mechanism is not influenced even by the initial reaction with oxygen which results in a small mass loss below 300 °C. The onset temperature of this initial reaction is essentially independent of hybrid composition, with a noticeable deviation to higher values only in case of G1E0 hybrid.

The second step, the combustion of products of the first degradation step, is much harder to define since in hybrids with increased quantity of GLYMO it consists of several overlapping effects, as can be clearly seen from DTG curves (Figure 11). This makes the onset temperature very difficult to determine accurately, so temperatures of maximal degradation rate are used instead. These temperatures are comparable for all hybrid materials, and are approximately 50 °C higher than for unmodified epoxy resin, which can be ascribed to inhibiting influence of inorganic phase [48]. Series 2 hybrids show the shift of degradation onset to lower temperatures in oxidative atmosphere as well. This effect is noticeable also in the second degradation step, as can be seen in Figure 13.

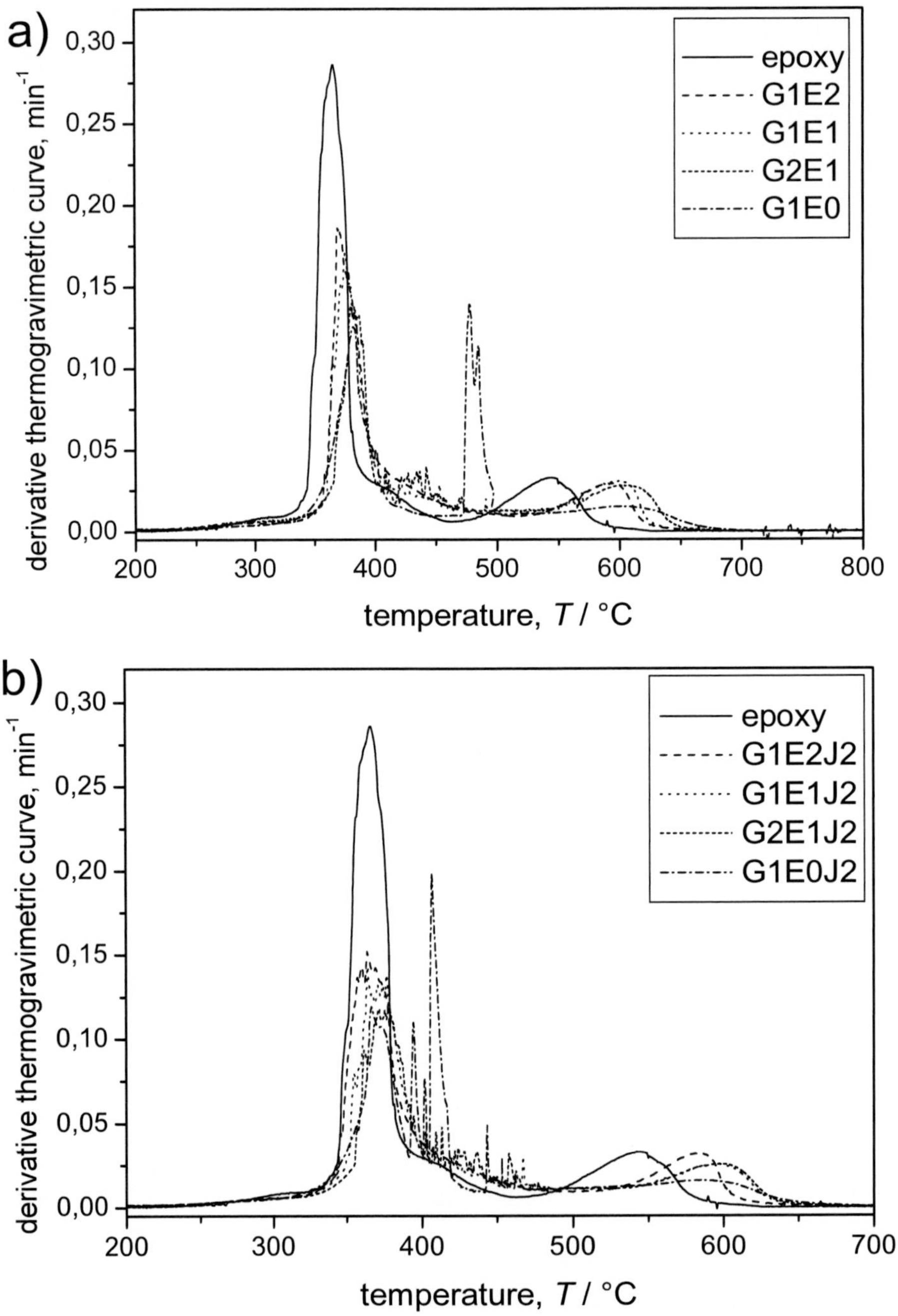

Figure 11. Derivative thermogravimetric curves in atmosphere of air: a) Series 1 hybrids and b) Series 2 hybrids, in comparison with unmodified epoxy.

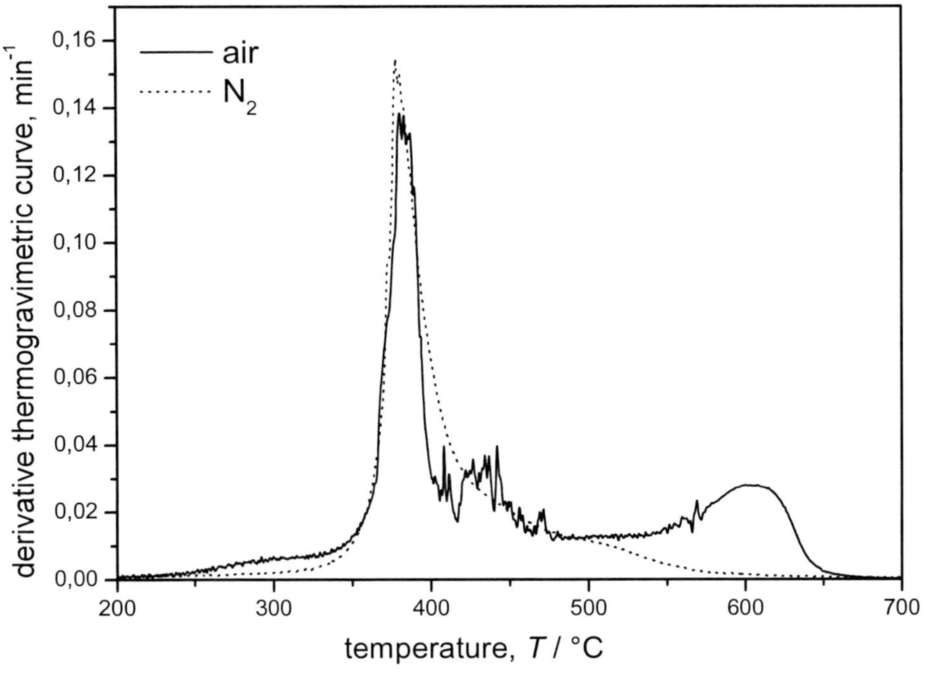

Figure 12. Comparison of derivative thermogravimetric curves for degradation of G2E1 hybrid in nitrogen and air atmosphere.

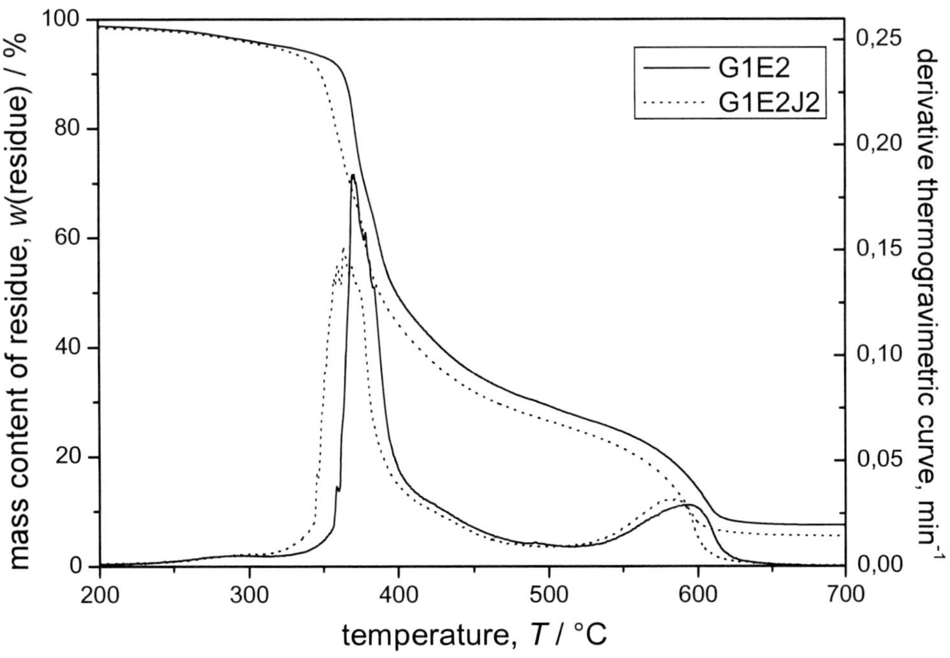

Figure 13. Comparison of thermogravimetric and derivative thermogravimetric curves for hybrids G1E2 and G1E2J2 in atmosphere of air.

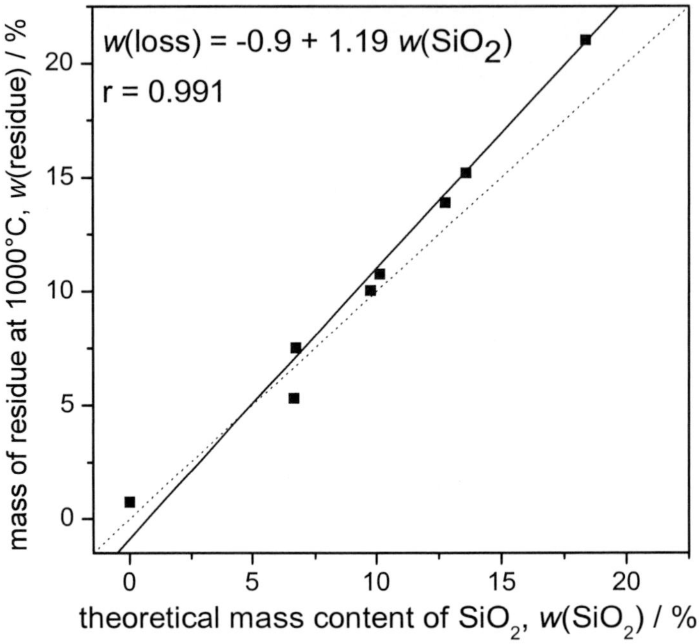

Figure 14. Linear dependence of total residue on theoretical mass content of SiO_2 (full line) and expected/theoretical dependence (dashed line).

Table 6. Theoretical content of SiO_2 in hybrid materials, $w(SiO_2)$, difference of experimentally determined total residue at 1000 °C and theoretical content of SiO_2, $\Delta w(SiO_2)$, as well as content of carbon char after the first step calculated from the equation (7), $w(C,1)$

Material	$w(SiO_2)$, %	$\Delta w(SiO_2)$, %	$w(C,1)$, %
Epoxy	0.00	0.73	23.86
G1E2	6.72	0.80	20.50
G1E1	10.14	0.61	23.30
G2E1	13.59	1.61	28.14
G1E0	20.54	4.11	23.86
G1E2J2	6.65	-1.35	19.90
G1E1J2	9.76	0.29	20.23
G2E1J2	12.75	1.13	21.25
G1E0J2	18.36	2.67	23.55

Mass residue at 1000 °C shows a linear dependence on theoretical SiO_2 content (Figure 14). It is assumed that the residue after thermo-oxidative degradation is completely inorganic (SiO_2) since all carbon char burns off, but there are some discrepancies with the theoretical (expected) dependence. The difference between the experimental and theoretical inorganic residue is given in Table 6. Increased difference between the two values for the hybrids with increased content of GLYMO (G2E1 and G2E1J2, and especially G1E0 and G1E0J2) is expected, and clearly follows the increase of theoretical inorganic content. In these hybrids the inorganic network prevents oxygen diffusion and thus protects carbon char from combustion. An interesting exception is the hybrid material G1E2J2, which shows a negative deviation from the theory. This deviation is too large to be ascribed to experimental error. It may be assumed that content of oxygen in SiO_2 for this material is substoichiometric, due to insufficient condensation of the inorganic network [2].

Content of carbon char after the first degradation step was calculated in the same manner as for the degradation in nitrogen (equation 4), and does not show a significant dependence on composition of hybrid materials. It varies at ~23 %, the same as for the unmodified resin.

Heats of degradation for all hybrids were determined from DTA curves (Figure 15) as previously described, and are considerably higher than was the case in inert atmosphere. Total area, position and relative contribution of the most intensive maximum are given in Table 7. Once more, a lack of regularity in total heat of reaction is obvious, as it decreases with the increase of combustible organic phase. On the other hand, the heat calculated from the most intensive maximum regularly decreases with increased content of inorganic phase (Figure 16), and shifts to higher temperatures as can be expected from inhibiting effect of the inorganic phase. The most intensive thermal effect overlaps the second step of mass loss (Figure 17) ascribed to combustion of organic char, since the combustion heat of carbon is quite high (~ 410 kJ/mol).

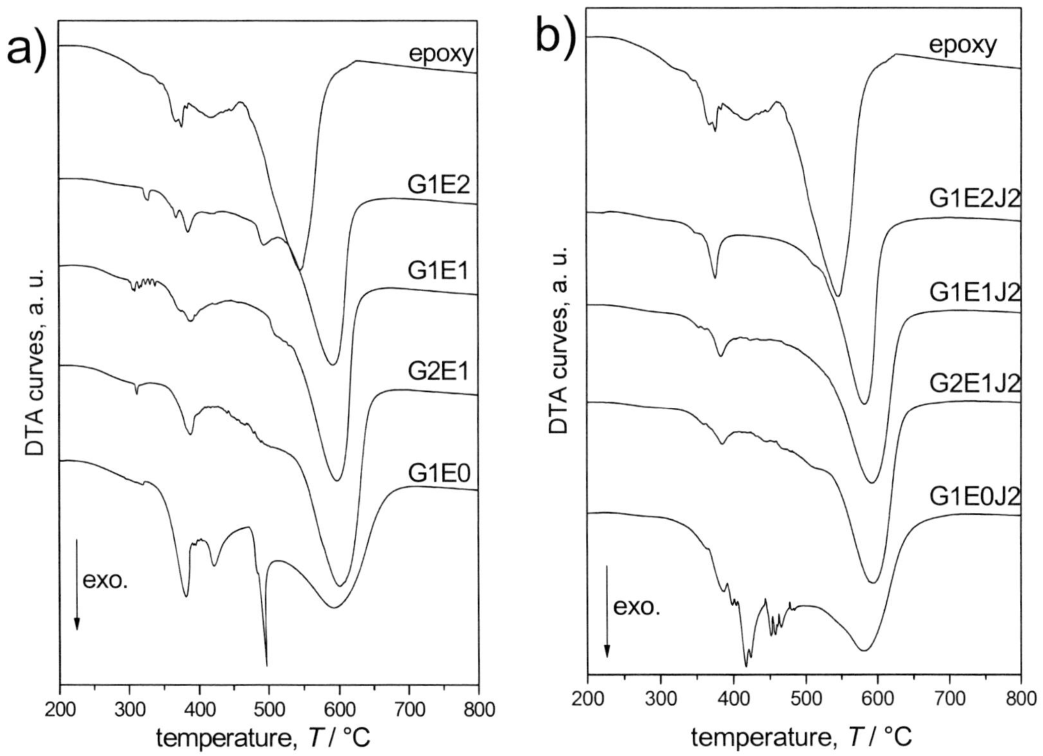

Figure 15. Differential thermal analysis curves in atmosphere of air for a) Series 1 hybrids and b) Series 2 hybrids, in comparison with unmodified epoxy.

Table 7. Area under differential thermal analysis curve, A_{DTA}, temperature of maximum degradation rate, T_{max}, and the relative area for the second degradation step, A_{rel}, as determined from DTA curves

Material	A_{DTA}, a. u./g	T_{max}, °C	A_{rel}, %
Epoxy	3400.9	545	66.05
G1E2	2610.4	591	65.90
G1E1	2975.1	597	63.08
G2E1	3444.2	601	60.99
G1E0	3844.9	592	48.96
G1E2J2	2459.8	582	64.71
G1E1J2	2638.5	592	64.45
G2E1J2	2788.7	594	60.77
G1E0J2	3312.7	580	49.57

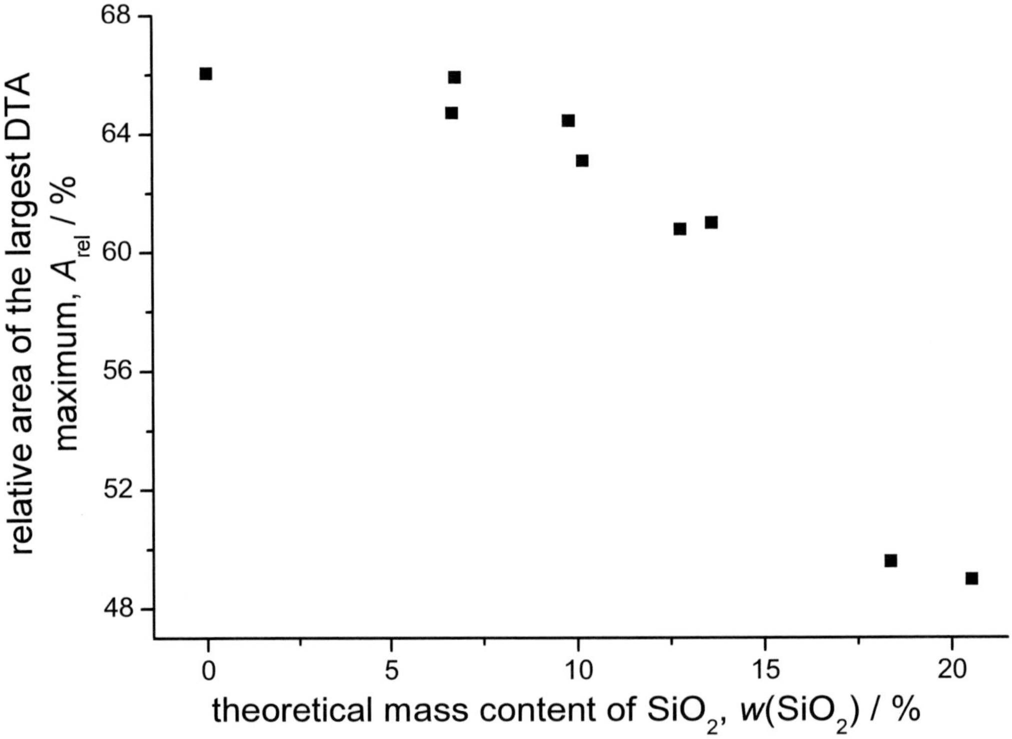

Figure 16. Dependence of relative area of the most intensive differential thermal analysis curve maximum on theoretical SiO_2 content.

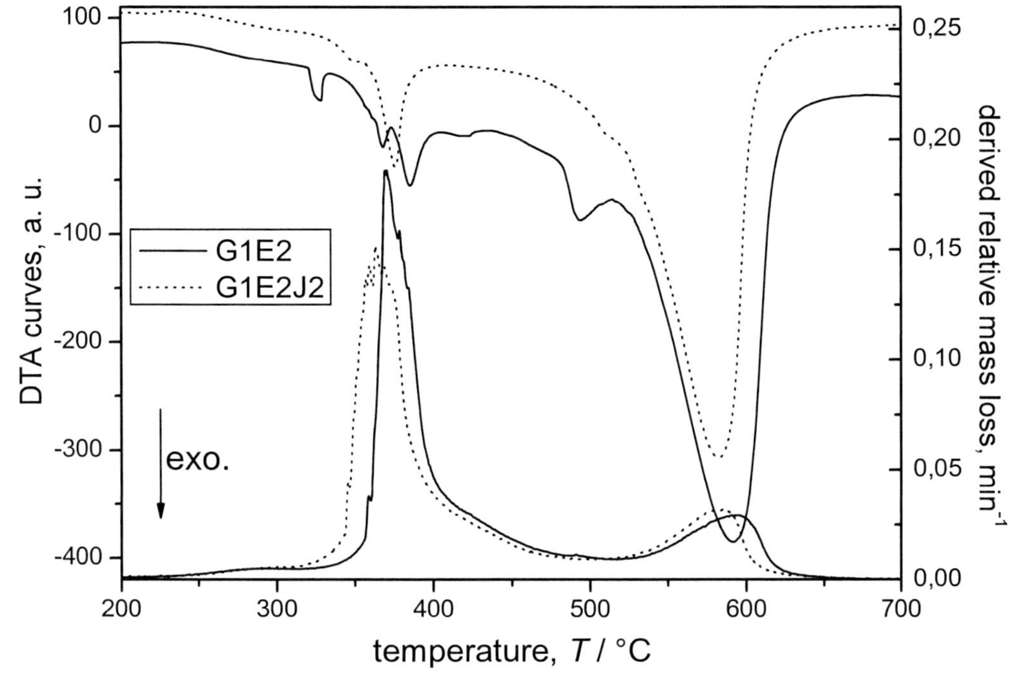

Figure 17. Comparison of differential thermal analysis curves and derivative curves for listed hybrids in oxidative atmosphere.

CONCLUSION

Organic-inorganic hybrid materials were prepared from diglycidyl ether of bisphenol A (DGEBA) and organically modified silicon alkoxide, 3-glycidyloxypropyltrimethoxysilane (GLYMO), making use of a poly(oxypropylene) diamine (Jeffamine D230) as both a curing agent for epoxy and a basic catalyst for GLYMO hydrolysis. Influence of D230 amount on thermal stability of hybrid materials was studied by combined differential thermal analysis and thermogravimetric analysis. It was determined that hybrid materials show improved thermal stability in comparison with unmodified epoxy resin, and that relative improvement is the greatest for inorganic phase (SiO_2) mass content < 10 %. Increased amine content has detrimental influence on thermal stability of hybrid materials. The main degradation step of hybrid materials does not depend on the atmosphere.

REFERENCES

[1] Judeinstein, P.; Sanchez, C. *J. Mater. Chem.* 1996, *6*, 511-525.

[2] Yoldas, B. E. *J. Sol-Gel Sci. Technol.* 1993, *1*, 65-77.

[3] Wojcik, A. B.; Klein, L. C. *J. Sol-Gel Sci. Technol.* 1994, *2*, 115-120.

[4] Mehrotra, R. C. In *Chemistry, spectroscopy and applications of sol-gel glasses*; Reisfeld, R.; Jørgensen, C. K.; Ed.; Springer-Verlag: Berlin - Heidelberg, 1992.

[5] Weng, W. H.; Chen, H.; Tsai, S. P.; Wu, J. C. *J. Appl. Polym. Sci.* 2004, *91*, 532-537.

[6] Kambour, R. P.; Ligon, W. V.; Russell, R. R. *J. Polym. Sci. Polym. Lett.* 1978, *16*, 327-333.

[7] Matějka, L.; Dukh, O.; Kolařík, J. *Polymer* 2000, *41*, 1449-1459.

[8] Matějka, L.; Pleštil, J.; Dušek, K. *J. Non-Cryst. Solids* 1998, *226*, 114-121.

[9] Matějka, L.; Dukh, O.; Brus, J.; Simonsick Jr, W. J.; Meissner, B. *J. Non-Cryst Solids* 2000, *270*, 34-47.

[10] Špírková, M.; Brus, J.; Hlavatá, D.; Kamišová, H.; Matějka, L.; Strachota, A. *J. Appl. Polym. Sci.* 2004, *92*, 937-950.

[11] Špírková, M.; Brus, J.; Hlavatá, D.; Kamišová, H.; Matějka, L.; Strachota, A. *Surf. Coat. Int. Part B Coat. Trans.* 2003, *86*, 187-193.

[12] Matějka, L.; Dušek, K.; Pleštil, J.; Kříž, J.; Lednický, F. *Polymer* 1999, *40*, 171-181.

[13] Bauer, J. B.; Liu, D.-W.; Jackson, C. L.; Barnes, J. D. *Polym. Adv. Technol.* 1996, *7*, 333-339.

[14] Qi, C. Z.; Gao, H.; Yan, F. Y.; Liu, W. M.; Bao, G. Q.; Sun, X. D.; Chen, J. M.; Zheng, X. M. *J. Appl. Polym. Sci.* 2005, *97*, 38-43.

[15] Weng, W. H.; Chen, H.; Tsai, S. P.; Wu, J. C. *J. Appl. Polym. Sci.* 2004, *91*, 532-537.

[16] Lu, S. R.; Zhang, H. L.; Zhao, C. X.; Wang, W. Y. *J. Mater. Sci.* 2005, *40*, 1079-1085.

[17] Huang, C. J.; Fu, S. Y.; Zhang, Y. H.; Lauke, B.; Li, L. F.; Ye, L. *Cryogenics* 2005, *45*, 450-454.

[18] Liu, Y. L.; Lin, Y. L.; Chen, C. P.; Jeng, R. J. *J. Appl. Polym. Sci.* 2003, *90*, 4047-4053.

[19] Liu, Y. L.; Wu, C. S.; Chiu, Y. S.; Ho, W. H. *J. Polym. Sci. Part A Polym. Chem.* 2003, *41*, 2354-2367.

[20] Mascia, L.; Tang, T. *J. Sol-Gel Sci. Technol.* 1998, *13*, 405-408.

[21] Mascia, L.; Prezzi, L.; Lavorgna, M. *Polym. Eng. Sci.* 2005, *45*, 1039-1048.

[22] Prezzi, L.; Mascia, L. *Adv. Polym. Technol.* 2005, *24*, 91-102.

[23] Landry, M. R.; Coltrain, B. K.; Landry, C. J. T.; O'Reilly, J. M. *J. Polym. Sci. Part B Polym. Phys.* 1995, *33*, 637-655.

[24] McCarthy, D. W.; Mark, J. E.; Schaefer, D. W. *J. Polym. Sci. Part B Polym. Phys.* 1998, *36*, 1167-1189.

[25] Mauri, A. N.; Riccardi, C. C.; Williams, R. J. J. *Macromol. Symp.* 2000, *151*, 331-340.

[26] Verganelakis, V.; Nicolaou, P. D.; Trapalis, C.; Kordas, G. *J. Non-Cryst. Solids* 2000, *265*, 265-275.

[27] Ochi, M.; Takahashi, R.; Terauchi, A. *Polymer* 2001, *42*, 5151-5158.

[28] Ochi, M.; Takahashi, R. *J. Polym. Sci. Part B Polym. Phys.* 2001, *39*, 1071-1084.

[29] Ochi, M.; Matsumura, T. *J. Polym. Sci. Part B Polym. Phys.* 2005, *43*, 1631-1639.

[30] Jensen, R. E.; Palmese, G. R.; McKnight, S. H. *Int. J. Adhes. Adhes.* 2006, *26*, 103-115.

[31] Macan, J.; Ivanković, H.; Ivanković, M.; Mencer, H. J. *J. Appl. Polym. Sci.* 2004, *92*, 498-505.

[32] Macan, J.; Ivanković, H.; Ivanković, M.; Mencer, H. J. *Thermochim. Acta* 2004, *414*, 219-225.

[33] Macan, J.; Brnardić, I.; Orlić, S.; Ivanković, H.; Ivanković, M. *Polym. Degrad. Stab.* 2006, *91*, 122-127.

[34] Lin, S. T.; Huang, S. K. *Eur. Polym. J.* 1997, *33*, 365-373.

[35] Hsiue, G. H.; Wei, H. F.; Shiao, S. J.; Kuo, W. J.; Sha, Y. A. *Polym. Degrad. Stab.* 2001, *73*, 309-318.
[36] Serier, A.; Pascault, J. P.; Lam, T. M. *J. Polym. Sci. Part A Polym. Chem.* 1991, *29*, 209-218.
[37] Penn, L. S.; Wang, H. In *Handbook of composites* (2nd edition); Peters, S. T.; Ed.; Chapman & Hall: London, 1998.
[38] Paterson-Jones, J. C. *J. Appl. Polym. Sci.* 1975, *19*, 1539-1547.
[39] Patterson-Jones, J. C.; Smith, D. A. *J. Appl. Polym. Sci.* 1968, *12*, 1601-1620.
[40] Grassie, N.; Guy, M. I.; Tennent, N. H. *Polym. Degrad. Stab.* 1986, *14*, 125-137.
[41] Lee, L.-H. *J. Polym. Sci. A* 1965, *3*, 859-882.
[42] Zakordonskiy, V. P.; Hnatyshin, S. Y.; Soltys, M. M. *Pol. J. Chem.* 1998, *72*, 2610-2620.
[43] Dyakonov, T.; Mann, P. J.; Chen, Y.; Stevenson, W. T. K. *Polym. Degrad. Stab.* 1996, *54*, 67-83.
[44] Rose, N.; Le Bras, M.; Delobel, R.; Costes, B.; Henry, Y. *Polym. Degrad. Stab.* 1993, *42*, 307-316.
[45] Lee, L.-H. *J. Appl. Polym. Sci.* 1965, *9*, 1981-1989.
[46] Damian, C.; Espuche, E.; Escoubes, M. *Polym. Degrad. Stab.* 2001, *72*, 447–458.
[47] Wu, C. S.; Liu, Y. L.; Chiu, Y. C.; Chiu, Y. S. *Polym. Degrad. Stab.* 2002, *78*, 41-48.
[48] Pospíšil, J.; Nešpurek, S. *Prog. Polym. Sci.* 2000, *25*, 1261-1335.
[49] Cardiano, P.; Sergi, S.; Lazzari, M.; Piraino, P. *Polymer* 2002, *43*, 6635-6640.
[50] Schmidt, H. *J. Non-Cryst. Solids* 1988, *100*, 51-64.

239In: Research in Hybrid Materials
Editors: Simon J. Brunner and Julian W. Egger
ISBN 978-1-60456-718-2

Chapter 9

SIMULTANEOUS HYBRID PROCESSES AS A WAY TO IMPROVE CHARACTERISTICS OF CULTIVATION OF MICRO-ORGANISMS AND BIODESTRUCTION OF POLLUTANTS

A. Ye. Kouznetsov*[3] *and S. V. Kalyonov
D.Mendeleyev University of Chemical Technology of Russia
Miusskaya sq., 9, 125047, Moscow, Russia

ABSTRACT

Some systems for investigation of the biosynthesis, biotransformation and biodestruction processes with a simultaneous chemical or photochemical reaction have been developed. It is expected that such hybrid processes will improve biotechnological performances and a variety of the biotechnological methods for pollutant's destruction.

Keywords: hybrid processes, cultivation, oxidation, biodestruction, waste water treatment, bioreactors, artificial peroxisome, hydrogen peroxide, UV-light, photolyase, high density culture.

We define a hybrid process as one where chemical or photochemical reactions and biocatalysis or biodestruction occur in the same volume and simultaneously proceed. They are typical for natural ecosystems. For the purposes of practical application such processes could improve bioavailability of resistant substrate, characteristics of the systems of biopurification and biodestruction, microbial growth and biosynthesis. For scientific research an evolution of these systems in conditions of autoselection has some interest as well.

Examples of various abiotic systems of transformation (oxidative, hydrolytic, reductive) suitable for hybrid processes of cultivation and biodestruction are:

[3] e-mail: aekuz@muctr.edu.ru

- Fenton reagent (oxidation of organic substances)

$Fe^{2+} + H_2O_2 \rightarrow Fe^{3+} + OH^- + OH^\cdot$

- Kostelfranco reagent (oxidation of organic substances)

ascorbate + Cu^{2+}

- Photochemical oxidation

hν 330-380 nm

$C_{org.} \rightarrow CO_2 + H_2O$

- Advanced photocatalytic oxidation

hν 330-380 nm

$TiO_2 + C_{org.} \rightarrow TiO_2 + CO_2 + H_2O$

- hydrolysis of organic substances, catalyzed by TiO_2, FeOOH, Al_2O_3
- dechlorination by smectites with Cu(II)
- redox processes with oxides Fe (II, III) and Mn (III,IV)etc.

For destruction of resistant chemical pollutants it would be more suitable to use more aggressive abiotic oxidants like hydrogen peroxide or photochemical ones. However there will be a need to solve the problem of maintainance of the phisiological activity of biological component in a hybrid process.

Using model systems with addition of hydrogen peroxide or Fenton reagent (H_2O_2 + Fe^{2+}) into nutrient medium before or during fermentation as well as ultraviolet irradiation (UVA and UVB subrange) of cell suspensions during cultivation several hybrid processes have been studied:

- biodestruction of ion exchange resins as potential waste from atomic power station,
- biodestruction of phenol by an isolate with bacteria and yeast,
- cultivation of yeast *Candida tropicalis* on sucrose,
- cultivation of *Azotobacter* strains,
- biodestruction of phenol by consortia of yeast and bacteria *Azotobacter* on nitrogen free medium.

The summary of the investigations are following:

- pre-treatment of ion exchange resins by Fenton reagent improves their bioavailability;
- phisiological conditions of microbial population for their functioning at the abiotic factors impact, especially being aggressive, are very important for the survival and high

activity of micro-organisms;

- at the cultivation in fed batch conditions and at the addition of H_2O_2 the amount of extracellular metabolites and residual concentration of substrate were lower than in the control (without addition of H_2O_2); there was not an accumulation of metabolites inhibiting growth of micro-organisms in the process of phenol biodestruction (fig. 1) and at the growth of *C. tropicalis* on sucrose (fig. 2); these peculiarities kept for a long time after ceasing of H_2O_2 addition;

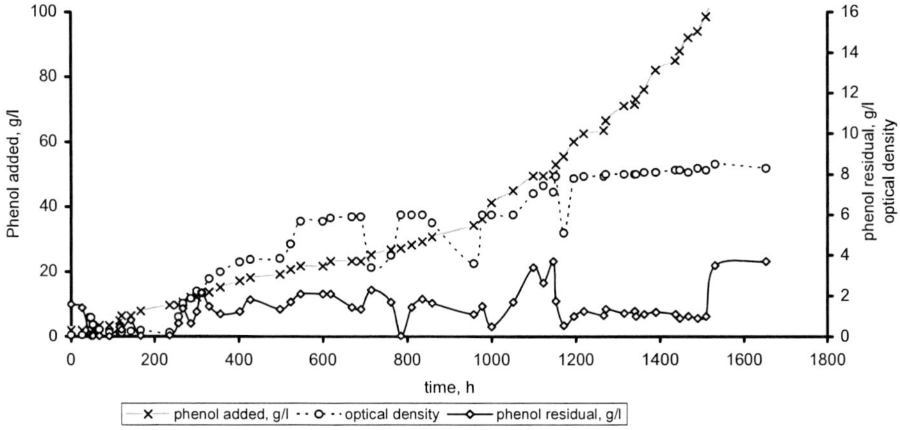

Figure1. Biooxidation of phenol in fed batch conditions with fractional addition of phenol and small amounts of H_2O_2. (since 1 to 1000 h after start of the experiment) into the biooxidation zone and using yeast biocenosis preadapted to hydrogen peroxide.

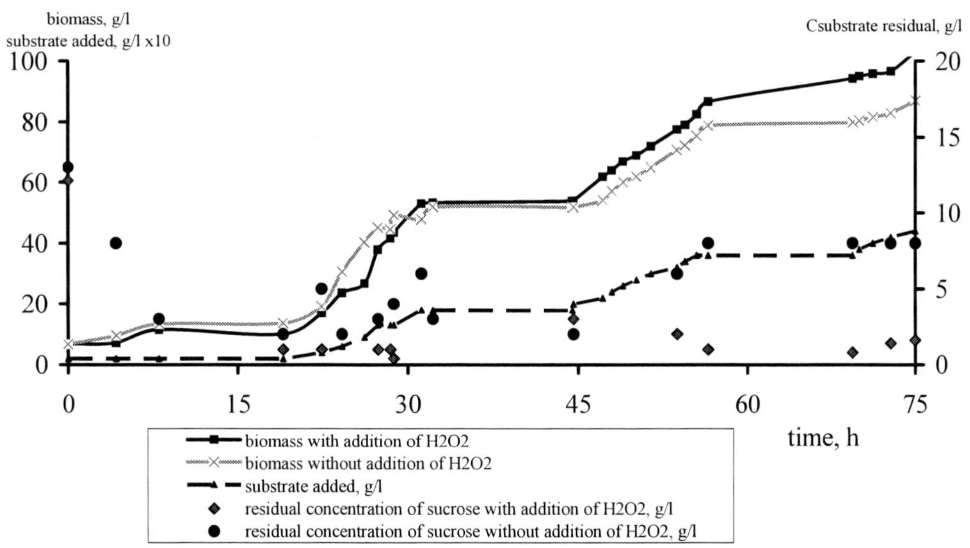

Figure 2. Growth of yeast *C. tropicalis* on sucrose in fed batch conditions with and without addition of H_2O_2.

- in case of high density cultivation with fractional addition of dry sucrose and periodical addition of H_2O_2 the yeast cells of *C. tropicalis* accumulated up to the maximal level of 150-170 g of dry weight per one liter of culture liquid;
- the rate of oxidation of substrate in fed batch conditions was limited only by oxygen mass-transfer (for biodestruction of phenol and for the growth of *C. tropicalis*);
- diauxy in the consumption of sucrose and extracellular intermediates (acetic acid, ethanol) temporarily accumulated in the culture liquid disappeared (for the growth of *C. tropicalis* on sucrose);
- there was not a dying phisiological activity of cells in fed batch conditions (for biodestruction of phenol and for the growth of *C. tropicalis*);
- cells are more resistant to H_2O_2 at the active stage of the growth (phenol biodestruction, growth of *C. tropicalis*);
- cells pre-adaptated to H_2O_2 were more resistant to various stresses:

    - their activity recovered better after substrate starvation;
    - they could grow at higher running concentration of substrate – up to 5 g of phenol per liter of medium – for phenol biodestruction; up to 700 g of sucrose per liter of medium – for C. tropicalis growth;
    - they were high resistant to hydrogen peroxide: population of phenol biodestructors survived after addition of 2% of H2O2; population of yeast C. tropicalis survived after one-time addition of 16.5% H2O2 (perhydrol diluted 2-fold and burning skin); the last one could be charged by addition up to 15 g of 100% H2O2 per 1 L of medium per 1 hour in continuous condition and could use H2O2 as a source of oxygen;
    - addition of H2O2 to the populations with Azotobacter strains stimulated in some cases their growth on nitrogen free medium with sucrose;
    - there was observed a biodestruction of phenol by consortia of resistant to H2O2 phenol-degrading yeast and bacteria Azotobacter on nitrogen free medium;
    - under simultaneous action of visible and UV-light the yield of yeast C. tropicalis on medium with sucrose was 10-15% above the control (fig 3).

The data obtained in the experiments with simultaneous action of biological and aggressive oxidizing factors (H_2O_2, UV-light) indicate: i) the importance of the supporting optimal red-ox status in the microbial population stipulated, in particular, by concentration of active forms of oxigen; ii) a possibility to achive best output results at the controllable cultivation of micro-organisms at the conditions of active functioning the reparation mechanisms.

For instance, the primary chemical responses initiated by UV-light, H_2O_2 or other forms of active oxygen are accompanied by generation of particles (free radicals etc.), which participate in regulation of physiological state of cells, activate anti-stress systems of reparation and initiate the processes of oxidation of biosubstrates.

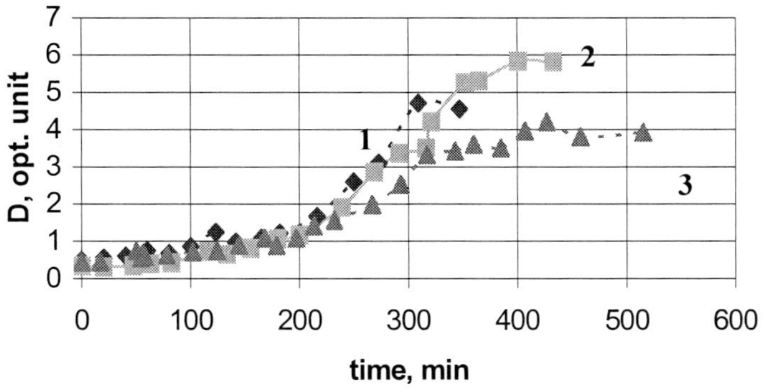

Figure 3. Growth of yeast *Candida tropicalis* under irradiation of UV- and visible light. 1 – control without UV-irradiation, 2 – with UV- and visible light irradiation, 3 – with UV-irradiation in the dark.

Among the mechanisms of reparation in cells there is photoreparation by photolyase after UV exposure. The last one functionates together with other ways of response to stress factors. For anti-stress systems of reparation some intersection reactions are possible as well. The system of reparation with photolyase functioning in response to UV irradiation has a maximal sensitivity to light at two wavelength at visible and UV ranges of spectrum. Our preliminary results obtained in the experiments with yeast *Candida* (fig. 3) has shown that one of the ways to improve the efficiency of the hybrid photodestruction and biodestruction processes could be through the simultaneous influence of visible and UV light.

Based on the results obtained three novel methods for biodestruction of pollutants were proposed by us:

- closed intensive microbiological system for the high concentrated waste treatment (fig. 4);
- artificial peroxisome (fig. 5);
- "Solar" photobioreactor (fig. 6).

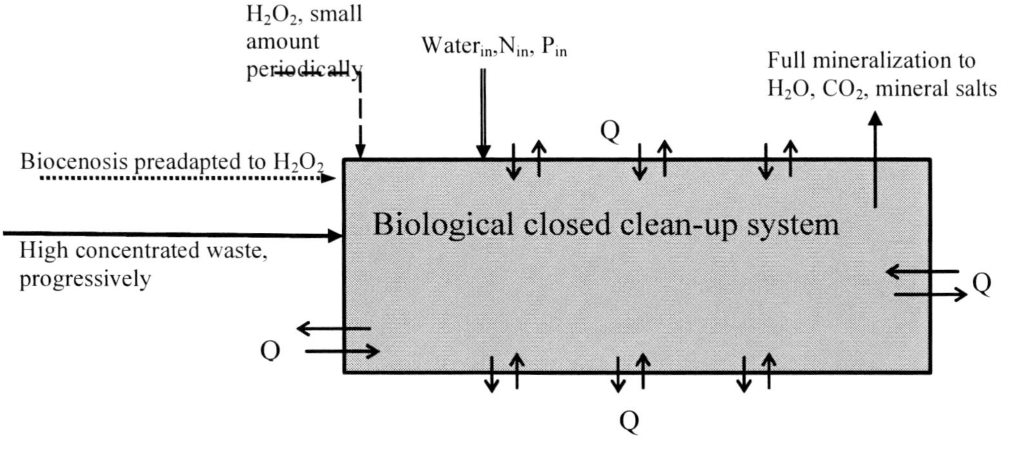

Figure 4. Closed intensive microbiological system for the high concentrated waste treatment.

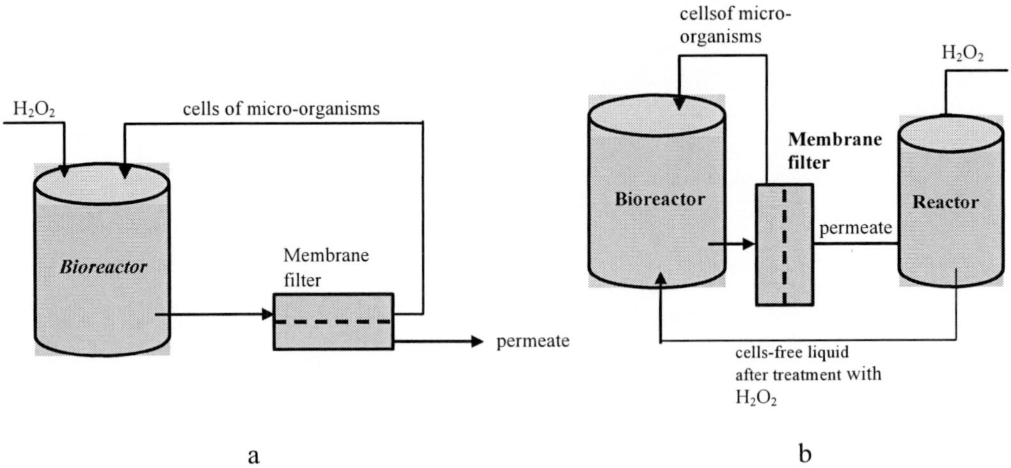

Figure 5. Artificial peroxisome.

A possibility to protect cells against oxidative stress, induced by addition of H_2O_2 or UV-irradiation, through photoreparation way by photolyase is being under study.

A mathematical model of growth of cells under simultaneous irradiation of visible and UV-light taking into account photoinactivation, photodynamic action, photoreparation and photostimulation will be developed.

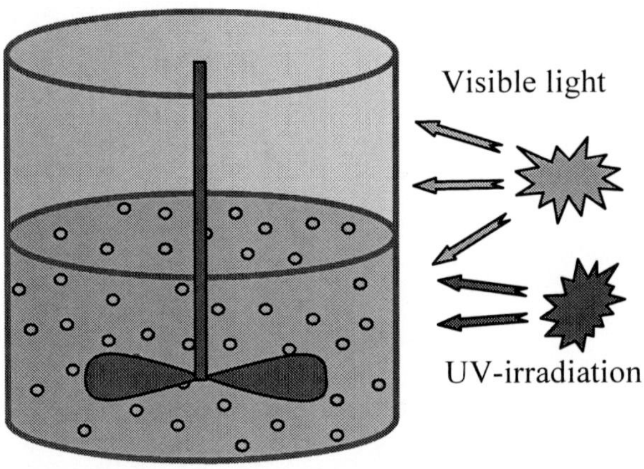

Figure 6. "Solar" photobioreactor.Possible effects: - oxidative stress; - photoreparation; - sinchronisation of cell division; - enhancing substrate bioavailability; - increase of rate of oxidative processes; - high-density culture.

245In: Research in Hybrid Materials
Editors: Simon J. Brunner and Julian W. Egger

ISBN 978-1-60456-718-2

Chapter 10

OXIDATION AND AGEING OF NANOCOMPOSITES AND HYBRIDS

Radu Setnescu[a b], Silviu Jipa[a b], Tanța Setnescu[a] and Traian Zaharescu[a]

[a] National Research Institute on Electrical Engineering (INCDIE ICPE – CA), 313 Splaiul Unirii, Bucharest 030138, Romania

[b] Valahia University of Targoviste, 1 Bd. Castanilor, Targoviste 130 024, Romania

ABSTRACT

Oxidation behavior of various copolymers and hybrid materials is discussed on the basis of recent references in the field. Structural and functional changes induced by oxidation ageing are discussed in some significant cases. The influences of chemical structure, additives or impurities are also discussed as possible factors influencing the photo-oxidation susceptibility of nanocomposite materials. The study revealed that the oxidation and ageing of nanocomposites and hybrids is tentatively explored, in spite of the interest regarding long term exploitation of these materials.

ABBREVIATION LIST

EPDM:	ETHYLENE-PROPYLENE-DYENE MONOMER RUBBER
EPDM-G-MA:	EPDM GRAFTED WITH MALEIC ANHYDRIDE
2M2HT:	DIMETHYL DEHYDROGENATED TALLOW AMMONIUM
DTG:	DIFFERENTIAL THERMOGRAVIMETRY
HALS:	HINDERED AMINE LIGHT STABILIZER
MA:	MALEIC ANHYDRIDE
MMT:	MONTMORILLONITE
MT2ETOH:	METHYL TALLOW DI-HYDROXYETHYL AMMONIUM

ABBREVIATION LIST (CONTINUED)

PMMA:	POLY(METHYLMETHARCYLATE)
P(MMA-GMA):	POLY(METHYLMETHACRYLATE *CO* GLYCIDYL METHACRYLATE)
P(MMA-MA):	POLY(METHYLMETHACRYLATE *CO* MALEIC ANHYDRIDE)
P(MMA-MA5-70)-S IO_2:	IS A SAMPLE CODE IN PAPER [28] (SEE FOOTNOTE 1)
P(MMA-GMA5-70) -SIO_2:	SIMILAR AS ABOVE, EXCEPTING MA WHICH IS REPLACED BY GMA
PP:	POLYPROPYLENE
PP-G-MA:	POLYPROPYLENE GRAFTED MALEIC ANHYDRIDE
PP- OD3M:	PP-G-MALEIC ANHYDRIDE/ AMINE MODIFIED MONTMORILLONITE (SEE OD3MA) NANOCOMPOSITE
PP-2M2HT:	PP-G-MALEIC ANHYDRIDE/ AMINE MODIFIED MONTMORILLONITE (SEE 2M2HT) NANOCOMPOSITE
PP-MT2ETOH:	PP-G-MALEIC ANHYDRIDE/ AMINE MODIFIED MONTMORILLONITE (SEE MT2ETOH)
PP/PP-G-MA:	BLEND OF PP AND POLYPROPYLENE GRAFTED MALEIC ANHYDRIDE
PS:	POLYSTYRENE
OD3MA:	OCTADECYL TRIMETHYL AMINE,
OLSB:	ORGANICALLY MODIFIED SILICATES
TEOS:	TETRAETHOXYSILANE
TGA:	THERMOGRAVIMETRIC ANALYSIS
TGA-FTIR-MS:	THERMOGRAVIMETRIC ANALYSIS COUPLED FOURIER TRANSFORM INFRARED SPECTROSCOPY AND
XRD:	X RAY DIFFRACTION

INTRODUCTION

Organic-inorganic *hybrids* and *nanocomposites* are novel materials containing an inorganic phase dispersed in a polymer matrix at molecular or nano-level. In fact, organic and inorganic moieties seem to be chemically linked to each other in the case of hybrids, while physical interactions seem to occur between organic (polymer) phase and inorganic components in nanocomposites. The properties of nanocomposites and hybrids are expected to be different than the behavior of classical polymer composites.

Beside the applications in which improved fire resistance is relevant [1,2], hybrid materials can be used in various other fields. For example, it can be mentioned the rhodium or iridium complexes with 2,2'-bipyridine grafted onto *in situ* generated titanium dioxide particles, which exhibited very good catalytic activity in ketone hydrogenation. Other major advantages consist in re-usability of the catalyst and no-significant Rh leaching [3].

Nanocomposites belong to a new class of composites, firstly developed by Toyota group [4 - 6]. At least one dimension of the dispersed particles of these materials is in the nanometer range [7]. The nanoscale fillers bring about sharp difference in the physical characteristics of basic materials comparative to the behavior of conventional microscale counterparts [4]. For instance, increased Young's modulus and tensile strength, improved gas barrier properties [8], reduced thermal expansion coefficients, and reduced flammability [9-12] are observed with only a few percent of nanofiller content (≤ 5%) thus maintaining polymer processability, cost and clarity.

Montmorillonite (MMT) is now the most used layered silicate to obtain polymer nanocomposites [13]. Changeable cations present in the region between two MMT layers can be substituted by organic cations to enhance the interlayer spacing and to induce organophilic properties in the silicate layer [14]. Alkyl ammonium salts containing long hydrocarbon chains are usually applied as surfactants [15,16], but their main problem is low thermal stability [17, 18]. Therefore, has been tried the use of other salts, exhibiting higher thermal stability, such as imidazolium salts [14], to replace ammonium salts as *surfactants* for nanocomposites preparation. In the case of polymers containing non-polar groups (such as polypropylene), a *compatibilizing agent* presenting hydrophilic groups grafted on a non-polar chain {[(e.g. polypropylene grafted maleic anhydride or maleic anhydride grafted on poly(propylene-ethylene) [13]} is used during processing to facilitate the homogeneous dispersion of MMT.

There are two types of nanocomposites which exhibit different dispersion degrees of silicate layers. One kind of nanocomposites presents *intercalated morphology*, that is represented by multilayer-structured and highly ordered materials, in which polymer chains are inserted into 2-3 nm spaced galleries between parallel silicate layers. The second group is *exfoliated* or *delaminated* type. The morphology reveals nanometric thick silicate monolayers homogeneous dispersed in polymer matrix. This kind of structure consists in extensive penetration of the polymer through silicate layers, causing an enhancement of the space between two crystallites. Delaminate nanocomposites seem to exhibit increased mechanical properties as compared to intercalated systems [14]. It was noted that both morphologies could coexist in a polymer matrix [13].

After an initial euphoric period caused by detailed investigation of nanocomposite properties and hybrid materials, several problems appeared in relation to their oxidation. Thus, coloration or worsening of some valuable properties was observed as a result of oxidation induced at elevated temperatures or by light, which promotes material degradation [19]. It was observed that the behavior of these materials is different to that known for pure polymers or classical composites. In spite of extensive literature concerning the synthesis and properties of both nanocomposites and hybrids, there are relative few papers concerning their oxidation and ageing under various stress conditions [17,20-26]. In this chapter, we try to review recent results in the above mentioned field of oxidative degradation and ageing of hybrid and nanocomposite materials.

THERMALLY INDUCED OXIDATION

In a recent paper [14], the effect of some imidazolium derivative surfactants on thermal strength and thermal oxidation stability of PS/MMT nanocomposites has been assessed. While one or two unresolved stages (located on 400 – 460 °C range) were observed in DTG under nitrogen, two separate peaks (at 370 – 430 °C and respectively ca. 560 °C) were recorded under air static atmosphere (see Figure 1).

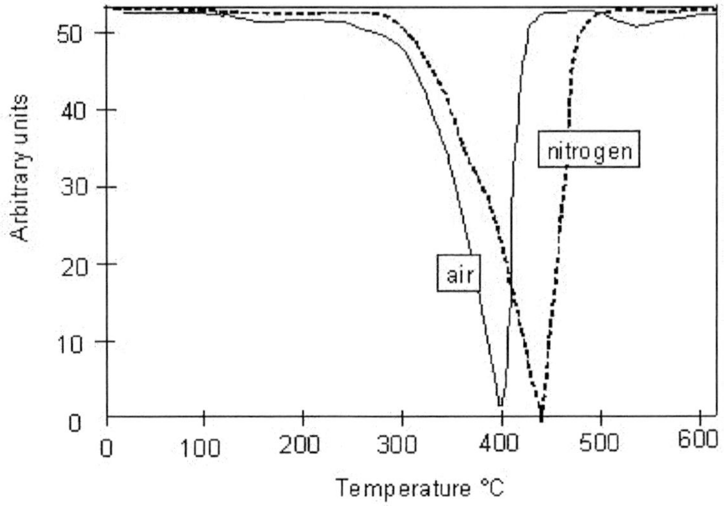

Figure 1. Typical DTG curves recorded in air or nitrogen from PS/MMT nanocomposites (Adaptation from paper [14])

It has been found that E_a values of thermal oxidation of PS/MMT nanocomposites, calculated by Kissinger method [27], decreased as the length of the hydrocarbon surfactant radical (see scheme 1) increases, namely PS < PS/MMT-C_{12} < PSMMT - C_{16} < PS/MMT - C_{18}. This behavior has been explained by increase in the percentage of exfoliation that occurs in the case of large interlayer spacing in organically modified MMT.

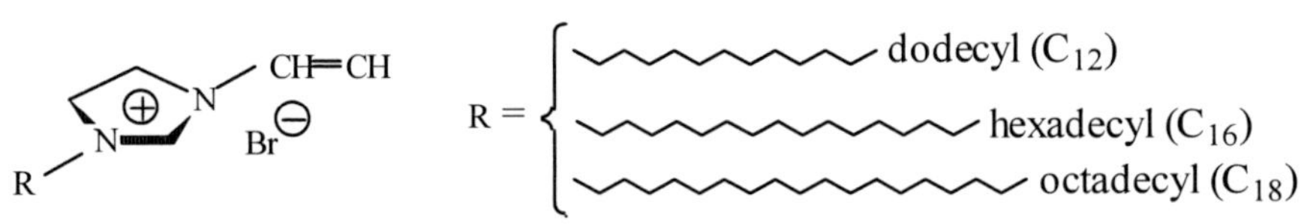

Scheme 1. Structure of imidazolium surfactants used in paper [14]

On the other hand, all PS/MMT nanocomposites studied by Abate et al. [14], exhibited lower values of both $T_{5\%}$ (temperature corresponding to 5% of degradation) and E_a (activation energy), when thermal degradation occurs under air atmosphere as compared to nitrogen atmosphere (Figure 2). The lower values in both cases were observed for PS. Oppositely, the higher figures for PS/MMT-C_{18} were observed, suggesting that MMT increased both thermal strength and thermo-oxidative stability in comparison to the basic polymer.

The thermal stability of organically modified layered silicates (OLS), which is a key factor in synthesis and processing of polymer/layered silicates nanocomposites, has been studied by TGA and TGA-FTIR-MS methods [17]. It has been reported that the onset temperature of OLS decomposition is about 180 °C. In addition, it has been found that the initial degradation of the surfactant (alkyl quaternary ammonium salts) follows a Hoffman elimination reaction. On the other hand, no influence on the above mentioned onset temperature of some experimental factors, such as the architecture of the surfactant molecule (i.e. trimethyl or dimethyl groups), chain length, surfactant formulation, modified ratio or preconditioning (washing) was observed. Catalytic sites occurring on the aluminosilicate layers were found to reduce the characteristic temperature of thermal stability of a certain fraction from the surfactants by an average of 15 – 25 °C.

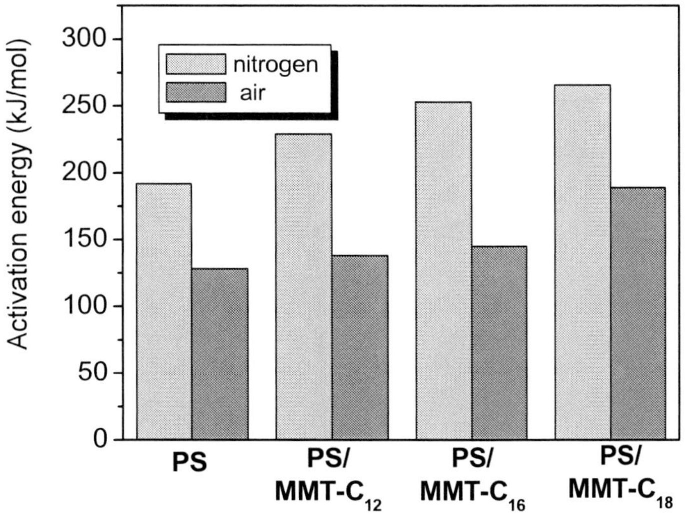

Figure 2. Activation energy of thermal decomposition and thermo-oxidative degradation of some PS-montmorillonite nanocomposites containing imidazolium salts surfactants (after Abate et al [14]).

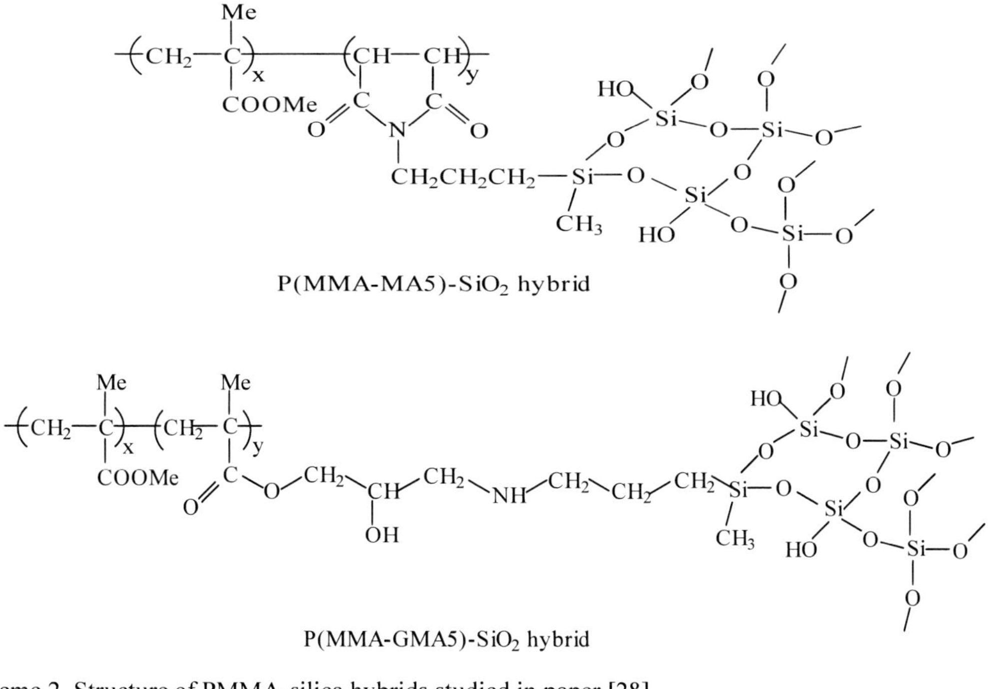

Scheme 2. Structure of PMMA-silica hybrids studied in paper [28]

A study on two series of PMMA-silica hybrids (see Scheme 2) [28] revealed that thermooxidation exhibits three stages on DTG curves, which can be assigned to the following processes:

- degradation of partly oxidized groups formed by oxygen trapping of the free radicals generated by scission of weak bonds [29,30];
- oxygen attack on the radicals generated by random scission within the polymer chains;
- degradation of SiO_2 network [31].

It was found also that the themooxidation stability, as it is described by $T_{5\%}$, is higher in the case of P(MMA-MA)-SiO_2 hybrids as compared to the copolymer P(MMA-MA), but it tends to decrease as increasing the SiO_2 content in the case of P(MMA-GMA)-SiO_2 hybrids (Figure 3). These results could be explained, in the case of P(MMA-MA)-SiO_2 hybrids, in the terms of retarding the initial thermooxidation of PMMA segments induced by SiO_2. Then, the same moieties of SiO_2 increase random scission as suggest the slight diminution in T_m values (see Figure 4). An enhancing in degradation rate of interfacial amine bond on early stages, that is induced by SiO_2, was supposed to explain the thermooxidation behavior of P(MMA-GMA)-SiO_2 hybrids, which exhibited the independence of T_m values on SiO_2 content (see Figure 4).

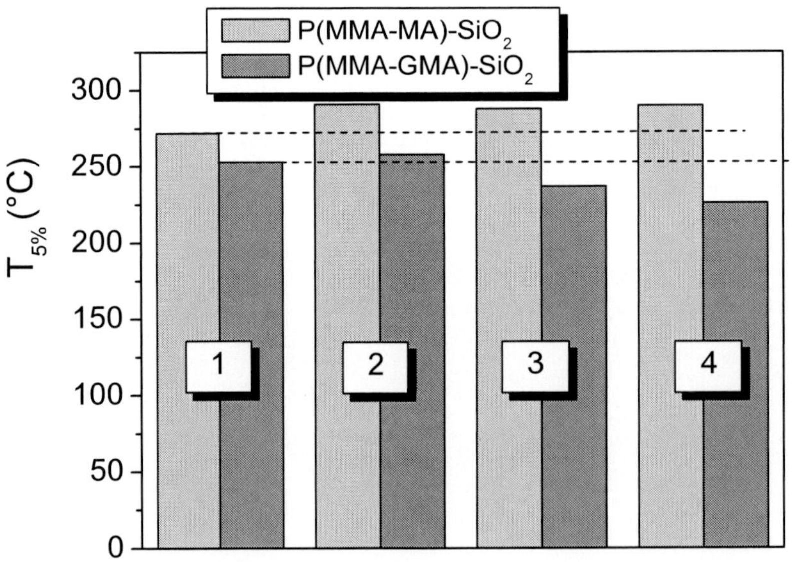

Figure 3. T5% values as a function of composition for two series of PMMA-SiO2 hybrids, respectively: 1-pure copolymers (MA5-100 and GMA5-100); 2 - MA5-90 and GMA 5-90; 3 - MA5-80 and GMA5-80; 4 - MA5-70 and GMA5-70 (Adaptation of data in paper [28]) (see footnote 1 for samples names).

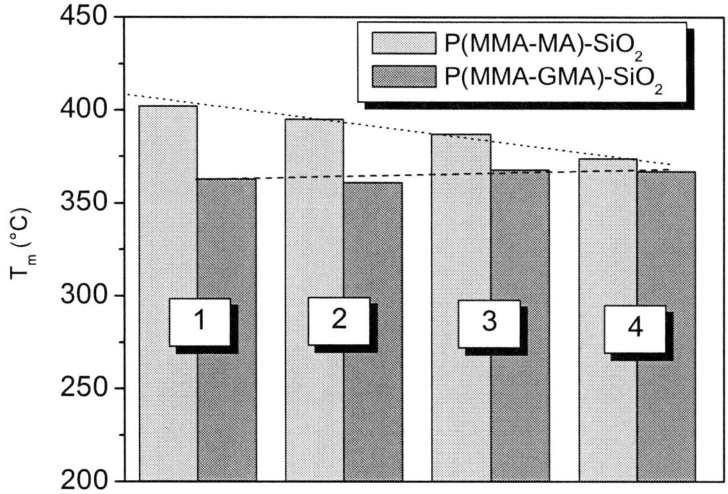

Figure 4. T_m values as a function of composition for two series of PMMA-SiO_2 hybrids, respectively: 1-pure copolymers (MA5-100 and GMA5-100); 2 - MA5-90 and GMA 5-90; 3 - MA5-80 and GMA5-80; 4 - MA5-70 and GMA5-70 (Adaptation of data in paper [28]). (see footnote 1 for samples names).

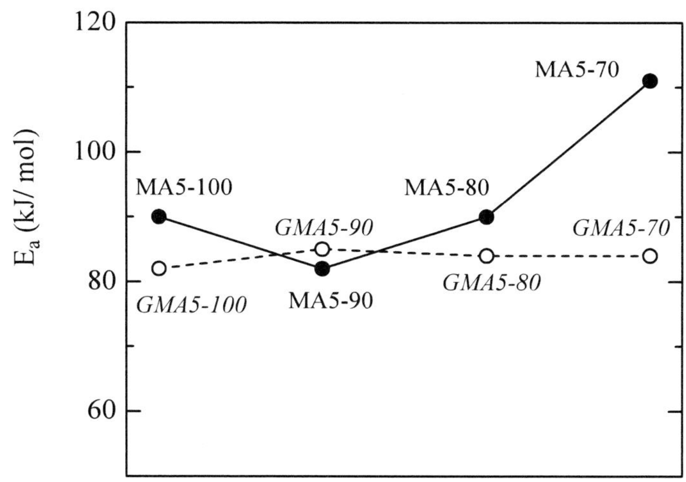

Figure 5. E_a values as a function of composition for thermal oxidation of two series of PMMA-SiO_2 hybrids (Adaptation of data in paper of Wang et al [28]); see footnote 1 for samples names.

The kinetic analysis of thermal oxidation applied to the two hybrid series mentioned above has shown that E_a values of random scission depend on SiO_2 concentration in a specific manner, which can be related to the nature of co-monomers. Thus, for P(MMA-MA5)-SiO_2[4] hybrids, E_a increased with increasing the SiO_2 content, but opposite trend has been observed

[4] P(MMA-MA5-xy) or P(MMA-GMA5-xy), xy being 70, 80, 90 or 100, are sample codes used in paper [28]. For example, hybrid MA5-70 has been prepared by using *70 wt%* of a mixture of 95% (mol/mol) of MMA and *5%* of diethoxysilyl functional groups which has been condensed with 30%wt of TEOS based on the weight of monomer mixture (MMA, MA and 3 aminopropyl methyl diethoxysilane, 2.0: 0.1: 0.2 weight parts)

under nitrogen. With P(MMA-GMA5)-SiO_2 hybrids, E_a is basically independent on both the SiO_2 content and surrounding atmosphere (nitrogen or air) (Figure 5). These results illustrate that the scission process in hybrids is more complex than in the case of pure copolymers. The reason for enlarged values of E_a under air is still unclear.

Nanofiller particles induced also improved thermal and thermo-oxidative properties, being found that inorganic component act as a shield for thermal oxidation, shifting the temperature range of volatilization towards that of thermal degradation in nitrogen. They bring also to a reduction of 55% in heat release rate during combustion in the case of PE-hydrotalcites nanocomposites [32]

Thermogravimetric studies on epoxy MMT nanocomposites showed also that the epoxy matrix is sheltered from the action of oxygen. Cone calorimeter tests indicated that the rate of heat release in the nanocomposite combustion is strongly reduced [32].

In the case of nanocomposites based on an ethylene vinyl acetate copolymer (27 wt% vinyl acetate) and various (organo-modified) clays, it was found that nanostructure and tensile properties mainly depend on the nature of the organic modifier while the delay in thermal volatilization of EVA during thermo-oxidation is mainly driven by the nature of the clay (mainly its aspect ratio), with no significant influence of the nanostructure and of the nature of organic modifier [33].

Photo-Induced Oxidation

Photooxidation studies on various polymers (such as PP [13,17,24,26], polyethylene [21], polycarbonate [22])-MMT nanocomposite systems lead to the conclusion that nanocomposites degrade easier than pure polymers. Such a behavior toward photo-ageing could be an important problem in outdoor applications of these materials [25].

An FTIR study on UV irradiated polypropylene - graft - maleic anhidride/clay nanocomposites [34] revealed that the initial band at 1780 cm^{-1} disappeared at the beginning of the exposure (after 12 hours). This process has been related to maleic anhydride conversion. Consequently, a weak band at ca. 1740 cm^{-1} appeared, which has been related to the formation of carbonyl species, including ester ones. As the irradiation time increased, progressive formation of the bands at 1715 cm^{-1} (ascribed to a mixture of carboxyl and ketone carbonyl groups) and 1780 cm^{-1} (shoulder) is observed. According to other papers [13], the positions of these bands are slightly different, namely 1713 cm^{-1} (assigned to carbonyl group of carboxylic acids in dimer form [35]), 1735 cm^{-1} (assigned to either carbonyl group of esters [36] or carbonyl vibration of carboxylic acids [37], and 1780 cm^{-1} (γ-lactones). Both above mentioned bands were related to the degradation of the polymer matrix, and the process seems to be similar to that occurring in the pure polymer, taking into account the evolution of the bands related to carbonyl and hydroxyl (ca. 3400 cm^{-1}) groups. However, the behavior of the studied samples has been different depending on their composition. Thus, the absorbance at 1715 cm^{-1} starts to increase after 20h for the samples PP-g-MA, while the same process starts significantly latter with pure PP. Thus, pure PP was less oxidizable than PP-g-MA copolymer. All nanocomposite samples exhibited higher values of the absorption at 1715 cm^{-1} as compared to PP or PP-g-MA. However, after ca. 40 h of irradiation, the absorption tends to level off in the case of nanocomposites while i-PP exhibited an autoacceleration. This behavior has been assigned to a prevention of oxidation reactions caused by reduction in molecular motions in the presence of silicate layers. Natural exposure tests shown similar behavior, without the initial decreasing in absorption at 1780 cm^{-1}. Acceleration factors, calculated by formula (1) indicate clearly that pure polymer is less susceptible to photo-oxidation as compared to studied nanocomposites (Figure 6).

$$acc.factor = \frac{\tau_n}{\tau_a} \qquad (1)$$

where τ_n and τ_a are the durations up to pre-established and attended value of a degradation parameter under natural and accelerated conditions, respectively.

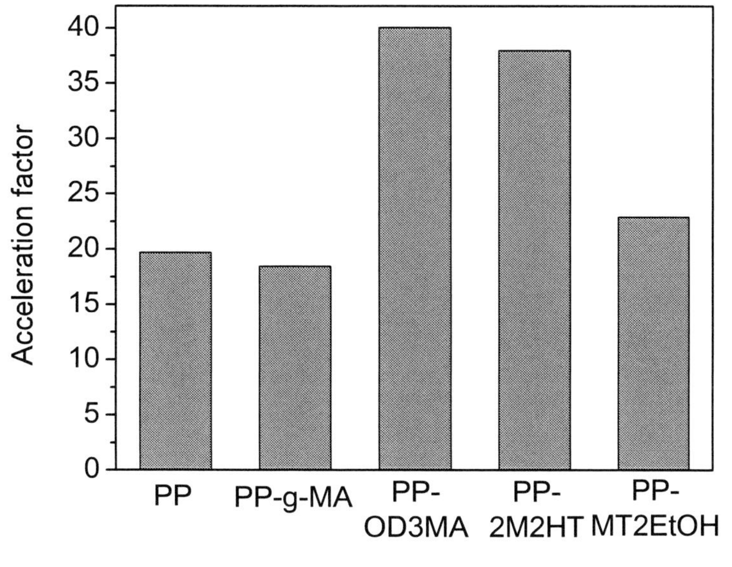

Figure 6. Acceleration factors of photo-induced oxidation of various polypropylene - modified montmorillonite nanocomposites (OD3MA = octadecyl trimethyl amine, 2M2HT = dimethyl dehydrogenated tallow ammonium; MT2EtOH = methyl tallow di-hydroxyethyl ammonium) (Adaptation of data in paper of Diagne et al [34]).

On the other hand, the clay type seems to influence on a high extent the oxidation susceptibility, rather than the mixing state of the PP-montmorillonite nanocomposites. Thus, the highest and the lowest values of the acceleration factors correspond (see Figure 6) to tactoid-exfoliated structures, while the intermediate value is exhibited by a highly exfoliated system. A different behavior in photo-oxidation and thermal oxidation has been observed in the case of the above mentioned PP-montmorillonite nanocomposites [34]. Thus, PP-OD3MA (see Figure 6) degraded without any induction period, while its counterparts remained undegraded even after 3000 hours of thermal ageing. The results were explained in terms of the importance of ion exchange that may play a role in thermooxidation mechanism.

The influence of some preparation factors on the photo-oxidation behavior has been emphasized in a series of papers [13,26,38]. Thus, in the paper [13] the effect of the preparation mode (injection vs extrusion) is discussed in the case of some PP-graft-maleic anhydride - MMT nanocomposites. XRD results revealed significant differences in the structure of the resulted nanocomposites, which can be assigned rather to the differences in composition that it is happened during the manufacture process.

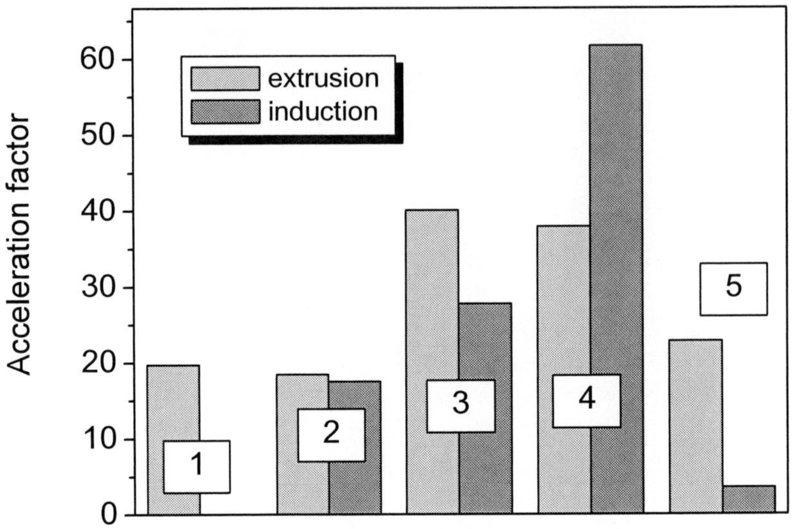

Figure 7. Acceleration factors for PP-g-maleic anhydride/ MMT nanocomposites exposed to photo-oxidation under natural and UV conditions: 1 - pure PP; 2 - PP-g-maleic anhydride; 3 - PP-OD3M; 4 - PP-2M2HT; 5 - PP-M2EtOH (Adaptation from paper [38]

However, similar composition materials, but differently processed, were found to exhibit unlike behavior regarding photo-induced oxidation, caused by their preparation mode. Thus, the samples prepared by injection molding were found to degrade more rapidly than their counterparts prepared by extrusion, the differences being clearer in the case of carbonyl groups, 1715 cm^{-1}. The increasing in CO content exhibited an induction period of ca. 40 h in the case of extrusion prepared samples, whereas no induction has been observed to the injection molded samples. As it can be seen in Fig. 7, the differences in photo-oxidation behavior of PP-MMT nanocomposites appear to originate in both type of ammonium salt used to modify MMT and in preparation mode. It was found that both parameters influenced FTIR absorption bands (even in the case of initial, unexposed samples) suggesting differences in the reaction sequence involved in photo-oxidation process. Thus, less carboxyl and ketone groups, but more ester ones, were found in the case of extrusion prepared samples as compared to injection molded counterparts.

A coherent explanation of the lower photo-stability of nanocmposites has been done in a series of papers by Morlat et al. [13, 25, 26]. Three main factors that explain the observed decreasing in photo-stability of nanocomposites as compared to base copolymers are supposed:

- processing antioxidant (stabilizer) adsorption on silicate layers, or antioxidant intercalation between layers;
- catalytic effect of transitional metals (iron mainly) in oxidation processes;

- decomposition of surfactant molecules and increasing in photo-sensitivity induced by decomposition products.

The effect of antioxidants on the oxidation of nanocomposites has been discussed in the papers of Morlat et al [13,5,26]. The accumulation of oxidation products, as it has been displayed by FTIR data, has been found to be different in the case of nanocomposite and of pure polymer, due to the interaction antioxidant - nanofiller (i.e. adsorption of antioxidant on MMT layers). Thus, in the case of some PP based systems, it has been observed that both absorbance at 3400 cm^{-1} and 1713 cm^{-1} increased easier in the case of samples containing MMT as compared to the materials free of MMT. The increasing in ROOH exhibited long induction period with PP and PP/PP-g-MA (80/ 20, wt%) samples, while considerably shorter induction periods were observed with MMT containing samples (Figure 8). During the induction time of photo-treatment, the antioxidant acts as a sacrificial additive [13], i.e. it is consumed rapidly (being designed for thermal oxidation protection, not for photo-stabilization). After the antioxidant depletion, the absorbance in the oxidation assigned bands increased steadily as increasing the irradiation time. The accumulation rates of oxidation products as they have been detected by FTIR spectroscopy were found to be basically the same in the case of both base polymer and nanocomposite materials [13, 26]. The effect of reduction in induction period of oxidation in the presence of MMT has been assigned to antioxidant adsorption on MMT layer [13]. Similar results were reported with a phenol antioxidant (Irganox 1010) or a HALS [25], even their action mechanism was different (i.e. radical trapping in the case of the phenol and redox, in the latter case).

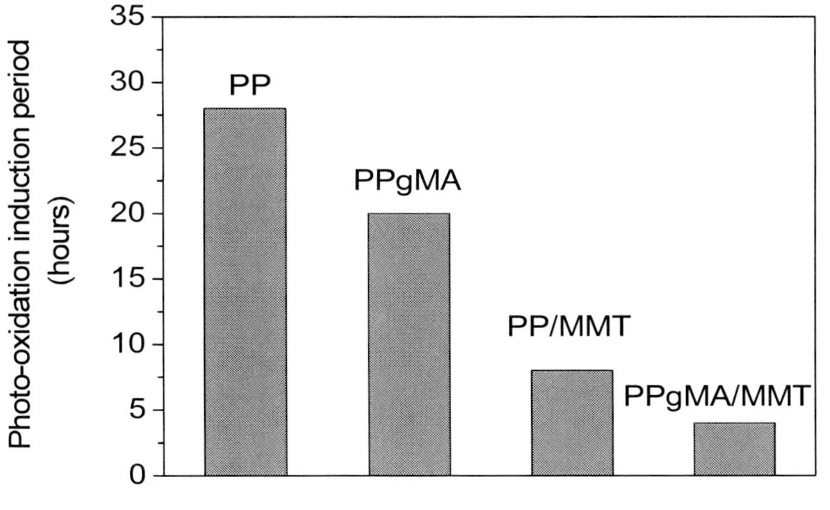

Figure 8. Induction period of photo-oxidation ($\lambda \geq 300$ nm, 60 °C) of various PP based materials. (Adaptation form paper of Morlat et al [13])

In the case of EPDM/MMT nanocomposites similar conclusions were drawn, but furthermore, the effect of MMT dispersion on the photo-stability has been clearly pointed out [26]. Thus, it was observed the lack of induction period in increasing of the absorption band at 1713 cm^{-1} occurred in the investigation on nanocomposites based on EPDM. Both pure EPDM-g-MA and microcomposite exhibited induction periods (of about 6h), when oxidation was not detected. This effect of the adsorption process of antioxidant molecules on the MMT surface has been assumed to explain this observation. Such a process, leading to the decrease in stabilizer effectiveness, becomes more significant, when the accessible surface of MMT is highly enhanced, similarly noted for nanocomposites. To illustrate this enhancement, it can be noted the spacing values, between MMT layers (d_{001}) (evaluated from XRD data), which were found to be 1.34 nm and 1.43 nm for MMT powder and EPDM/MMT microcompozite, respectively [26]. These values increased significantly to 4.60 nm in intercalate nanocomposite (low content of organic ammonium salt, i.e. octadecyltrimethyl ammonium bromide) to attend infinite value (no XRD peak at low angle values has been observed) in the case of exfoliated materials (high content in surfactant).

The adsorption of antioxidants on clay particles has been already reported [39,40], and it could be pointed out by the experimentally observed decrease in antioxidative activity. This decreasing can be assigned to reducing in both effective stabilizer concentration in polymer matrix and stabilizer mobility.

However, there are also some doubts concerning the role of adsorption processes in decreasing the photo-stability of nanocomposites. Thus, stabilizer adsorption or intercalation are possible processes in nanocomposites, if we take into consideration the differences in the polarity of the active groups in antioxidants or hindered amines, on one side and polymers and organo-clay, on the other side. These differences can explain the occurrence of some relative interphase insolubility phenomena. The organically modified clays as well as surfactant molecules still contain enough hydrophilic sites able to facilitate some interactions with stabilizer molecules. They could result in different distributions of stabilizer added in nanocomposite as compared to base polymer. Such interactions would lead to the swelling of the structured MMT layers, but no experimental evidence has been provided in this direction [25] (the interlayer spacing values, as they were evaluated from X-ray diffraction, exhibited similar values in the case of organically modified MMT and stabilizers containing organo-MMT). On the other hand, the interaction stabilizer-clay seems to be rather weak as it was suggested by the experimental investigations according to which stabilizers were easily removed by solvent extraction [25].

Antagonistic effects were also reported in the presence of HALS and talc in formulated materials, as well as in the case of HALS and some transitional metals pairs, suggesting that adsorption could be accompanied by other factors in increased photo-susceptibility of nanocomposites [25]. Taking into account the natural origin of various nanofillers used in preparation of nanocomposites, the existence of iron ions can be assumed in all the cases. It is well established that, at low concentrations, iron ions act as pro-oxidants in photo-oxidation processes, due to their capability in the initiation of oxidation by redox reactions involving hydroperoxides [41,42]. It has been shown that the catalytic effect of such ions is characterized by shortening of the induction period, but the influence on steady state oxidation has been not observed. Thus, the observed decrease in induction periods and similarly, in oxidation rates of nanocomposites by comparison to base polymers could be explained to be influenced by iron ions. The behavior of some EPDM/MMT nanocomposites prepared by iron free hydrothermally synthesized MMT has been studied by Morlat-Thérias et al. [26]. The results obtained in photo-oxidative degradation (under UV irradiation, $\lambda \geq 300$ nm; T = 60 °C) suggest similar mechanisms [as it has been suggested by similar rates of increasing in absorption band at 1713 cm^{-1} (carbonyl groups)], irrespective of material morphology and composition. The photo-stability of nanocomposites (both intercalated and exfoliated materials) has been lower than in the case of basic polymer (EPDM-g-MA), and no induction period has been observed. This aspect suggests that the role of iron ions in the photo-degradation of MMT nanocomposites should be supplementary clarified. Other data suggested that MMTs may initiate themselves the polymer degradation due to the presence of some structural elements, such as Fe^{2+} or Fe^{3+} ions, in a similar manner to that observed in the case of other products, such as pesticides [43].

On the other hand, the organic component of the modified MMT could be involved in the degradation of nanocomposites, the photo-oxidation process being more pronounced in the case of exfoliated EPDM/MMT nanocomposite samples. The influence of higher content of organic cations on the progress in thermal degradation of EPDM-g-MA materials is illustrated in Figure 9.

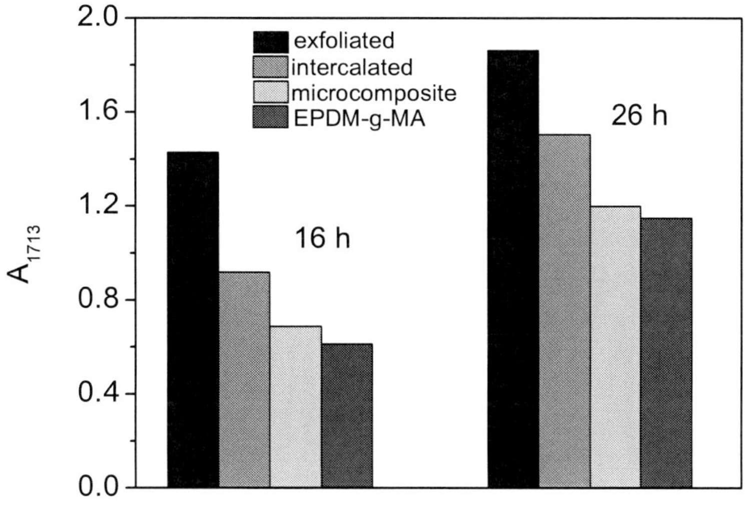

Figure 9. Comparative values of absorbance in 1713 cm^{-1} region of UV exposed EPDM-g-MA based materials (Adaptation from paper [25])

Thermal degradation of the organic modifiers has been early reported [10,44], being observed that the organic-treated clay could influence the degradation reactions of the polymer during extrusion. Thus, tetraalkylammonium ions can undergo Hofmann elimination, leading to amines and unsaturated end groups in organic moieties from ions. These compounds may undergo photo-oxidation, where degradation products may results via peroxyradicals.

There are several factors influencing the nanocomposites oxidation. Among them, metal ions content, added antioxidants, the presence of nanofiller dispersion and antioxidant adsorption capacity, organic ions of the surfactants and the compatibilizing agent are the most important. Their simultaneous action leads to a general decreasing in material stability comparative to pristine polymers.

CONCLUSION

In spite of few available data regarding the thermal or photo-oxidation of nanocomposite and hybrid materials, different behavior as compared to classic composites or pure polymers has been revealed. This behavior seems to be due to specific interactions between organic and inorganic nanoscale moieties and could result in improved thermooxidation strength. However, poorer stability to photo-oxidation of nanocomposites has been surprisingly observed, and several factors which could induce this sensitivity are considered. The knowledge of efficient ways to the improvement in these materials durability remains an attractive goal to be reached.

REFERENCES

[1] Lin, S.-T.; Huang, S. K. *Eur. Polym. J.* 1997, *33*, 365-373.

[2] Hsiue, G.-H.; Wei, H.-F. *Polym. Degrad. Stab.* 2001, *73*, 309-318.

[3] Maillet, C.; Janvier, P.; Pipelier M.; Praveen, T.; Andres, Y.; Bujoli, B. *Chem. Mater.*, 2001, *13*, 2879-2884..

[4] Kojima, Y.; Usuki, A.; Kawasumi, M.; Okada, A.; Fukushima, Y.; Kurauchi, T.; Kamigaito, O. *J. Mater. Res.*, 1993, *8*, 1185-1189.

[5] Kojima, Y.; Usuki, A.; Kawasumi, M.; Okada, A.; Fukushima, Y.; Kurauchi, T.; Kamigaito, O. *J. Mater. Res.*, 1993, *8*, 1179-1185

[6] Kojima, Y.; Usuki, A.; Kawasumi, M.; Okada, A.; Fukushima, Y.; Kurauchi, T.; Kamigaito, O. *J. Polym. Sci. Part. A Polym. Chem.* 1993, *31*, 983-986.

[7] Zanetti, M.; Kashiwagi, T.; Falqui, L.; Camino, G.; *Chem. Mater.* 2002, 14, 881-887.

[8] Manias, E.; Touni, A.; Wu, L; Strawhecker, K.; Lu, B.; Chung, T. C. *Chem. Mater.* 2001, *13*, 3516-3523.

[9] Zhu, J.; Morgan, A. B.; Lamelas, F. J.; Wilkie, C. A.; *Chem. Mater.* 2001, *13*, 3774-3780.

[10] Gilman, G. W.; Jackson, C. L.; Morgan, A. B.; Harris, R. Jr. *Chem. Mater.* 2000, *12*, 1866-1873.

[11] Gilman, J. W. *Appl. Clay. Sc.* 1999, *15*, 31-49.

[12] Alexandre, M.; Dubois, Ph. *Mater. Sci. Eng. R.* 2000, *28*, 1-63.

[13] Morlat, S.; Mailhot, B.; Gonzales, D.; Gardette, J. L. *Chem. Mater.* 2004, *16*, 377-383.

[14] Abate 2004 - Abate, L.; Blanco, I.; Bottino, F. A.; Di Pasquale G.; Orestano A.; Pollicino A. „Kinetics of The Thermal Degradation of PS/MMt Nanocomposites Prepared with Imidazolium Surfactants", 3rd Modest Conference Lyon 2004, France Aug. 29 - Sept. 2, 2004, http://modest.univ-lyon1.fr/Page_progr.shtml; paper on conference's CD.

[15] Bala, B.; Samantary, B. K.; Srivastava, S. K. *Mater. Res. Bull.* 2000, *35*, 1717-1724.

[16] Giannelis, E. P.; Krishnamoorti, R. K.; Manias, E. *Adv. Polym. Sci.* 1998, *138*, 107-147.

[17] Xie, W.; Gao, Z.; Pan, W. P.; Hunter, D.; Singh, A.; Vaia, R. *Chem. Mater.* 2001, *13*, 2979-2990.

[18] Xie, W.; Gao, Z.; Liu, K.; Pan, W. P.; Vaia, R.; Hunter, D.; Singh, A. *Thermochim. Acta* 2001, *367*, 339-350.

[19] Tidjani, A.; Wilkie, C. A. *Polym. Degrad. Stab.* 2001, *74*, 33-37.

[20] Leroux, F.; Meddar, L.; Mailhot, B.; Thérias-Morlat, S.; Gardette, J.-L. *Polymer* 2005, *46*, 3571 – 3578.

[21] Qin, H.; Zhao, C.; Zhang, S.; Chen, G.; Yang, M. *Polym. Degrad. Stab.* 2003, *81*, 497-500.

[22] Sloan, J. M.; Patterson, P.; Hsieh, A. *Polym. Mater. Sci. Eng.* 2003, *88*, 354-359.

[23] Lee, H. C.; Lee, T. W.; Lym, Y. T.; Park, O. O. *Appl. Clay. Sci.* 2002, 21, 287-293.

[24] Mailhot, B.; Morlat, S.; Gardette, J. L; Boucard, S.; Duchet, J.; Gerard, J. F. *Polym. Degrad. Stab.* 2003, *82*,163-167.

[25] Morlat-Thérias, S.; Mailhot, B.; Gonzales, D.; Gardette, J-L. *Chem. Mater.* 2005, *17*, 1072-1078.

[26] Morlat-Thérias, S.; Mailhot, B.; Gardette, J-L.; D Silva, C.; Haidar, B.; Vidal, A. *Polym. Degrad. Stab.* 2005, *90*, 75-85.

[27] Kissinger, H. E., *Anal. Chem.*, 1957, *29*, 1702-1706.

[28] Wang, Y. T.; Chang, T. C.;Hong, Y. S.; Chen, H. B. *Thermochimica Acta* 2002, *397*, 219-226.

[29] Kashiwagi, T.; Hirata, T.; Brown, J. E. *Macromolecules* 1985, *18*, 131-138.

[30] Marning, L. E.; Sogah, D. Y.; Cohen, G. M. *Macromoecules* 1989, *22*, 4652-4654.

[31] Chang, T. C.; Wang, Y. T.; Hong, Y. S.; Chiu, Y. S. *J. Polym. Sci. Polym. Chem.* 2000, *38*, 1972-1980.

[32] Costantino, U.; Gallipoli, A.; Nocchetti, M.; Camino, G.; Bellucci, F.; Frache, A. 2005, *90*, 586-590.

[33] Peeterbroeck, S.; Alexandre, M.; Jérôme, R.; Dubois, Ph. *Polym. Degrad. Stab.* 2005, *90*, 288-294.

[34] Diagne, M.; Dasilva, A.; Tidjani, A.; Photo- and thermo-oxidation of nanocomposites of polypropylene-graft-maleic anhydride/clay 3rd Modest Conference Lyon, France Aug. 29-Sept. 2, 2004 http://modest.univ-lyon1.fr/Page_progr.shtml; paper on conference's CD.

[35] Philippart, J. L.; Sinturel, C.; Arnaud, R; Gardette, J. L. *Polym. Degrad. Stab.* 1999, *64*, 213-225.

[36] Geuskens, G.; Kabamba, M. S. *Polym. Degrad. Stab.* 1982, *4*, 69-76.

[37] Swern, D.; Witnauer, L. P.; Eddy, C. R.; Parker, W. E. *J. Am. Chem. Soc.* 1955, *77*, 5537-5541.

[38] Gueye 2004 - Guèye, M.; Diagne, M.; Vidal, L; Tidjani, A. „Influence of the mode of preparation on the photo-aging behaviour of polypropylene-graft-maleic anhydride/clay nanocomposites" 3rd Modest Conference Lyon, France Aug. 29 - Sept. 2, 2004 http://modest.univ-lyon1.fr/Page_progr.shtml; paper on conference's CD.

[39] Boyd, S. A.; Shaobai, S.; Lee, J. F.; Mortland, M. *Clays Clay. Miner.* 1988, *36*, 125-130.

[40] Mortland, M. M.; Shaobai, S.; Boyd, S. A. *Clays Clay Miner.* 1986, *34*, 581-585.

[41] Scott, G.; Sheena, H. H.; Hariman, A. M. *Eur. Polym. J.* 1978, *14*, 1071-1077.

[42] Verdu, J. *Macromol. Symp.* 2001, *169*, 275-287.

[43] Cervini-Silva, J.; Larson, R. A.; Wu, J.; Stucki, J. W. *Environ. Sci. Technol.* 2001, *35*, 805-809.

[44] Davis, R. D.; Gilman, J. W.; Vanderhart, D. L. *Polym. Degrad. Stab.* 2003, *79*, 111-121.

INDEX

A

B

C

D

G

H

I

J

K

L

M

N

O

T

U

V

W

X

Y

Z